Geothermal Energy

Ingrid Stober · Kurt Bucher

Geothermal Energy

From Theoretical Models to Exploration and Development

 Springer

Ingrid Stober
Applied Geosciences
Karlsruhe Institute of Technology KIT
Karlsruhe
Germany

Kurt Bucher
University of Freiburg
Mineralogy and Petrology
Freiburg
Germany

ISBN 978-3-662-50193-1   ISBN 978-3-642-13352-7   (eBook)
DOI 10.1007/978-3-642-13352-7
Springer Heidelberg New York Dordrecht London

© Springer-Verlag Berlin Heidelberg 2013
Softcover reprint of the hardcover 1st edition 2013
This work is subject to copyright. All rights are reserved by the Publisher, whether the whole or part of the material is concerned, specifically the rights of translation, reprinting, reuse of illustrations, recitation, broadcasting, reproduction on microfilms or in any other physical way, and transmission or information storage and retrieval, electronic adaptation, computer software, or by similar or dissimilar methodology now known or hereafter developed. Exempted from this legal reservation are brief excerpts in connection with reviews or scholarly analysis or material supplied specifically for the purpose of being entered and executed on a computer system, for exclusive use by the purchaser of the work. Duplication of this publication or parts thereof is permitted only under the provisions of the Copyright Law of the Publisher's location, in its current version, and permission for use must always be obtained from Springer. Permissions for use may be obtained through RightsLink at the Copyright Clearance Center. Violations are liable to prosecution under the respective Copyright Law.
The use of general descriptive names, registered names, trademarks, service marks, etc. in this publication does not imply, even in the absence of a specific statement, that such names are exempt from the relevant protective laws and regulations and therefore free for general use.
While the advice and information in this book are believed to be true and accurate at the date of publication, neither the authors nor the editors nor the publisher can accept any legal responsibility for any errors or omissions that may be made. The publisher makes no warranty, express or implied, with respect to the material contained herein.

Springer is part of Springer Science+Business Media (www.springer.com)

Preface

Geothermal energy is an inexhaustible source of thermal and electrical energy on a human time scale. Its utilization is friendly to the environment and supplies base-load energy. The energy source does not depend on weather, and the energy is supplied 24 h per day during 7 days a week. Utilization of geothermal energy increases the regional and local net product. It relieves dependence from fossil fuels and helps to conserve the valuable chemical resources for the future. Deep geothermal resources provide thermal and electrical (converted thermal) energy thus providing reliable energy for the future.

Utilization of deep geothermal resources extracts hot fluid from thermal reservoirs. These hot waters are reinjected to the reservoirs, thus maintaining the natural equilibrium and permitting a sustainable and sparing management of the resource. Geothermal installations and power plants are characterized by small land use. Visual impact on landscape is minimal (an aspect that is particularly critical in densely populated areas).

Electrical energy from geothermal resources can provide an important contribution to the base-load electrical energy supply and may replace large-scale power plants fired with fossil fuels.

The utilization of geothermal energy from shallow resources for the production of energy at low temperature for heating and cooling applications made tremendous progress in the past decade.

Geothermal energy from deep sources and reservoirs can contribute significant base-load energy. The necessary technology of Enhanced Geothermal Systems (EGS) can be installed nearly everywhere. However, the EGS technology needs further improvements and research. Successful demonstration projects would help to popularize EGS further.

Long-term concepts of energy politics integrate geothermal energy sources because it supplies base-load energy. Intelligent combination of geothermal systems with other sources of renewable energy can create sustainable synergy benefits. For residential homes, for example, combining ground source heat pump systems with solar thermal systems proved to be highly energy efficient. Geothermal power from a deep hydrothermal system can be combined with a biogas installation to improve energy efficiency.

This book aims to offer the reader a general overview over the many different aspects of utilization of geothermal energy. We are looking forward to the further rapid development of this fascinating source of energy in the years to come. We wish all of us a reliable, save, and environmentally friendly supply of thermal and electrical power. We hope to contribute to the sustainable use of energy with this book.

Ingrid Stober
Kurt Bucher

Contents

1	**Thermal Structure of the Earth**	1
	1.1 Renewable Energies, Global Aspects	2
	1.2 Internal Structure of the Earth	2
	1.3 Energy Budget of the Planet	7
	1.4 Heat Transport and Thermal Parameters	8
	1.5 Brief Outline of Methods for Measuring Thermal Parameters	13
2	**History of Geothermal Energy Use**	15
	2.1 Early Utilization of Geothermal Energy	16
	2.2 History of Utilization of Geothermal Energy in the Last 150 years	21
3	**Geothermal Energy Resources**	25
	3.1 Energy	26
	3.2 Significance of "Renewable" Energies	28
	3.3 Status of Geothermal Energy Utilization	29
	3.4 Geothermal Energy Sources	31
4	**Applications of Geothermal Energy**	35
	4.1 Near Surface Geothermal Systems	37
	4.2 Deep Geothermal Systems	42
	4.3 Efficiency of Geothermal Systems	53
	4.4 Major Geothermal Fields, High Enthalpy Fields	55
5	**Potential Perspectives of Geothermal Energy Utilization**	61
6	**Geothermal Probes**	65
	6.1 Planning Principles	66
	6.2 Construction of Ground Source Heat Exchangers	66
	6.3 Dimensioning and Design of Geothermal Probes	72
	6.3.1 Heat Pumps	73
	6.3.2 Thermal Parameters and Computer Programs for the System Design of Ground Source Heat Pump Systems	77

	6.4	Drilling Methods for Borehole Heat Exchangers	84
		6.4.1 Rotary Drilling	87
		6.4.2 Down-the-Hole Hammer Methods	89
		6.4.3 Concluding Remarks, Technical Drilling Risks	90
	6.5	Backfill and Grouting of Geothermal Probes	94
	6.6	Construction of Deep Geothermal Probes	98
	6.7	Operating Geothermal Probes: Potential Risks, Malfunctions and Damages	99
	6.8	Special Systems and Further Developments	101
		6.8.1 Geothermal Probe Fields	102
		6.8.2 Cooling with Geothermal Probes	103
		6.8.3 Combined Solar Thermal: Geothermal Systems	104
		6.8.4 Geothermal Probe: Performance and Quality Control	105
		6.8.5 Geothermal Probes Operating with Phase Changes	110
7	**Geothermal Well Systems**		**115**
	7.1	Building Geothermal Well Systems	116
	7.2	Chemical Aspects of Two-Well Systems	119
	7.3	Thermal Range of Influence, Numerical Models	120
8	**Hydrothermal Systems, Geothermal Doublets**		**125**
	8.1	Geologic and Tectonic Structure of the Underground	126
	8.2	Thermal and Hydraulic Properties of the Target Aquifer	129
	8.3	Hydraulic and Thermal Range of Hydrothermal Doublets	137
	8.4	Hydrochemistry of Hot Waters from Great Depth	141
	8.5	Reservoir-Improving Measures, Efficiency-Boosting Measures, Stimulation	144
	8.6	Productivity Risk, Exploration Risk, Economic Efficiency	146
		8.6.1 Exploration Risks	148
	8.7	Some Site Examples of Hydrothermal Systems	153
		8.7.1 High-Enthalpy Hydrothermal Systems	153
		8.7.2 Low-Enthalpy Hydrothermal Systems	153
	8.8	Project Planning of Hydrothermal Power Systems	160
		8.8.1 Phase 1: Preliminary Study	161
		8.8.2 Phase 2: Feasibility Study	162
		8.8.3 Phase 3: Exploration	162
		8.8.4 Phase 4: Development	162
9	**Enhanced-Geothermal-Systems, Hot-Dry-Rock Systems, Deep-Heat-Mining**		**165**
	9.1	Techniques, Procedures, Strategies, Aims	167
	9.2	Historical Development of the Hydraulic Fracturing Technology, Early HDR Sites	169
	9.3	Stimulation Procedures	170
	9.4	Experience and Dealing with Micro-Seismicity	176
	9.5	Recommendations, Notes	177

10	**Environmental Issues Related to Deep Geothermal Systems**		183
	10.1	Seismicity Related to EGS Projects	185
		10.1.1 Induced Earthquakes	187
		10.1.2 Quantifying Seismic Events	189
		10.1.3 The Basel Incident	190
		10.1.4 Observed Seismicity at Other EGS Projects	193
		10.1.5 Conclusions and Recommendations Regarding Seismicity Control in Hydrothermal and Petrothermal (EGS) Projects	196
	10.2	Interaction Between Geothermal System Operation and the Subsurface	198
	10.3	Environmental Issues Related to Surface Installations and Operation	200
11	**Drilling Techniques for Deep Wellbores**		203
12	**Geophysical Methods, Exploration and Analysis**		221
	12.1	Geophysical Pre-drilling Exploration, Seismic Investigations	222
	12.2	Geophysical Well Logging and Data Interpretation	227
13	**Testing the Hydraulic Properties of the Drilled Formations**		233
	13.1	Principles of Hydraulic Testing	234
	13.2	Types of Tests, Planning and Implementation, Evaluation Procedures	243
	13.3	Tracer Experiments	249
	13.4	Temperature Evaluation Methods	252
14	**The Chemical Composition of Deep Geothermal Waters and Its Consequences for Planning and Operating a Geothermal Power Plant**		255
	14.1	Sampling and Laboratory Analyses	256
	14.2	Deep Geothermal Waters, Data and Interpretation	258
	14.3	Mineral Scales and Materials Corrosion	271
References			279

Chapter 1
Thermal Structure of the Earth

Island of Vulcano, southern Italy

1.1 Renewable Energies, Global Aspects

The term "renewable energy" is used for a source of energy from a reservoir that can be restored on a "short time scale" (in human time scales). Renewable energy includes geothermal energy and several forms of solar energy such as bio-energy (bio-fuel), hydroelectric, wind-energy, photovoltaic and solar-thermal energy. These sources of energy are converted to heat or electricity for utilization. An example: The "renewable" aspect of burning firewood in a cooking stove lies in the relatively short period of time required to re-grow chopped down forests with solar energy and the process of photosynthesis. In contrast, it will take much more time to "renew" coal beds when burning coal for the same purpose, although geological processes will eventually form new coal beds. The "renewable" aspect of geothermal energy will be explained and discussed in detail in this chapter.

The International Geothermal Association (IGA) wrote in a recent press release on "Renewable Energy Policy Network for the 21st Century" that the global production of renewable energies increased by 16 % from 2007 to 2008 and amounts to a total of 280 GW (IGA 2009; www.geothermal-energy.org). The growth in renewable energy consumption is larger than the increase in fossil fuel consumption in Europe and the US. Political and financial programs support the development and use of renewable energies in more than 60 countries.

Large hydroelectric systems had the largest share in electricity production from renewable energy sources in 2008 with 860 GW_{el} followed by wind energy with 121 GW_{el}, small hydroelectric systems with 85 GW_{el} and biomass conversion with 52 GW_{el}. Photovoltaic systems (13 GW_{el}) and geothermal systems (10 GW_{el}) follow with a large gap. Thermal energy production from renewable sources is dominated by biomass (250 GW_{th}), followed by solar thermal systems (145 GW_{th}) and geothermal systems (50 GW_{th}) (Bertani 2007, 2010).

Geothermal energy has the potential to become a significant source of energy in the future because it is available everywhere and withdrawals are continuously replenished. From a human perspective the resource is essentially unlimited. Heat and electricity can be continuously produced and therefore it is a base load resource. The utilization is friendly to the environment and the land consumption for the surface installations is small. The coming years will show how the optimistic expectations and the positive perception of geothermal energy utilization will succeed in regions with low-enthalpy geothermal resources.

1.2 Internal Structure of the Earth

Geothermal energy is the thermal energy stored in the Earth body, geothermal energy is underground heat. 99 % of the Earth is hotter than 1,000 °C and only 0.1 % is colder than 100 °C. The average temperature at the Earth surface is 14 °C. The surface temperature of the sun is about 5,800 °C, which corresponds to the temperature at the center of the Earth (Fig. 1.1).

1.2 Internal Structure of the Earth

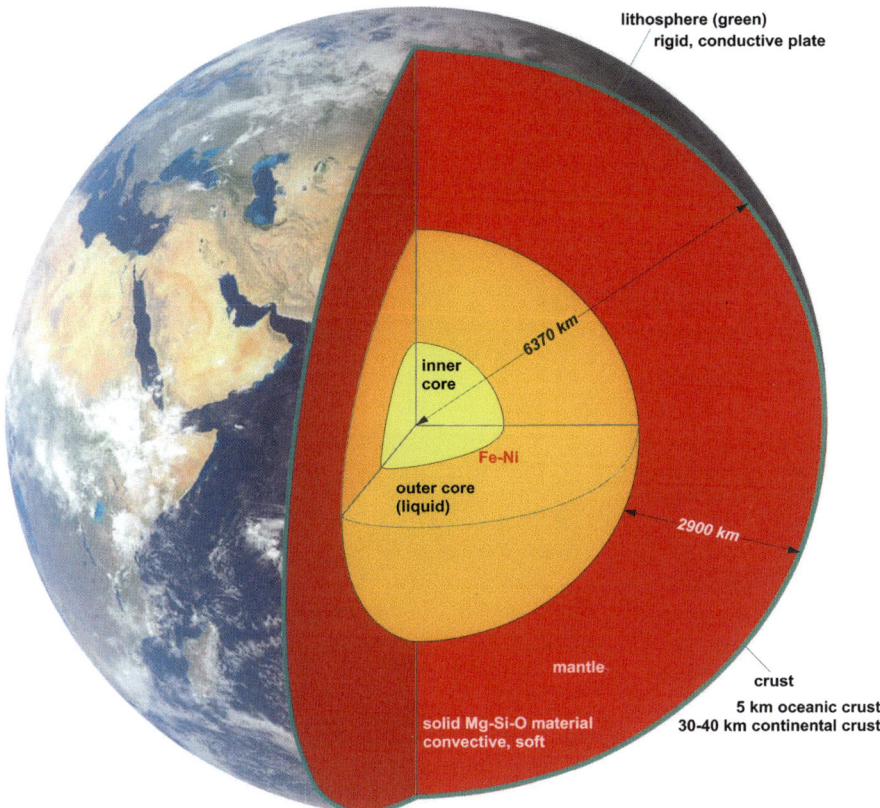

Fig. 1.1 Internal structure of the Earth

The Earth has a layered internal structure (Fig. 1.1) with a solid core of high-density material, an iron-nickel alloy surrounded by an outer core of the same material in a low-viscosity state. A thick internally layered, viscous magnesium silicate mantle encloses the core. The surface zone of the planet is build up of a thin rigid crust, whose composition is different on continents and oceans. This layered structure developed from a more homogeneous system by gravitational compaction and differentiation during the earliest history of the planet.

The total thickness of the core (Fig. 1.1) exceeds the thickness of the mantle. However, the core represents only about 16 % of the volume of the Earth and, because of its high density about 32 % of the mass of the planet.

At 6,000 km depth the inner core temperatures are above 5,000 °C and the pressure is about 400 GPa. Iron meteorites that arrive at the Earth surface occasionally from space consist of material similar to the Ni–Fe alloy of the core (Fig. 1.2). The molten Fe–Ni metal outer core (about 2,900 °C) is together with the rotational movement of the planet responsible for the Earth's magnetic field. The core—mantle boundary is a zone of dramatic changes in composition and

Fig. 1.2 Widmanstätten figures made visible on an iron meteorite. The structure results from the intergrowth of the two minerals kamacite and taenite with different Fe/Ni ratios

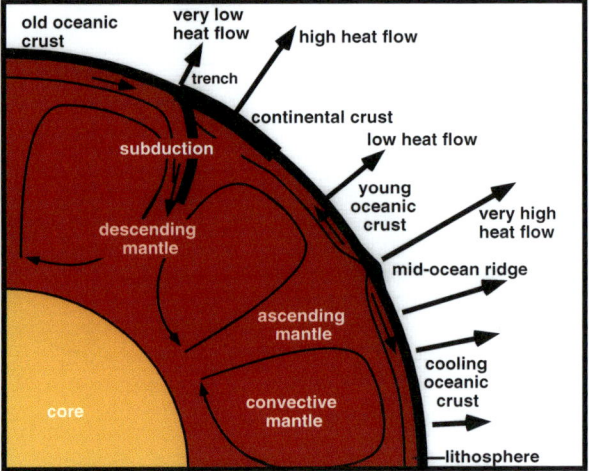

Fig. 1.3 Thermal convection currents in the viscous mantle drive plate tectonics (movement of rigid lithospheric plates) and controls large-scale heat flow (*black arrows*)

density where molten metal from the outer core and solid mantle silicate minerals mix. Beneath the lithosphere, the upper mantle reaches to a depth of about 1,000 km. The boundary layer between lithospheric and convective mantle at 100–150 km depth is rheologically soft and melt may be present locally facilitating movement of the lithospheric plates. The solid but soft mantle is in a very slow convective motion driven by the heat of the core and transmitted through the core—mantle boundary (hot plate). A part of the heat given off to the mantle arises from the enthalpy of crystallization at the boundary between inner solid and outer liquid core in an overall environment setting of a cooling planet. The motor for mantle convection operates since the formation of the Earth.

The lithosphere is the rigid lid of the planet that is subdivided into a series of mobile plates that move individually as a result of pull and drag forces exerted by the convecting mantle (Fig. 1.3). The lithospheric mantle is separated from the crust by the petrographic MOHO and consists of the same rock types as the mantle

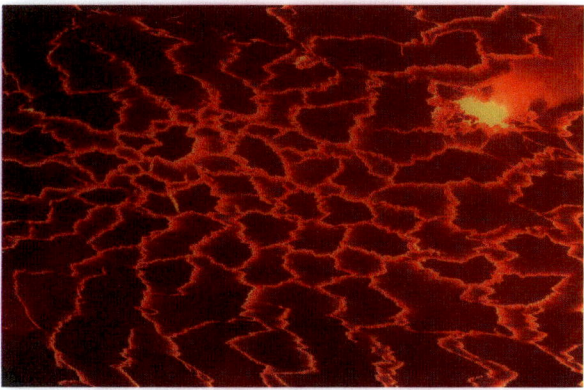

Fig. 1.4 Nyiragongo lava lake near Goma 2011 (Photograph: Spiegel Online)

as a whole. The convecting mantle creates distinct thermal regimes at the Earth surface resulting from upwelling and subsiding hot mantel material and from the mechanical and thermal response of the lithosphere.

Convergent plate motions may create mountain belts such as the Alps and the Himalayas. The dense oceanic lithosphere of two convergent plates may be subducted and recycled into the mantle. Melting and release of H_2O from the subducting slab can generate massive amounts of melts in the overriding plate and the transfer of heat to shallow levels of the crust. Examples are the volcanic chain of the Cascades, parts of the Andes, Aleutian Islands, Japan, Philippines, Indonesia, North Island of New Zealand and many other volcanic areas of the world.

Extending lithosphere creates rift and graben structures typically with a pronounced thermal response at the surface. Examples of this setting include the East African rift valley and the Basin and Range Province in the western US (Fig. 1.4). Extensional oceanic plate margins are mid-ocean ridges and the sites of the most prominent volcanic activity of the planet. The Mid-Atlantic Ridge and the East Pacific Rise are examples of these settings.

Particularly spectacular large scale geologic structures are tied to the linear upwelling systems, so called mantle diapirs or "hot spots". A hot spot under oceanic lithosphere is causing intraplate volcanism on the Hawaii islands, a hot spot under continental lithosphere is causing extremely dangerous rhyolite volcanism in the Yellowstone area (USA) with are all associated forms of hydrothermal activity such as geysers, mud volcanoes, gas vents and others (Fig. 1.5). The massive Yellowstone eruptions, that have devastated the whole Earth, have a periodicity of about 600 Ka. The last one occurred about 0.6 Ma ago. The crown of the hot spots is located underneath Iceland where it coincides with the extension of the mid Atlantic ridge causing an abnormally high volcanic activity and a massive heat transfer to very shallow levels of the crust. The Italian volcanoes are classic examples of volcanism and associated hydrothermal and degassing phenomena (Fig. 1.6).

Fig. 1.5 Different stages of an eruption of Echinus Geyser in the Norris Geyser Basin, Yellowstone National park, Wyoming, USA

Fig. 1.6 Volcanic phenomena on the island of Vulcano (Italy): **a** Volcanic steam degassing from crater flank; **b** volcanic gasses bubble from hot water pond; **c** Steam degassing on the crater ridge; **d** Sulfur crusts on the crater ridge; **e** One of the authors submerged in poisonous volcanic gasses (photograph taken by the other author); **f** Sulfur crystals deposited from the oxidation of primary H_2S gas with atmospheric oxygen ($2H_2S + O_2 = S_2 + 2H_2O$)

1.3 Energy Budget of the Planet

The average temperature at the Earth surface is 14 °C, at the core—mantle boundary the temperature is in the range of 3,000 °C. This temperature difference between the surface and the interior is the driving force for heat flow, which tries to eliminate ΔT. The process is known as so-called Fourier conduction. Heat is continuously transported from the hot interior to the surface. The terrestrial heat flow is the amount of energy (J) transferred through a unit surface area of 1 m² per unit time (s) and is referred to as heat flow density (q). In its general form, the Fourier equation is:

$$q = -\lambda \nabla T \quad [J\ s^{-1} m^{-2}] \quad (1.1a)$$

where λ is a material constant explained below. The general form can be rewritten for the case of one dimensional and constant temperature gradient as:

$$q = -\lambda \Delta T/\Delta z \quad [J\ s^{-1} m^{-2}] \quad (1.1b)$$

where $\Delta T/\Delta z$ is a constant temperature gradient in vertical (z) direction.

The average global surface heat flow density is about 65×10^{-3} W/m² (65 mW/m²). The planet looses heat because of this heat transfer from the interior to the surface. On the other hand, the planet gains some energy by capturing solar radiation. Electromagnetic solar radiation is created in the sun by nuclear fusion reactions that are ultimately converted to other forms of energy on the planet Earth such as coal, oil, gas, wind, hydroelectric, biomass (crop, wood), photovoltaic and solar thermal. The average global solar energy received by the Earth is 170 W/m², 2,600 times the amount lost by heat flow from the interior. This corresponds to 5.4 GJ per year per 1 m² surface area, which is approximately the energy that can be extracted from one barrel of oil, 200 kg of coal, or 140 m³ of natural gas (source: World Energy Council). The total integrated heat flow of the planet corresponds to the impressive thermal power of 40 terra Watt (4×10^{13} W).

The measured surface heat flow density has several contributions. Only a small part of it is related to the Fourier heat flow from core and mantle as described above (about 30 %). 70 % is caused by heat generated by the decay of radioactive elements in the crust, mostly in the continental "granitic" crust. Specifically uranium (^{238}U, ^{235}U), thorium (^{232}Th) and potassium (^{40}K) in the continental crust produce ~900 EJ/a (9×10^{20} J/a). Together with the contribution of the interior of $\sim 3 \times 10^{20}$ J/a, the planet looses 1.2×10^{21} J/a (1.2 ZJ/a) thermal energy to the space. Most of it is restored in the crust continuously.

Heat production in the crust is thermal energy produced per time and volume (J/s m³). The crust is composed very differently and its thickness differs considerably. Continental crust is typically thick, granitic and rich in radioactive elements, oceanic crust is thin, basaltic and poor in radioactive elements. Therefor heat production of crustal rocks differs over a wide range (Table 1.2). The total global radioactive heat production is estimated to be on the order of 27.5 TW (Ahrens 1995).

Surface heat flow q (W/m^2) composed of the heat flow from the interior and the heat production in the crust varies within a surprisingly narrow range of 40–120 mW/m^2. This is only a factor of 3. The global average of 65 mW/m^2 corresponds to an average temperature increase in the upper part of the Earth crust of about 3 °C per 100 m depth increase. Departures from this average value are designated to heat flow anomalies or thermal anomalies. The variation is caused by the different large-scale geological settings as outlined above and by the diverse composition of the crust. Negative anomalies, colder than average, are related to old continental shields, deep sedimentary basins and oceanic crust away from the spreading ridges. Positive anomalies, that are hotter than normal geotherms, are the prime targets and the major interest of geothermal exploration. Extreme heat flow anomalies are related to volcanic fields and to mid ocean ridges. In low enthalpy areas heat flow anomalies are often related to upwelling fluids (upwelling groundwater). The advective fluid flow also transports thermal energy to near surface environments.

Average heat flow density is 65 mW/m² at the surface of continents (see above) and 101 mW/m² from oceanic crust. The global average of 87 mW/m² corresponds to a global heat loss of 44.2×10^{12} W (Pollack et al. 1993). A net heat loss of 1.4×10^{12} W (Clauser 2009) of the planet results from the difference between the heat lost to space and the heat production due to radioactive decay and other internal sources. The cooling process of the planet is very slow however. During the last 3 Ga (from a total of 4.6 Ga) the average mantle cooled 300–350 °C. The heat loss by thermal radiation from the interior is minimal (by a factor of 4,000) compared with the thermal energy gained by solar radiation.

The total amount of heat (thermal energy) stored by the planet is about 12.6×10^{24} MJ (Armstead 1983). Therefore, the geothermal energy resources of the planet are truly enormous and omnipresent. Geothermal energy is everywhere available and can be extracted at any spot of the planet. Geothermal energy is friendly to the environment and it is available 24 h a day 365 days per year anywhere on the planet. Today it is used insufficiently but geothermal energy has a hot future.

1.4 Heat Transport and Thermal Parameters

A prerequisite for the design of geothermal installations is availability of data and information on the physical properties of rocks. Rock properties are required at sites of shallow geothermal installations and deep geothermal systems for heat and electricity production alike. Particularly needed are rock properties that relate to transport and storage of heat and fluids in the subsurface. Thermal properties include thermal conductivity, heat capacity and heat production; hydraulic properties embrace for example porosity and permeability. Important properties of deep fluids are their density and compressibility.

Geothermal heat can be transported by two basic mechanisms: (1) by heat conduction through the rocks and (2) by a moving fluid (groundwater, gasses), a mechanism referred to as advection. Conductive heat flow can be described by the

1.4 Heat Transport and Thermal Parameters

empirical transport equation: $q = -\lambda \, \Delta T$ (Fourier law). It expresses that the heat flux (Watt per unit area of cross section) is caused by a temperature gradient ΔT between different parts of a geologic system and that it is proportional to a material property λ called thermal conductivity [J s^{-1} m^{-1} K^{-1}]. Thermal conductivity λ depicts the ability of rocks to transport heat. It varies considerably between different types of rock (Table 1.1). Rocks of the crystalline basement such as granites and gneisses conduct heat 2–3 times better than unconsolidated material (gravel, sand). Measured thermal conductivities for the same rock type may vary over wide ranges (Table 1.1) because of variations in the modal composition of rocks, different degrees of compaction, cementation or alteration, but also because of anisotropy caused by layering and other structures of the rocks. The thermal conductivity of stratified, layered or foliated rocks depends on its direction. It is generally anisotropic. In schists, for instance λ vertical to the schistosity can be only a third or less than λ parallel to the schistosity. Thick schist formations hamper vertical heat flux from the interior to the surface and thus have an insulating effect. The positive thermal anomaly at Bad Urach (SW Germany), for example, has been associated with the presence of thick shale series in the section (Schädel and Stober 1984).

All rocks contain a certain amount of voids in the form of pores and fractures. It is crucial for the heat transport properties of the rocks if the voids are filled with a liquid fluid (water) or gas (air). Air is an isolator with a very low λ value (Table 1.1). This is why in shallow geothermal systems the position and variation of the water table has a profound effect on the thermal conductance of unconsolidated rocks.

Table 1.1 Thermal conductivity and heat capacity of various materials

Rocks/fluids	Thermal conductivity λ [J s^{-1} m^{-1} K^{-1}]	Specific heat capacity [kJ kg^{-1} K^{-1}]
Gravel, sand dry	0.3–0.8	0.50–0.59
Gravel, sand wet	1.7–5.0	0.85–1.90
Clay, loam moist	0.9–2.3	0.80–2.30
Limestone	2.5–4.0	0.80–1.00
Dolomite	1.6–5.5	0.92–1.06
Marble	1.6–4.0	0.86–0.92
Sandstone	1.3–5.1	0.82–1.00
Shale	0.6–4.0	0.82–1.18
Granite	2.1–4.1	0.75–1.22
Gneiss	1.9–4.0	0.75–0.90
Basalt	1.3–2.3	0.72–1.00
Quartzite	3.6–6.6	0.78–0.92
Rocksalt	5.4	0.84
Air	0.02	1.0054
Water	0.59	4.12

Data for 25 °C 1 bar. *Source* VDI 4640, Schön (2004), Kappelmeyer and Haenel (1974), Landolt-Börnstein (1992)

Thermal conductivity λ of air is 100 times smaller and the one of water is 2–5 smaller than that of rocks (Table 1.1). As a result the thermal conductivity of dry, air filled gravel and sand is about 0.4 J s^{-1} m^{-1} K^{-1}, however, for wet, water saturated gravel the thermal conductivity may be 2.1 s^{-1} m^{-1} K^{-1} or higher. Knowing the water table and its temporal variation is critically important for determining the heat extraction capacity of a geothermal probe (Sect. 6.3.2). This is extremely so in strongly karstified rocks.

The thermal conductivity (k) controls the supply of thermal energy for a given temperature gradient. The heat capacity (C) is a rock parameter that portrays the amount of heat that can be stored in the subsurface. It is the amount of heat ΔQ (thermal energy J) that is taken up or given off by a rock upon a temperature change ΔT of one Kelvin:

$$C = \Delta Q / \Delta T \qquad [\text{J K}^{-1}] \tag{1.2a}$$

The specific heat capacity (c) also simply specific heat of rocks (material) is the heat capacity per unit mass. It characterizes the amount of heat ΔQ that is taken up per mass (m) of rock per temperature increase ΔT:

$$c = \Delta Q / (m \Delta T) \qquad [\text{J kg}^{-1} \text{ K}^{-1}] \tag{1.2b}$$

If C is normalized to a constant volume (V) rather then mass, it is designated volumetric heat capacity also volume-specific heat capacity (s):

$$s = \Delta Q / (V \Delta T) \qquad [\text{J m}^{-3} \text{K}^{-1}] \tag{1.2c}$$

The two parameters are connected by the equation $(c = s/\rho)$, where ρ is the density (kg m^{-3}). Heat capacity and thermal conductivity depend on pressure and temperature. Both parameters decrease with increasing depth in the crust. As a consequence, for a specific material the temperature rises as depth decreases.

Table 1.1 lists specific heat capacities of common rocks. For solid rocks c typically varies between 0.75 and 1.00 kJ kg^{-1} K^{-1}. The heat capacity of water c = 4.19 kJ kg^{-1} K^{-1} is 4–6 times higher than c of solid rocks. Water stores many times more heat than rocks. Referred to the volumetric heat capacity water stores about twice the amount of heat than rocks. Consequently, highly porous aquifers of unconsolidated rock store more thermal energy than low-porosity aquifers with poor hydraulic conductivity consisting of dense rocks.

Heat flow density (q) and thermal conductivity (λ) reflect the temperature distribution at depth. The temperature gradient is the temperature increase per depth increment (grad T or ΔT) at a specified depth. Equation 1.3 shows that T at a specific given depth (for constant one dimensional gradients) is given by the heat flow density and the thermal conductivity:

1.4 Heat Transport and Thermal Parameters

$$\Delta T/\Delta z = q/\lambda \quad [K\ m^{-1}] \tag{1.3}$$

For example: With the average continental surface q = 0.065 (W m^{-2}), λ = 2.2 [J s^{-1} m^{-1} K^{-1}] for typical granite and gneiss (Table 1.1) a constant $\Delta T/\Delta z$ = 0.03 [K m^{-1}] or 3 °C per 100 m depth increase follows from Eq. 1.3. The temperature increases in the upper kilometers of the central European continental crust with 2.8–3.0 °C per 100 m of Δz, consistent with the typical mean λ-values of crustal hard rock material (Table 1.1) and the typical measured surface heat flow density of 65 mW m^{-2}. Vice versa, Eqs. 1.1b or 1.3 can be used to roughly calculate q for given T-gradients and rock material.

Temperature gradients, heat flow density and hence the temperature distribution in the subsurface is not uniform. If the deviation from average values is significant the features are termed positive or negative temperature (thermal) anomaly. There are numerous geologic causes of positive thermal anomalies including active volcanism (as described above) and upwelling hot deep waters in hydrothermal systems. Upwelling thermal waters are typically related to deep permeable fault structures often in connection with graben or basin structures or boundary fault systems of mountain chains. Hydrothermal waters commonly reach the surface and discharge as hot springs. Positive anomalies can also be caused by the presence of large volumes of rock with a high thermal conductivity such as rock salt deposits. Salt diapirs preferentially conduct more heat to the surface than other surrounding sedimentary rocks. So that high heat flow is channelized in the salt diapirs. Thick insulating strata in sedimentary sequences such as shales with low thermal conductivity (often strongly anisotropic as discussed above) may retard heat transfer to the surface. Unusually high local geochemical or biogeochemical heat production can also be a reason of heat anomalies. Positive anomalies are prime target areas for geothermal projects because their exploration and development require smaller drilling depth (Chap. 5).

All rocks contain a certain measurable amount of radioactive elements. The energy liberated by the decay of unstable nuclei is given off as ionizing radiation and then absorbed and transformed to heat. In common rocks the heat production of the decay chains of the nuclei ^{238}U, ^{235}U and ^{232}Th and the isotope ^{40}K in potassium are the only significant contributions. Uranium and thorium occur in accessory minerals, mainly zircon and monazite, in common rocks such as granite and gneiss. Potassium is a major element in common rock forming minerals including K-feldspar and mica.

Total radioactive heat production of a rock can be estimated from the concentrations of uranium c_U (ppm), thorium c_{Th} (ppm) and potassium c_K (wt %) (Landolt-Börnstein 1992):

$$A = 10^{-5}\rho(9.52c_U + 2.56c_{Th} + 3.48c_K) \quad [\mu J\ s^{-1} m^{-3}] \tag{1.4}$$

where ρ is the density of the rock [kg m^{-3}]. Some typical values for radiogenic heat production of selected representative rocks are listed in Table 1.2.

Table 1.2 Typical radiogenic heat production of selected rocks

Rock type	Heat Production [$\mu J\ s^{-1}\ m^{-3}$]
Granite	3.0 (<1–7)
Gabbro	0.46
Granodiorite	1.5 (0.8–2.1)
Diorite	1.1
Gneiss	4.0 (<1–7)
Amphibolite	0.5 (0.1–1.5)
Serpentinite	0.01
Sandstone	1.5 (0.2–2.3)
Shale	1.8

Source (Kappelmeyer and Haenel 1974; Rybach 1976)

Because radiogenic heat production is related to the amount of K-bearing minerals and zircon in a rock, granite and other felsic rocks, they produce more heat than gabbros and mafic rocks (Sect. 1.3). Mantle peridotite and its hydration product serpentinite produce less than 0.01 $\mu W\ m^{-3}$ (Table 1.2). A part of the radioactive elements can be mobilized by water–rock interaction and dissolve in hydrothermal fluids. Some thermal waters contain a considerable amount of radioactive components and are thus radioactive (Sect. 10.2).

The heat equation describes the variation of temperature in a rock in space and time (Carlslaw and Jaeger 1959). Solutions to the equation depict the distribution of heat in the subsurface and its variation with time. The partial differential heat equation can be written as:

$$\partial(\rho cT)/\partial t = \nabla(\lambda \nabla T) + A - v\nabla T + \alpha g T/c \qquad (1.5)$$

where the first term on the right hand side of the equation describes the heat conduction (see also Eq. 1.1a), A stands for the depth and material dependent internal heat production [$J\ s^{-1}\ m^{-3}$], the third term describes advective heat transfer (generally mass transfer) and the last term expresses the pressure effect with the density ρ (kg m^{-3}), v velocity (m s^{-1}), g acceleration due to gravity [m s^{-2}] and α [K^{-1}] the volumetric linear coefficient of thermal expansion defined by $\alpha = (1/V)\partial V/\partial T$. For most rocks $\alpha = 5$–$25\ \mu K^{-1}$.

The analytical solution of Eq. 1.5 for one dimensional heat transport (along depth coordinate z), constant thermal conductivity (λ), constant radiogenic heat production (A), for a homogeneous isotropic volume of rock, no heat transport by mass flux and ignoring the pressure dependence is:

$$T(z) = T_0 + 1/\lambda q_0 \Delta z - A/(2\lambda)\Delta z^2 \qquad (1.6)$$

where T_0 is the temperature at z_0 the top of the considered volume of rock, q_0 the heat flow density at z_0 and Δz the thickness of the considered rock volume. This simplified heat equation (Eq. 1.6) can be used to construct a thermal profile through the crust by adding layer by layer for the case of conductive heat transport and radiogenic heat production in the individual layers.

1.5 Brief Outline of Methods for Measuring Thermal Parameters

Thermal conductivity of rocks can be measured on drillcores in the laboratory or in situ in boreholes directly. There are different methods and types of devices for measuring the thermal conductivity of rocks and soils on the market. All are based on the same principle: the sample is exposed to defined and controlled local heating and temperature sensors measure the temperature response to heating in space and time. The transient line source method is widely used in needle-type measuring instruments. A long and thin heating source is brought in contact with the sample and is heated with constant power, while simultaneously the temperature of the source is registered. The slower the source temperature rises, the higher is the thermal conductivity of the sample material.

Probably the most commonly used method for thermal conductivity measurements in geology is the use of so-called divided bar instruments. The instruments are commercially available also as portable electronic divided bar machines. Portable divided bars apply thermal gradient across a sample along with a substance of known thermal conductivity used as standard. Thermal conductivity of the sample is measured by the device relative to the standard.

Thermal conductivity measuring bars have differential temperature adjustment provisions and provide accurate results with a variance of only 2 %. Portable divided bars can be easily calibrated and weigh only 8 kg facilitating easy transport. Divided bar systems also generate less noise and can be used to measure thermal conductivity of fresh core samples even during remote drilling operations. Additionally, these rock thermal conductivity-measuring bars can provide readings for varying temperatures over a range of 20 °C. Portable thermal conduction measuring devices are very useful in geothermal energy explorations (web page: Hot Dry Rock, Australia).

The heat capacity of rocks is measured with a calorimeter in the laboratory. There are a large variety of calorimeters and the various instruments are used for very different purposes. The parameter C defined in Eq. 1.2a is measured with instruments that add or remove a defined amount of heat to the calorimetric system (sample plus embedding material, usually a liquid) and monitor the temperature response of the process.

The density of rocks in the form of drillcores is measured using Archimedes's principle. This means the mass of an irregularly shaped body like a piece of rock, is first measured by a balance. Then the mass of the body is submerged in a liquid of known density (e.g. water 1,000 kg m^{-3} at about 25 °C and 1 bar) and measured by the balance. The volume of the sample follows from the difference of the two measurements, thus the density $\rho = m/V$ can be calculated from the data. The density of cuttings is measured with a pycnometer, a simple laboratory device for measuring densities of liquids and solids.

Chapter 2
History of Geothermal Energy Use

Huaqin Hot Springs near Xi'an, China

Geothermal energy, heat from the interior of the planet Earth, has been utilized by mankind since its existence. Hot springs and hot pools have been used for bathing and health treatment, but also for cooking or heating. The resource has also been used for producing salts from hot brines. For the early man the Earth internal heat and hot springs had religious and mythical connotation meaning. They were the places of the Gods, represented Gods or were endowed with divine powers. In many modern societies bathing in hot spring spas has still preserved the meaning of a divine ceremony.

Natural springs, where water emerges from the underground, have been symbols of life and power in all religions and civilizations. The mythical significance of springs producing hot and highly mineralized water from which minerals precipitate and form sinter, crusts and unusual mineral deposits was and still is immense.

Thermal springs had a religious and social function from early on. Godly healing power has been attributed to hot springs, where gods were near. Thermal springs and spas were centers of cultural and civilization development. In the Roman Empire, the middle Chinese Dynasties and the Ottoman Empires spas have been centers of balneological use of hot springs, where physical health and hygiene (modern term: wellness) have been combined with cultural and political conversation and progress of the time.

Natural hot springs (onsen) are numerous and highly popular across Japan. Every region of the country has its share of hot springs and resort towns, which come with them. There are many types of hot springs, distinguished by the minerals dissolved in the water. Different minerals provide different health benefits and all hot springs are supposed to have a relaxing effect on your body and mind. Hot spring baths come in many varieties, indoors and outdoors, gender separated and mixed, developed and undeveloped. Many hot spring baths belong to a ryokan, while others are public bath houses. An overnight stay at a hot spring ryokan is a highly recommended experience to any visitor of Japan.

Hot springs have been (and still can be) regarded as godly messengers of the immense energies stored in the subsurface of planet Earth.

2.1 Early Utilization of Geothermal Energy

Archeological finds prove that North American Indians utilized geothermal springs several thousands of years ago. Hot springs of South Dakota (USA) have been battlegrounds among Sioux and Cheyenne tribes. Healing powers from the deep interior of the Earth have been attributed to the hot waters from the springs. A bathtub carved into the rocks at the springs witnesses the use of the waters by the Indians for therapeutical bathing. They also drank hot spring water to cure gastro-intestinal health problems. Later, white settlers started to use the hot springs for balneological purposes commercially. Today, the hot water is utilized for cooling and heating purposes with the assistance of heat pumps. Similar "Indian Hot Springs" are found along Rio Grande in Texas and in Mexico. The natives of North America have also used them for therapeutical purposes and for bathing in rock pools since time immemorial. Several thousand thermal springs are known in the USA.

Fig. 2.1 Fishing Cone Geyser in Yellowstone Lake (Yellowstone National Park, USA), (Photograph: US gov)

A peculiarity is Fishing Cone Geyser submerged in water near the Shore of Yellowstone Lake, which has been used for cooking fish by fishermen (Fig. 2.1). The small crater had been above water surface of the lake for some time and the fishermen held the rods with the still flouncing fish for cooking into the boiling and steaming small crater either from the boat or from the beach. Today Fishing Cone Geyser submerged in the lake water and the hot water eruptions stopped.

Historical written documents by the Romans, Japanese, Turks, Icelander, also from Maori in New Zealand describe the occurrence and utilization of hot springs for cooking, bathing and house heating. About 2,000 years ago, bathing and treatment centers have been erected at the hot springs Huaquingchi and Ziaotangshan near Beijing in China.

About 3,000 years ago, gods of Greek civilization have been associated with thermal and mineral waters and their healing power. In the 3rd to the 1st century B.C. Celts worshipped springs with healing power, e.g. the thermal springs of Teplice in Northern Bohemia. Bath in Southern England is associated with the cure of king Bladud, the father of King Lear, from leprosy in 863 B.C. Bath are the waters of Sul, the god of wisdom.

The Celts and then particularly the Romans demonstrably extensively utilized thermal springs in central Europe. Already more than 2,000 years ago, the Romans heated their baths with geothermal energy. It is proven that the Romans settled preferably in the vicinity of thermal springs from 2nd century B.C. Examples

Fig. 2.2 Ruins of the Roman bathing facility at the thermal springs of Badenweiler in the Rhine rift valley (southern Germany)

are Aix-en-Provence (Aquae Sextiae), Bagnière de Luchon in the Pyrenees, Wiesbaden Germany (Aquae Mattiacorum), Baden–Baden Germany (Aquae Aureliae), Badenweiler (Aqua Villae) (Fig. 2.2) and many other places. No other epoch of the western civilization celebrated bathing and bathing culture with more delight than the classic Roman period. "Sanus per aquam" healthiness through water was the motto of the Romans. Bathing was the most important pastime of the Romans. Wellness has been a central aspect of their lifestyle; bathing was a feast for all senses. The bath was the place for social gathering, used for business affairs and for sports.

In Roman times established spas offered regular bathing programs, which were fundamentally related to the believe in gods that were responsible for health. Liable for the success of a treatment were primarily the gods of the local springs such the Celtic-Roman god Apollo-Grannus and not so much the well-trained balneologists. In Roman spas cured patients donated sanctified platelets to express gratitude for the celestial accomplishment.

The hot springs of Badenweiler in the Black Forest (Germany), as an example, have been used by the Celts (known from coin finds). Shortly after the Roman conquest of the lands East of the Rhine river at the end of the first century A.C. The invaders raised a civil settlement and a bathing house (Fig. 2.2). During Roman times the water must have been significantly warmer than today's 26.4 °C, because the Romans built the large bathing halls without heating systems (Cataldi 1992). Also the mineralization of the water was probably higher than today even in the year 1560 according to the "spa travel guide" (Badenfahrtbüchlein) of

2.1 Early Utilization of Geothermal Energy

Fig. 2.3 Thermal spa Baden–Baden, Roman soldier spa, underfloor heating system, first geothermal heating system

Georgius Pictorius. After the withdrawal of the Romans the spa sunk into oblivion. It was rediscovered and unearthed in 1784.

The roman settlement Baden–Baden, Aquae Aureliae, in the foothills of northern Black Forest can be traced back to the first century. It developed into an important administrative town during the 2nd and 3rd Century. Aquae Aureliae was a flourishing town in the Roman province Germania Superior. The roman city centered on the curative thermal springs, which were the source of the economic success and importance. The luxury imperial spa built by order of the roman emperor Caracalla is located underneath today's market square of Baden–Baden. The spa was destroyed in the year 260. The distinctly more frugal soldier spa is situated at some distance from the imperial spa. The extremely comfortable roman spas were technically highly sophisticated and very cultivated institutions. The spas were built with a so-called hypocaust system (hypocaustum) of central and underfloor heating, in other words with a geothermal heating system (Fig. 2.3). The Romans used the spas wearing wooden sandals protecting them from the hot floors.

Many of the spas have been abandoned after the retreat of the Romans from large areas of Europe. The early Christians preferred to build the first churches close to curative hot springs that have been used from ancient times. In central

Europe of the Middle Ages thermal springs had such an enormous importance that e.g. Charlemagne (Charles the Great) expanded the imperial seat in Aachen to his palatinate and in the year 794 declared it to his permanent residence. The thermal springs of Aachen have already been used by the Celts and the Romans but have fallen into oblivion for several hundred years. The legend says Charlemagne was on a hunting trip in the vicinity of Aachen, in midst of overgrown remains of Roman times. The horse of the sovereign got stuck in a swamp. Charlemagne realized the sludgy water was hot and that steam emerged from the soil. Charlemagne has re-discovered the hot springs of Aachen.

The thermal spas southeast of Oradea in medieval Transylvania have been established at the hot springs of Peta River. The waters of Peta have later also been used as "defrost liquid" by directing them to the castle moat around the fortress of Oradea to prevent the water from freezing and to maintain the functionality of the moat.

In Chaudes-Aigues in central France, construction of the first district heating system, still functioning today, has been commenced in the 14th century (Lund 2007).

Most of the old roman spas were re-discovered in the 13th and 14th century. The big boom of the European thermal spas, however, started not before the 18th century. The spas developed to meeting places of the nobility, aristocracy and the rising bourgeoisie. The first scientific studies on the therapeutic use of thermal spas and the chemical composition of the waters have been written by the monk Savonarola and by the anatomist Fallopio in the 15th and 16th century.

The first reports from China on thermal springs including therapeutic instructions and farming guides go back as far as the 4th to the 6th century. For example, the diversion of thermal water to the fields for rice crop permits the first harvest already in March and allowed for three harvests in the year. The pharmacologist Li Shizhen has written the first scientific review of mineral and thermal waters in China in the 16th century. In his book, "Compendium of Materia Medica" he classified the waters on the basis of chemical and genetic criteria.

In 1560 Georgius Pictorius published an account of the spas of southern Germany ("Badenfahrtbüchlein") and instructions how to use them. It represents a first balneological treatise. Georgius Pictorius studied medicine at the University of Freiburg and was later regionally well known for his medical essays. He had studied all relevant experts on therapeutical bathing of the Antique and the Middle Ages. In his "Badenfahrtbüchlein" he described all classic spas in southwestern Germany that all are still in use today one by one.

Early experience with geothermal phenomena has also been reported from the mining industry. Agricola realized in 1530 already that the temperature in underground mines increases with depth. The first reported temperature measurements with a thermometer are probably those by De Gensanne in 1740 in a mine near Belfort France. Alexander von Humboldt measured a temperature increase of 3.8 °C per 100 m depth increase in the mining district of Freiberg, Saxsony, in the year 1791. This was the first report on the concept of the geothermal gradient, a fundamental parameter in geothermal energy exploitation. Its existence and variation was rapidly confirmed by data from Central and South-America. In Germany, temperature measurements in deep drill holes up to 1,000 m depth were carried out in the years 1831–1863. A few years later, measurements down to 1,700 m followed. An average temperature increase of 3 °C per 100 m emerged from the rapidly increasing volume of data, which is known today as the normal temperature gradient. The first measurement of the surface heat flow density has been achieved by Benfield (1939).

A surprisingly high temperature of 38.7 °C has been measured at the bottom hole of the 342 m deep drillhole Neuffen in Southern Germany in the year 1839. This corresponds to a geothermal gradient of 9 °C per 100 m. The first large geothermal temperature anomaly had been discovered.

2.2 History of Utilization of Geothermal Energy in the Last 150 years

Using thermal water for energy conversion did not start before the second half of the 19th century related to the rapid development of thermodynamics. Thermodynamics helped to efficiently convert energy from hot steam first in mechanical energy and then into electrical energy with the help of turbines and generators.

The development of geothermal power generation is clearly associated with the Larderello region of Tuscany in northern Italy (Tiwari and Ghosal 2005). Until the early 19th century the thermal springs near Larderello have been used for the production of boron and other substances dissolved in the thermal water. In 1827 Francesco Larderel, the founder of the boron industry, installed the first plant for geothermal energy conversion. One of the hot water ponds has been covered with a brick cupola. The construction was the first low-pressure steam boiler heated naturally with geothermal water. It produced the heat needed to evaporate the boron-rich water for the production of boron and additionally also powered pumps and other machines. The installation saved large amounts of firewood and the deforestation of the region could be brought to an end. In the year 1904 the first electrical power was produced from a geothermal energy source by coupling a steam engine to a generator in Larderello (Fig. 2.4).

When the first Larderello power plant went into operation in 1913 it already had an electrical power of 250 kW. In 1915 the power station had power of 15 MW and was driven by saturated steam. From the year 1931 on, new deep drillholes produced superheated steam for the electrical power plant with a temperature of 200 °C. Superheated steam did not contain constituents that cause corrosion and scale formation in contrast to saturated steam. The installation of heat exchanger systems was therefore not necessary. In 1939 the total installed power of all Larderello power plants was up to 66 MW. The Italian geothermal fields were destroyed at the end of WWII but rebuild after the war. Today 545 MW electrical power is installed at the Larderello plants, 1.6 % of the total electrical energy production in Italy (2010).

Fig. 2.4 Lardarello 1904: The picture shows Principe Piero Ginori-Conti with his apparatus that converted geothermal to electrical energy for the first time in history. The installation had the power to light fife light bulbs (Photograph: Unione Geotermica Italiana 2010)

The Larderello geothermal fields are caused by shallow level igneous intrusions at the convergent plate margin of the Apulian and Eurasian plates beneath Tuscany. Extremely high geothermal gradients result from the shallow magma chambers.

In 1890, early systematic geothermal heat utilization was accomplished in Boise, Idaho, USA by completing a district heating system. This system was copied in 1900 by Klamath Falls, Oregon, USA. Later, in 1926, Klamath Falls started to use a geothermal well to heat greenhouses. The first private homes were geothermally heated from separate wells in Klamath Falls in 1930.

The utilization of thermal water for heating homes and greenhouses started in the Reykjavik, Iceland, on a large scale in the 1920s. The name Reykjavik, steaming bay, was given by the Vikings because of the visibly steaming thermal springs. The first wells were drilled into hot water reservoirs for heating buildings as early as in the middle of the 19th century. Geothermal heating of public buildings and entire city districts followed.

Today, Iceland is clearly number one in utilization of geothermal energy in the world. 79,700 TJ or 53 % of primary energy is supplied by geothermal sources. Geothermal and hydroelectric energy provide 99.9 % of the country's electrical energy demand. Low-enthalpy geothermal fields near Reykjavik supply water with temperatures of up to 150 °C which can be used in house heating systems. More than half of the Iceland's population lives in the area. Geothermal fields provide heat and hot water for 90 % of the Icelandic households. The high-enthalpy fields are located along the active volcanic belt that crosses the island. Typical temperatures are 200 °C, but these waters are highly mineralized and gas-rich and cannot

Fig. 2.5 Nesjavellir Power Plant on Iceland; 120 MW electrical power plus 380 MW thermal power from 83 °C water in 2010, reservoir temperatures up to 380 °C (*sources* http://de.academic.ru/dic.nsf/dewiki/1010201 and http://www.or.is/media/PDF/Nesjavellir%20ENS%2007.pdf)

be used directly. The diverse power plants produce typically some 10 MW electrical power in steam turbines. The Nesjavellir plant in the southwest of Iceland is the largest electrical power plant on the island. It produces about 120 MW power. It uses the volcanic heat of the central volcano Hengill as well as the heat from springs and drilled wells (Fig. 2.5). The thermal waters of Iceland are employed in many different branches of the industry.

Following the development in Italy and on Iceland, in 1958 New Zealand erected its first geothermal plant in Wairakei; in 1959 an experimental facility started in Pathe, Mexico, and in 1960 northern California initiated the project The Geysers. Today, The Geysers comprise 21 power stations with a total installed capacity of 750 MW electrical power. It is the largest geothermal installation in the world. The produced electricity is sufficient to supply a city of the size of San Francisco.

However, severe setbacks occurred too. The profitability of geothermal energy production is subject to general economic conditions, such as demand, supply and price of other forms of energy for instance crude oil. Changing laws and environmental regulations may cause increasing efforts and costs (Chap. 10). Greece and Argentina, for instance, shut down existing geothermal installations due to environmental and economic reasons. Germany's deep wells for geothermal installations were drilled in the 1980s of the last century following increased oil and gas prices. The further development of deep geothermal systems came to a halt during the economic crisis and the associated collapse of the oil price. Resumption of geothermal energy projects follows the price of dwindling fossil fuel resources.

In the year 2003 the first electrical energy production from a geothermal source in Germany started in Neustadt-Glewe. 2007 the geothermal wells Landau and 2009 the wells in Bruchsal which were drilled in the 1980s already, started to produce electrical energy.

The earliest documented drilling for ground source heat pump systems in central Europe has been completed in the late summer of 1974 in Schönaich, southern Germany. For retrofitting an existing building (from 1965) with a ground source heat pump as the exclusive heating system five ground loops of 50–55 m depth have been installed with a distance of 4–5 m between the wellbores arranged in a linear array of five coaxial probes with thick-walled steel tubing (60 × 5 mm) and a coaxial plastic hose. The probes were loaded with a water-glycol mixture. Grouting of the annulus with a cement-bentonite suspension, a standard procedure today, had not been carried out at that time. Supply water temperatures in the probes were -3 to -4 °C for peak load periods (continuous outside temperatures of -15 to -20 °C during several weeks); return temperature was about $+1$ °C. The system was in operation for 30 years. One of the probes failed in 2005 probably because of a corrosion damage. Now the system runs with four probes and an oil-fired boiler.

In 1852, Lord Kelvin has invented the heat pump, a crucial piece of equipment for utilizing near surface geothermal energy. Heinrich Zoelly filed a patent application in 1912 to use a heat pump for extracting heat from the subsurface. The first successful implementation of a ground source heat pump system occurred not before the 1940s. These first ground source heat pumps (GSHP) in Indianapolis, Philadelphia and Toronto had ground collectors that have been emplaced close to the surface. An experimental installation of the Union Electric Company in St. Louis used spiral pipes in 5–7 m deep drill holes as heat exchangers. Other early systems such as in an administrative building in Zurich 1938 and the Equitable Building in Portland in 1948 used river- or groundwater as a heat source, thus they are not utilizing geothermal energy in a strict sense.

The US Department of Energy published in 2010 a comprehensive series of four books downloadable as PDFs from their website on the history of geothermal energy development in the USA in the Geothermal Technologies Program: "A History of Geothermal Research and Development in the United States". The series covers the years 1976–2006. A brief history of geothermal energy in the US can be found at http://www1.eere.energy.gov/geothermal/history.html.

Chapter 3
Geothermal Energy Resources

Natural thermal water spring Da Qaidam, China

3.1 Energy

In physics, energy is the ability of a physical system to do work on other physical systems. There are many different forms of energy including mechanical (potential, kinetic), thermal, electric, chemical and nuclear energy. Thermal energy can be understood as the random motion of atoms and molecules.

The different forms of energy can be converted from one to another. E.g., chemical energy is converted to mechanical energy in a combustion engine. Solar heat radiation is converted in a photovoltaic system into electrical current.

The distinction between renewable and non-renewable forms of energy resulted from the increasing insight that natural energy resources are limited. Non-renewable kinds of energy, also called fossil energy resources, include coal, oil, gas and nuclear fuels (e.g. uranium). These forms of energy renew on time scales that are not interesting for the present day human economic system.

Solar energy is considered a typical representative of renewable energy. Radiation from the sun produced by solar nuclear processes is unlimited and last forever from a human perspective although the processes on the sun will stop when all nuclear fuel has been used up. The radiation energy reaching the Earth can be transformed into electricity (photovoltaic) or heat (solarthermic). Wind energy, hydroelectric energy and biomass (wood, energy plants) are also ultimately derived from solar energy. These forms of energy are limited only by the amount of radiation reaching the Earth from the sun provided they are used in a sustainable manner. The energy is renewable in the sense that the sun replenishes the energy consumed in the form of, for example fire wood and the potential energy of water in a reservoir every day and for "quite some time" (billions of years, although humans will not be able to enjoy this). Note that also the fossil "non-renewable" forms of energy such as coal and oil represent stored solar energy. Also these energy forms are renewed although at time scales not interesting to short lived human economic processes.

Geothermal energy, the heat in the interior of the Earth is an energy that is not related to the solar energy but ultimately has been created by gravitational energy and radioactive decay of unstable atoms. It is renewable in the sense that there is a very large amount of heat stored in the body of the planet and the human consumption cannot deplete the energy reservoir. Consumed geothermal energy is renewed and replenished from the internal planetary reservoir and is unlimited from a human perspective if used sustainably. Using renewable energy resources sustainably means that the rate of consumption is equal or smaller to the rate of the renewing process. Renewable energy is characterized by renewing processes that are fast in human timescales.

From the laws of physics (law of thermodynamics) we know that energy cannot be created or destroyed. Energy can only be transformed from one form into one other. The total amount of energy remains constant and nothing is lost. Merely the usability value of one form of energy can be reduced during transformation and transport.

3.1 Energy

Many forms of energy cannot be directly used. For practical applications, they must be transformed into usable forms. Chemical energy, nuclear energy and radiation energy must be converted into mechanical, thermal or electrical energy.

Energy can be stored and it can be transported. Energy sources are stored in specific storage media. For example, typical fossil energy sources of the Earth including coal, oil and gas ultimately can be regarded as storage media for solar energy. In engineering the purpose of storing energy is to make it retrievable on demand and transportable. For example, chemical energy can be stored in transportable batteries and converted to electrical energy to drive an appliance when and where needed.

Heat that has been generated e.g. in a solar installation can be stored in a heat accumulator and utilized when the sun temporarily does not shine. Heat storage media are normally liquids (often water) or solids (typically rocks). Water is a preferred storage material because of its relatively high heat capacity (Sect. 1.4). In order to prevent unwanted heat loss and rapid cooling the storage device needs to be thermally insulated. In addition to the conventional heat accumulators that operate at elevated temperature (sensible heat accumulators), latent heat accumulators utilize the latent heat associated with phase transitions to store thermal energy. The accumulator material begins to (e.g.) melt at the temperature of the phase transition. Further heating, however, continues to melt the substance but the temperature does not change until all solid is transferred to liquid (equilibrium melting). Latent heat accumulators store considerably more energy than sensible heat accumulators do. The property is referred to as a higher energy density.

For example, the energy required to heat water from 0 to 80 °C is equivalent to melting ice of 0 °C. The energy that must be invested to convert water to steam at 100 °C corresponds to 5.4 times the energy to heat the water from 0 °C to the boiling point of 100 °C at 1 bar.

Primary energy must be converted to net energy before it can be consumed by the user. For the production of 1 KWh electrical energy typically 3 KWh primary energy such as coal and petroleum is needed. A part of the losses during energy conversion is intrinsic and cannot be avoided. The ratio of usable net energy to spent primary energy characterizes the efficiency of the energy conversion (Sect. 4.2).

Conversion losses arise, for example, in industrial conversion of chemical energy into thermal energy such as in a coal fired power plant or a domestic oil heating system from incomplete reaction due to practical and technical conditions. Thermal energy is lost from a heat reservoir because of conductive heat flow. Therefore, heat cannot be stored in a home or a building for long periods. In this context, the thermal insulation of a house is a crucial construction feature. A typical home in central Europe without special thermal insulation requires more than 20 L heating fuel per m^2 per year. In insulated low-energy buildings heating fuel consumption reduces to 7 L/m^2 per year. A state-of-the-art passive house burns only 1.5 L fuel per m^2 and year for covering its heat demand.

Entropy and the second law of thermodynamics have the consequence that the directions of energy conversion processes are not equivalent. For example, mechanical kinetic energy can be completely converted into thermal energy. In the opposite direction, the conversion is always incomplete; the conversion of thermal energy into mechanical energy has an intrinsic conversion efficiency smaller than one. Energy conversion processes are anisotropic or asymmetrical.

Geothermal energy or geothermal heat is the thermal energy stored below the surface of solid earth. Geothermal heat recovered from different depths below the surface provides unique and different possibilities of utilization. Consequently, geothermal heat use is subdivided into near-surface geothermal systems and deep geothermal energy systems.

3.2 Significance of "Renewable" Energies

The economic development of the energy sector depends also on factors that are not immediately discerned as relevant to the energy industry. The factors include population development, number of households, general economic trends, structural change and technological advance. In addition, economic parameters, legal framework and political environment further influence the growth of energy consumption and set guidelines for general development of energy use and consumption.

The known global reserves of conventional energy commodities such as fossil fuels and nuclear fuels are probably in the order of 35,000 EJ (Exajoules = 10^{18} J; 35 ZJ = 35×10^{21} J; IEA 2012; EIA 2012). This amount corresponds to 90 times the total global consumption of primary energy in 2007. Coal and lignite account for about 60 % of the reserves. Dividing the reserves by the current production results provides the statistical range of a particular energy source. The estimated ranges using available data from 2007 are: 42 years for crude oil, 61 years for natural gas, 129 years for coal and 286 years for lignite (EIA 2012). The statistical range for nuclear fuel (uranium) based on estimated reserves in 2005 is about 70 years (EIA 2012).

The ranges given above represent snapshots and are, although afflicted by large uncertainties, meaningful only at the time of the estimate. The actual "true" range of a resource mostly depends on the price the consumer is willing to pay for it. It is a question of the market influenced by the price of competing resources and technologies among many other factors. The number of years until all oil etc. is "used up", however, indicates that the commodity has a finite range because it is a fossil fuel that is formed by geological processes that are much slower than production and consumption by humans.

Total global energy consumption increased dramatically during the last 50 years parallel to the exponential growth of world population. Many competent sources predict an increase by three times in global energy consumption and an increase in world population from 6 to 10 billion people in the next 50 years. In 2003, the total annual energy consumption per capita in the USA has been 330 GJ (330×10^9 J; World Resources Institute). If the future 10 billion inhabitants of this planet were allowed to use the same amount of energy as the US citizens used in 2003 then the 2005 reserves would be "used up" in about 10 years. This little exercise shows that "renewable" energy utilization must be dramatically developed in the near future if a high standard of living is a goal for all inhabitants of the planet. Parallel to the development of "renewable" energy, energy must be used more efficient, conversion losses must be reduced, energy saving must be made attractive. Energy efficiency requires new inventive technologies. Research and development of this knowledge calls for time and money.

In recent years, the change of global climate and the associated increase of the mean annual surface temperature has become a major concern. It is evident that the major cause of global warming is linked to the burning of fossil fuels and the release of carbon dioxide and other greenhouse gases to the atmosphere produced in the process. The greenhouse effects of anthropogenic CO_2 emissions contribute 50 % of the observed global warming. There are many unwanted and on a long run very costly effects of global warming. The effects include displacement of vegetation zones, thawing of permafrost, melting of continental ice caps and associated sea level increase, melting of glacier ice in Alpine mountain ranges and associated effects on water and energy supply for large areas in mountain forelands, and predicted increase of extreme weather conditions.

The operation of electrical power plants and power producing systems utilizing "renewable" energies, such as photovoltaic, hydroelectric and geothermal systems is completely or nearly free of green house gas emissions. Thus, "renewable" energies are not only of vital importance for saving fossil fuel resources but also for preserving environmental conditions.

The declared goals of environmental politics of many countries include a significant increase in "renewable" energy utilization for electrical power production and in the total energy consumption (electrical power, heat, mobility). At the same time, major efforts are made to improve the energy efficiency of existing power plants. This leads to a reduction of imported energy and energy commodities, an improved flexibility of the energy supply system and an increased security of supply.

3.3 Status of Geothermal Energy Utilization

The USA have an installed capacity of 3.1 GW_{el} from geothermal sources and it is the leading country in geothermal energy utilization for electrical power production. In all Europe, about 1.4 GW_{el} electricity is produced from geothermal systems (Antics and Sanner 2007; IEA-GIA 2012). The situation differs from country to country depending on the available natural resources and the technology to use these resources. Deployed deep geothermal systems range from steam production from high enthalpy heat reservoirs, for example in Iceland and Italy, to direct use of hydrothermal reservoirs in deep sedimentary basins. Near surface geothermal energy systems, usually borehole heat exchangers or geothermal energy probes or ground source probes can be installed practically everywhere (Chap. 4).

In the year 2008, geothermal energy has been used for the production of electrical power in 24 different countries. The total capacity installed amounted to 10.9 GW$_e$ (Bertani 2007, 2010; IEA 2009; IEA-GIA 2012). For the most part, dry steam and flash steam systems produce this electrical power from high enthalpy reservoirs that are characterized by high temperatures at shallow depth (very large geothermal gradients). These open systems use the produced geothermal aqueous fluid directly to drive a turbine for electrical power generation. Less common are closed systems that utilize heat from produced geothermal fluid to drive a turbine in a secondary loop in low-temperature reservoirs.

Dry steam and flash steam systems are for example installed in the following countries: USA, Philippines, Mexico, Indonesia, Italy, Iceland, Russia (Kamchatka, Kuril Islands), Turkey, Portugal (Azores) and France (Guadeloupe).

Electrical power production from low-temperature geothermal reservoirs using binary systems like organic ranking cycle plants (ORC) or Kalina plants started few years ago and is operative at a relatively small number of sites. The systems have a substantial potential and many locations are well suited for binary low-T systems. At present, many projects are in the development stage. If consistently extended, these systems have great ability to contribute significantly to the electrical power and heat production in the future.

After the year 1975, the utilization of geothermal energy continuously and markedly increased (Fig. 3.1). In the period from 1980 to 2005, the global installed capacity increased continuously by about 200 MWe per year. After 2005, the annual growth boosted to 500 MW$_e$. In 2008, the USA had with 3,040 MW$_e$ the highest installed geothermal capacity worldwide. It was followed by Indonesia (992 MW$_e$),

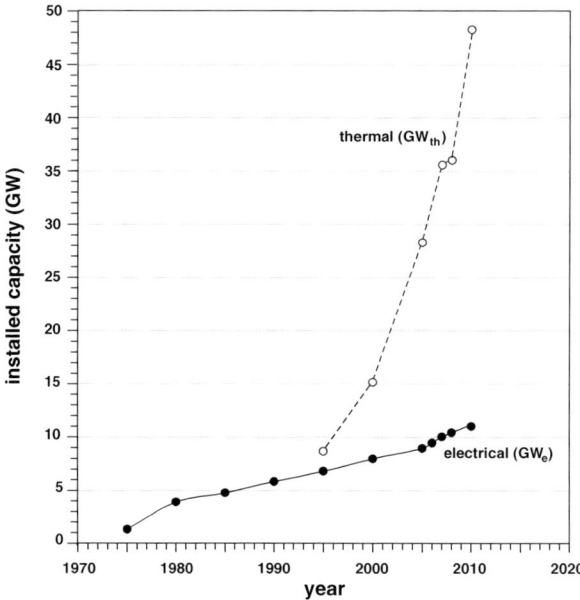

Fig. 3.1 Worldwide installed electrical and thermal power capacity from geothermal energy sources since 1975 (IEA 2009)

3.3 Status of Geothermal Energy Utilization

Mexico (958 MW$_e$), Italy (811 MW$_e$), New Zealand (632 MW$_e$), Iceland (575 MW$_e$) and Japan (535 MW$_e$). The 2012 number for the US is 3187 MW$_e$ (GEA 2012a, b).

In 2005, geothermal energy was utilized for heating purposes in 72 different countries. An estimated 45.8 GWth has been the total installed thermal capacity in 2010 (IEA 2009; IEA-GIA 2012). The global capacity doubles every five years (Fig. 3.1).

These numbers vary between different countries and strongly depend on geological conditions. For example, utilization of deep geothermal energy is in its infancy in Germany compared with Iceland, USA or New Zealand, because of the absence of high enthalpy fields in central Europe. Nevertheless, also in Germany and other geologically less favorable countries geothermal energy is increasingly utilized and has a promising marked. German governmental estimates expect that geothermal energy contribute 0.5 % to the electrical power production and 0.9 % to the useful heat production by the year 2020 and that these portions rise to 3.1–7.7 % respectively until 2050.

3.4 Geothermal Energy Sources

Subsurface temperatures in the uppermost meters of the crust are mainly controlled by climate. In wintertime, the ground can be frozen to one meter depth in moderate climate zones and considerably warmed up during summer. Heat input

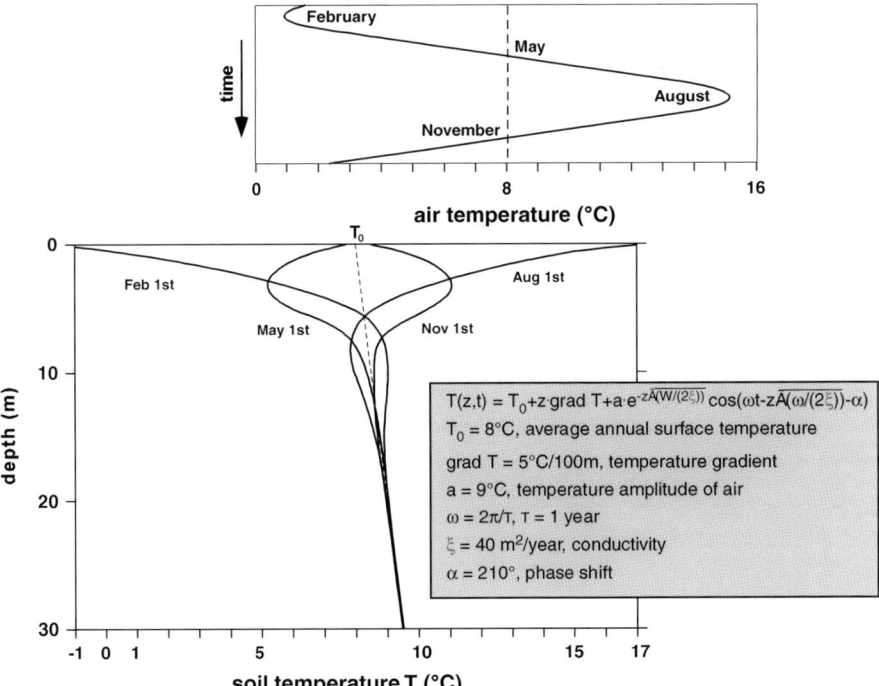

Fig. 3.2 Annual air and ground temperature variation in temperate zones (Lemmelä et al. 1981)

occurs directly by solar radiation and indirectly by heat exchange with air and infiltrated precipitation water.

The seasonal ground temperature variations decrease with depth. In moderate climate zones, the annual cycle disappears at 10–20 m depth. At this depth, temperature is constant throughout the year and its value corresponds closely to the local long-term average surface temperature (Fig. 3.2). Climatic effects at extended time scales such as ice ages are perceptible to greater depths (e.g. 200 m in central Europe). The consequences of the ice ages for the local geothermal gradients are still visible today. With increasing depth, temperature increases because of terrestrial heat flow and according to the local geothermal gradient. A large portion of the geothermal energy stored in these increasingly hot rocks is generated in the crust itself (Sect. 1.5).

Geothermal industry distinguishes between near-surface and deep geothermal energy utilization (Chap. 4). A notional boundary at 400 m depth and 20 °C separates the two quite different fields of geothermal energy uses. Deep geothermal energy utilization distinguishes furthermore between high enthalpy and low enthalpy reservoirs. The thermodynamic potential enthalpy reflects the heat content of material. Its symbol H stands for heat content (unit: Joule J). The distinction between the two types of reservoirs is at an imaginary dividing temperature of 200 °C.

Electrical power can be produced with high efficiency directly from steam turbines where steam is produced from high enthalpy reservoirs (in high enthalpy fields). High temperatures of more than 200 °C are required for the necessary steam pressure using water as heat transfer material. Producing electrical power from low enthalpy reservoirs is only possible with heat transfer substances with higher vapor pressure. Organic Rankine Cycle (ORC) plants use e.g. pentane and Kalina cycle plants ammonia-water mixtures as heat transfer substances (Sect. 4.2). The electrical efficiency of such plants varies between 10 and 15 % depending on transfer material and operation temperature.

The high enthalpy fields of the planet are typically located along volcanic belts related to tectonic plate boundaries (Sect. 1.2) but also to intraplate volcanic fields related to mantle plumes (or combinations thereof like on Iceland). Some high enthalpy fields are also related to hydrothermal convection linked to shallow level magma chambers and near-surface igneous intrusions in the crust. Many plutonic rocks indicate crystallization pressures of 50–100 MPa corresponding to intrusion depths of 1.5–3 km and crystallization temperatures in excess of 650 °C. Geothermal gradients in such regions can be extremely high and temperatures of up to 400 °C can be reached at very shallow depth of a few hundred meters below surface. In high enthalpy regions, the production of electrical power from geothermal sources is mature technology and well established. The electrical power consumption of San Francisco is supplied by geothermal power plants by nearly 100 %. On Iceland the electrical power production from geothermal sources exceeds local consumption leading to the establishment of new power consuming industries. Even export of electricity from Iceland to Europe via subsea cables is considered a viable project.

3.4 Geothermal Energy Sources

In deep geothermal systems, high and low enthalpy fields, the geothermal fluid that transports heat from the reservoir to the surface is natural liquid water or steam depending on the temperature and pressure conditions. The water is usually rich in dissolved solids and gasses (e.g. CO_2, H_2S). In high enthalpy fields, the aqueous fluid can be in a state of vigorous convection because of strong density contrasts caused by the very high temperature gradients. The convection cells are characterized by zones of upwelling hot water and descending cooler water.

Deep geothermal systems can be water-dominated (liquid-dominated) or gas-dominated (H_2O steam). In water-dominated systems liquid water is the pressure controlling fluid phase, although it may contain some dissolved gas, however below the saturation condition. Such systems are very common in a temperature range of 125–225 °C. These systems produce hot liquid water, a two-phase mixture of liquid water and steam, wet steam or occasionally also dry steam depending on prevailing pressure and temperature conditions. In gas-dominated systems, most commonly liquid water and steam coexist in a two-phase system with a gas (steam) continuum and gas as pressure controlling phase. Such geothermal systems are less common (e.g. Larderello Italy, Geysers USA) than liquid water dominated systems. Gas dominated systems are characteristic of high enthalpy fields and produce dry superheated steam, that is steam at temperatures considerably higher than the condensation point (on the boiling curve).

In regions with normal or slightly elevated geothermal gradients, low enthalpy systems produce warm or hot water depending on the depth of the borehole that can be used for heat or electrical power supply. If permeability of the reservoir is too low for fluid extraction heat can be extracted directly at depth by deep ground source probes (Chap. 4).

Chapter 4
Applications of Geothermal Energy

Enhanced geothermal system Soultz-sous-Forets, Alsace, France

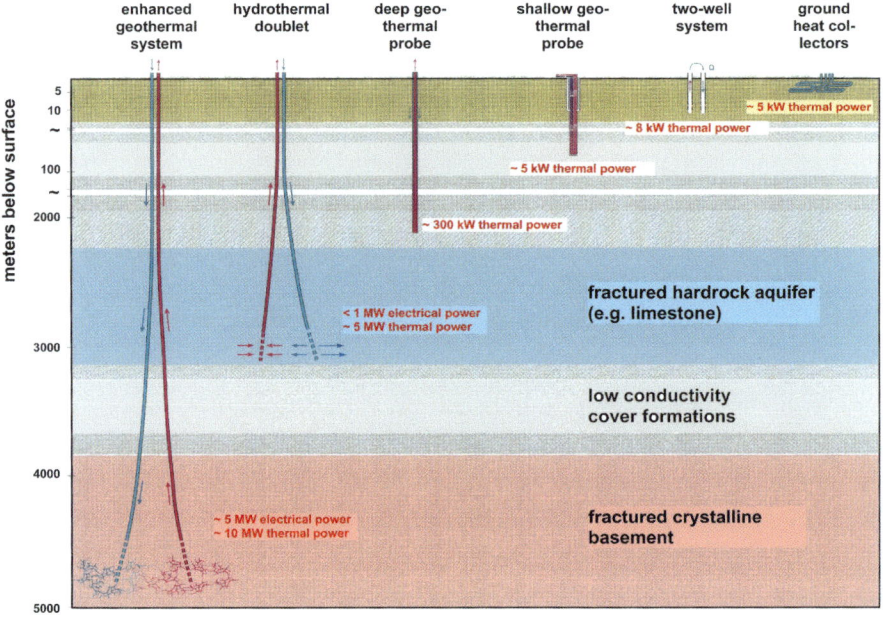

Fig. 4.1 Schematic illustration of different geothermal systems and their characteristic power output

The distinction between near surface and deep geothermal systems follows from the different depth levels of the geothermal reservoirs and different techniques of utilization (Fig. 4.1). Yet, the transition between the two worlds is smooth. Distinguishing the two main fields of geothermal energy utilization is useful, because their specific techniques for energy production require different geological and geophysical parameters for the description of the systems.

Deep geothermal systems exploit geothermal energy by means of deep boreholes. The mined thermal energy can be used directly and does not require further transformation.

Near surface geothermal systems, extract thermal energy from the uppermost layer of the earth crust. In most cases a depth of about 150 m is of interest. It may extend to a maximum of 400 m. Typical systems include: ground heat collectors, borehole heat exchangers, boreholes into groundwater, and geothermal energy piles. The exploitation is indirect and requires conversion with e.g. heat pumps. Direct use in the very low temperature range via heat pipes is under development. Railroad switch heaters and deicing of roads are typical potential applications.

With this definition of the boundary between shallow and deep systems, deep geothermal methods are employed at depth of 400 m and below. However, deep geothermal low-enthalpy systems in the proper and real sense are those at depth more than 1,000 m and above 60 °C. One needs to keep in mind, however, that in high-enthalpy fields high temperature fluids can be produced from boreholes in the range of hundreds of meters rather than thousands of meters as in the low-enthalpy deep geothermal fields.

4.1 Near Surface Geothermal Systems

Near surface geothermal techniques distinguish between open and closed systems with respect to the surrounding ground. The systems range from a few meters depth to some 10th of meters, rarely more than 150 m deep boreholes. Therefore, the temperature normally does not exceed about 25 °C.

Typical systems include: ground heat collectors, borehole heat exchangers, boreholes into the groundwater, and geothermal energy piles. At suitable temperatures, the utilization of waste water, mine water and tunnel waters also belong to near surface geothermal energy uses. In Switzerland, a number of road and rail tunnels produce warm water that is used for heating purposes. Examples include the Furka rail tunnel, the Gotthard road tunnel and the rail tunnel Ricken (Table 4.1). Utilization is made possible by means of heat pumps (www.geothermie.ch). A well-known geothermal system that uses waste water is the Olympic Village in Beijing, China. Waste water pumps heat and cool a total living space of 410,000 m^2.

Ground heat collectors consist of numerous horizontally installed plastic pipes of several hundred meters length at about 1–2 m depth (Fig. 4.2). The pipes must be mounted below the maximum penetration depth of winter frost. Also, the system needs to be above the level of solar regeneration in the summer. In the pipe system a circulating fluid (liquid) extracts heat from the ground. Strictly, ground heat collectors utilize not geothermal but rather solar radiation heat.

The most significant parameters controlling the thermal extraction output of such systems are the heat conductivity and heat capacity of the ground. The water and air content of the pore space and the ground temperature are important as well because of their effect on the key parameters heat conductivity and heat capacity. High porosity and void content of the ground typically reduce the heat conductivity.

If the groundwater table is low and the ground is in the vadose zone instead of the saturated zone, the voids are filled with air instead of water and the heat conductivity of the entire system is considerably lower (Sect. 1.4). Consequently, highly permeable sand and grit and water tables below 2 m below surface are problematic with respect to the efficiency of ground heat collectors.

The land required for ground heat collectors is large. The collector field cannot be overbuilt or covered because the system uses the solar heat input to the ground. If the groundwater table is temporarily low, irrigation of the ground heat collector

Table 4.1 Swiss tunnel water uses (Rybach et al. 2003)

Tunnel	Discharge (l/s)	Temperature (°C)	Thermal power (kW)
Gotthard	7,200	17	4,520
Furka	5,400	16	3,756
Grenchenberg	18,000	10	11,693
Rawyl	1,200	24	1,503

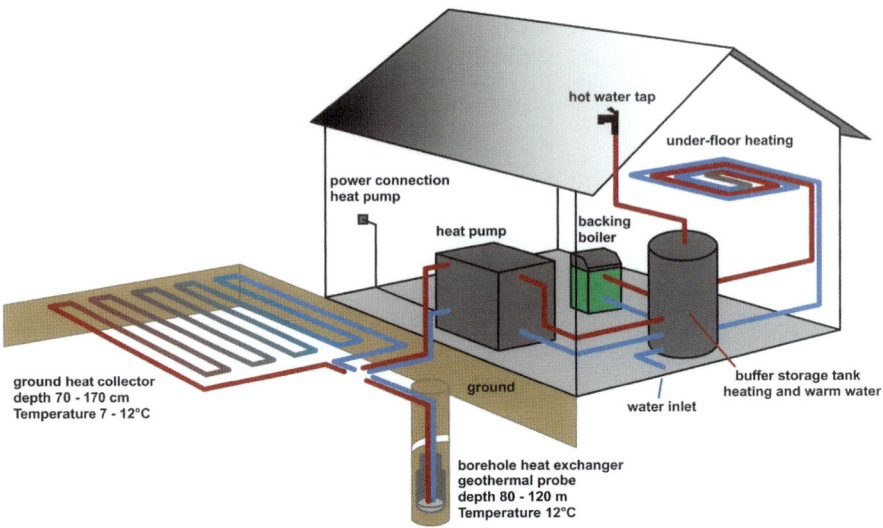

Fig. 4.2 Schematic illustration of ground heat collector and borehole heat exchanger for house heating

field may increase its efficiency. The systems require a considerable effort, which should not be undervalued, particularly if irrigation should be necessary.

Basis for planning for ground heat collectors are ground and soil maps and sections containing data on the structure of the near surface ground. These primary data are needed as input parameters for computer codes and techniques that model the heat conductivity structure of the ground as a function of soil compaction and water content (soil moisture). The computed models are essential for the final system design. Several computer models of variable complexity are presently available. However, no specific procedures for field tests have been developed so far. This is in marked contrast to thermal response tests for borehole heat exchangers. In addition, the computation tools cannot deal with heterogeneities of the ground. Furthermore, potential daily and annual variations of ground temperature and groundwater table are ignored in the system design. Nevertheless model computations are helpful and allow a generous dimensioning of the collector field and ensure that the spacing of the tubing is sufficiently wide, because of the potential for extensive icing of the ground.

Icing is an intrinsic system property of ground heat collectors; therefore, the systems cannot be operated with pure water. The system design must prevent massive freezing of the ground. The ground cools by operating the facility with the consequence of a retarded and shortened vegetation period. The biochemical activity of the soil biota including the production of humic and fulvic acids and other decomposition products of biomass may be altered. These chemical effects on the soil chemistry may trigger further chemical effects on the composition of seepage and groundwater.

4.1 Near Surface Geothermal Systems

A further reason why ground heat collectors cannot be run with pure water as a heat transport medium follows from its proximity to the surface. In wintertime, the system extracts heat from the ground at a low temperature level. Consequently, return temperatures commonly decrease to freezing conditions. Therefore, ground heat collectors need to be operated with special heat transfer liquids. For the approval and use of these liquids, detailed regulatory requirements must be obeyed, especially in groundwater protection fields.

Coiled tubes vertically installed in trenches and tube baskets are popular relatively recent new designs of ground heat collectors. Thermal power of the baskets ranges between 400 and 1,000 W depending on the size of the basket.

This technique of ground heat collectors is popular in e.g. Sweden and the USA where plots for family homes are typically larger than in e.g. densely populated central Europe and conform better to the area requirements for collector systems.

A further, relatively new type of near surface geothermal energy utilization uses building structures for heat exchange with the ground. Energy geostructures are elements of the foundation of a building in the ground that can be used for heating and cooling. Concrete is an ideal material for heat transfer because of its heat conductivity and heat storage capacity.

Foundation elements that function as geostructures are being equipped with plastic tubing for exchanging heat between the building and the ground for heating and cooling. The bundled tubing connects to one or several heat pumps. Proper hydraulic balancing increases the efficiency of the system. The foundation of the building serves as heat exchanger and geothermal system.

Energy piles or thermo-active piles are piles of reinforced concrete containing double or quadruple plastic U-tube heat exchanger or a network of polyethylene tubes. The tubes are completely embedded in concrete (Fig. 4.3). The heat transfer medium cycles between the pile and the heat pump in a closed loop. Depending on the energy demand of small or large industrial buildings, the installed thermal power of such systems ranges from 10 to 800 kW.

A very popular and widespread utilization of near surface geothermal energy is the use of borehole ground heat exchangers so called geothermal probes (Fig. 4.4). The heart of the system is a borehole of typically about 100 m depths. The deepest drill holes for geothermal probes reach 400 m. For many installations, more than one drill hole is used for energy exchange with the ground. In the borehole heat exchanger water or another heat transfer liquid such as water-anti freeze mixtures or also gases extract heat from the ground. The fluid circulates between a heat pump and the ground in a closed loop. The systems are technically mature and the installation is routine work for specialized commercial enterprises. Geothermal probes are also used for cooling in summer time. The geothermal borehole heat exchangers are particularly efficient in combination with solar-thermal installations. In Chap. 6, the combined systems will be presented in detail.

The geological structure and the ground properties are multifaceted and vary from place to place. The thermal properties of the ground differ from site to site accordingly. It is very important for the dimensioning of a geothermal installation to take the variability of the geological properties of the ground into account.

Fig. 4.3 Schematic illustration of an energy pile. In the indicated PE tubing a heat transfer liquid is cycled in a closed system

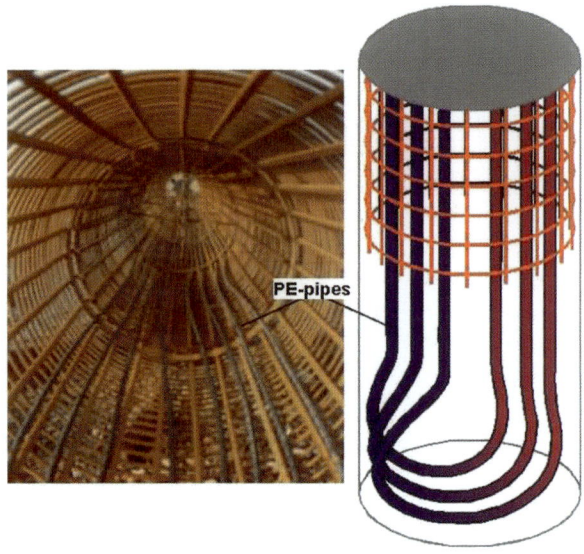

The thermal properties of some important types of rocks are compiled on Table 1.1. Highly permeable aquifers and aquifers with high groundwater flow velocities such as in karst areas are environmentally vulnerable. Drilling and casing of bores can be accompanied by mud losses, turbidity and chemical and microbial contamination and pollution of flowing groundwater. Drilling of a geothermal probe potentially intersects layers with different permeability, hydraulic situations and hydro-chemical properties. Tight annular void grouting and sealing of the annulus needs to conserve the layer separation. This is mandatory for any geothermal probe system. It is required by the matters of groundwater protection and it is essential for the efficiency and the economic lifetime of the installation.

Ideal sites are characterized by a uniform medium or low hydraulic conductivity. Areas with highly permeable karst aquifers or fractured hard rock aquifers are less favorable because of possible technical problems with drilling and casing. Drilling is often troubled by mud losses and can be associated with groundwater contamination. Moreover, it is often difficult to seal the annulus tight because of losses of cement slurry in the highly permeable voids. In such areas, higher costs must be expected for a professionally proper installed borehole heat exchanger. Occasionally the drilling is not successful and the wellbore must be abandoned and sealed.

In addition to the restricted potential of an area resulting from unfavorable geology, site-specific difficulties may trouble a geothermal energy project: Past losses, previous pollution, natural hazards, neighboring risks, adjacent bodies of water, protected areas, underground gas reservoirs, and others. Drilling into over- or under-pressured aquifers, or into layers of highly water-soluble minerals such as rock salt, gypsum or anhydrite may be potentially hazardous or cause technical drilling difficulties (Sect. 6.7).

4.1 Near Surface Geothermal Systems

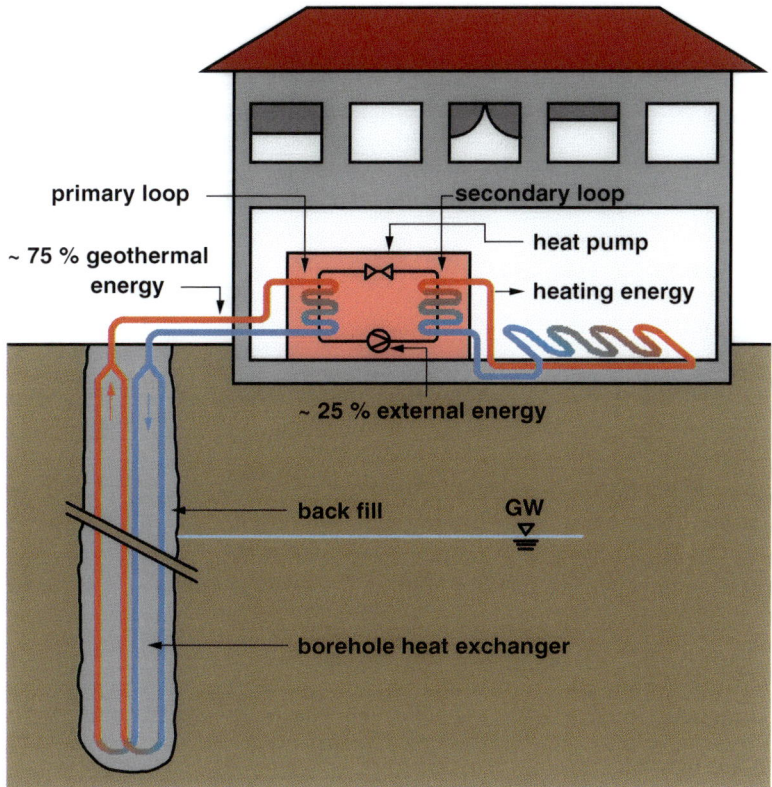

Fig. 4.4 General design of a borehole heat exchanger installation

Near surface geothermal energy can also be extracted directly from groundwater by means of an open two-well system. Heat is extracted from water of a production well and the cooled water re-injected and returned to the aquifer in a second well. It is important that the two wells do not influence each other thermally and hydraulically. The cool water must not be injected upstream from the production well. Furthermore, the chemical composition matters because many groundwaters have a disposition to precipitate scales. Detailed description of such systems is given in Chap. 7.

In order to utilize a near surface geothermal system for heating and heat production, the temperature of the circulated heat transfer fluid must be increased usually by making use of a heat pump. A heat pump is a device that transfers heat from a source at relatively low temperature to a heat sink at higher temperature using mechanical work. The mechanical work is provided by a pump driven typically with external electrical power. The device can be used for cooling (refrigerators and freezers are typical heat pump household devices) or heating (used in building space heating). Reversible cycle heat pumps are typically used for geothermal applications. The devices are equipped with a reversing valve so that the direction of heat flow may be reversed. The machines are evaporation–condensation systems and utilize the latent heat of condensation of a heat transfer fluid for space heating. The efficiency of a heat pump system using a specific heat source and operating at a particular temperature is characterized by the annual performance factor (APF). The ground source heat pump (GSHP) systems should operate at a minimum annual performance factor of four (Sect. 6.3). This means per unit of invested energy to drive the pump four units of energy should be extracted from the ground source. Investment costs, annual running costs, the primary energy requirement and the CO_2 emissions are the decisive criteria assessing the economical performance, the energy efficiency, and the environmental effects of ground source heat pump systems.

The legal and regulatory requirements for building and operating near surface geothermal systems vary from country to country (also between states and districts within countries). Normally they are based on groundwater and mining regulations. Typically, the authorities provide investors with guidelines and recommendations with detailed descriptions of all legal requirements for the building of the system of interest. Such guidelines also inform about existing restrictions for building a specific system. Potential restrictions include: groundwater protection areas, areas of unfavorable and difficult aquifer structure, drilling risks, and others. The guidelines may also assist developers and clients with recommendation procedures to follow in case of drilling into an artesian aquifer or under- or over-pressured confined aquifers, drilling into strata with gas over-pressure, drilling into large cavities or karst and into strata with soluble salts or with swelling minerals.

The utilization of geothermal energy for heating purposes requires significant initial investments. Prior to planning the system, potential and possible reductions of heating needs must be implemented. Highly recommended are thermal insulation measures, which directly reduce the need for heating. Included are masonry and façade insulation, thermally insulated high-quality windows and the like. Floor and wall heating systems significantly improve the economic viability of the heating system. Floor heating systems operating with supply temperatures of 35 °C or with concrete core temperatures of walls as low as 25 °C are far more economical than radiators running at 55 °C supplied by the heat pump. Economic and efficiency requirements also consider the hot water needs of a building (shower, washbasin, and the like). Expert advice and competent qualified planning of the total system assures an economic and environmentally friendly enduring operation.

4.2 Deep Geothermal Systems

Deep geothermal systems include hydro-geothermal low-enthalpy systems that use the heat stored in warm or hot water of deep aquifers (Fig. 4.5). The heat reservoir is exploited directly, generally employing a heat exchanger, occasionally also via a heat pump. The produced thermal water can be fed into the local and district heating grids or directly used in spas, heating of industrial complexes and heating of green houses. Conversion of the heat to electrical energy with supplementary technology such as Organic Rankine Cycle facilities or Kalina installations is possible above about 80 °C (Figs. 4.6a, b, c, and 4.7). However, economically feasible efficiency requires 120 °C or more.

Organic Rankine Cycle (ORC) plants work with an organic heat transfer fluid with a relatively low boiling temperature. The vapor phase of this fluid passes through a turbine thereby driving an electricity generator. Kalina installations use an ammonia–water mixture as a heat transfer fluid. The non-isothermal boiling of two-component fluid is a characteristic process of fluid mixtures (Kalina 1984; Ibrahim 1996).

The most popular kind-of-use of hydro-geothermal resources is the hydrothermal doublet (Chap. 8). The system is based on two wellbores drilled into a hot water aquifer, one of which is used as a production well where hot water is pumped from the aquifer to the surface, whereas the second well of the doublet is used for injecting the cooled water back into the subsurface reservoir. At the surface, the thermal energy of the hot water is transferred to a suitable fluid by means of a heat exchanger. The heat energy cannot be completely transferred and converted to electrical power. The hot water is typically cooled to about 55–80 °C only and, accordingly, much of the thermal energy remains in the thermal water. The residual heat has the potential to be utilized if appropriate customers and demand exist and the proper infrastructure can be installed. This also holds for enhanced geothermal systems (EGS) formerly known as hot-dry-rock (HDR) systems (Chap. 9). The economic success of a power plant depends much on selling the residual heat.

The cooled water with its residual heat is recycled to the aquifer from an injection well. The filter sections of the two wells of the doublet are at a exactly defined distance from each other (Fig. 4.5). Depending on the geological situation, injection may require a pump (Fig. 4.8). The need for recycling the produced hot water in a closed loop has several reasons. It is necessary to contribute to the recharge of the aquifer, because natural recharge of deep aquifers is a very slow process. Since a hydro-geothermal plant pumps large amounts of water it is simply necessary to make sure that the extracted water is replaced. Re-injection of cool water is also worthwhile for economical and practical reasons, because the waters contain typically high concentrations of dissolved solids and gases. For reasons of waste management, it is advantageous to dispose the waters in the original reservoir.

An example of a hydro-geothermal doublet is the plant of Riehen near Basel (Switzerland), which continuously supplies residential units in Switzerland and nearby Germany with thermal energy for heating (Fig. 4.8) since start-up in 1994. The two wells located at a distance of 1 km tap thermal water from the Muschelkalk aquifer at 1,547–1,247 m depth, respectively (Fig. 4.9).

Production and injection well of a hydro-geothermal doublet can be bored from one drilling site as inclined bores (Figs. 4.5, 4.10). This greatly reduces the area requirement of the surface installation of the plant. In the subsurface, bottom hole of the bores in the hot water reservoir are typically 1,000–2,000 m away from each other. The optimal distance of the wells must be determined pre-drilling

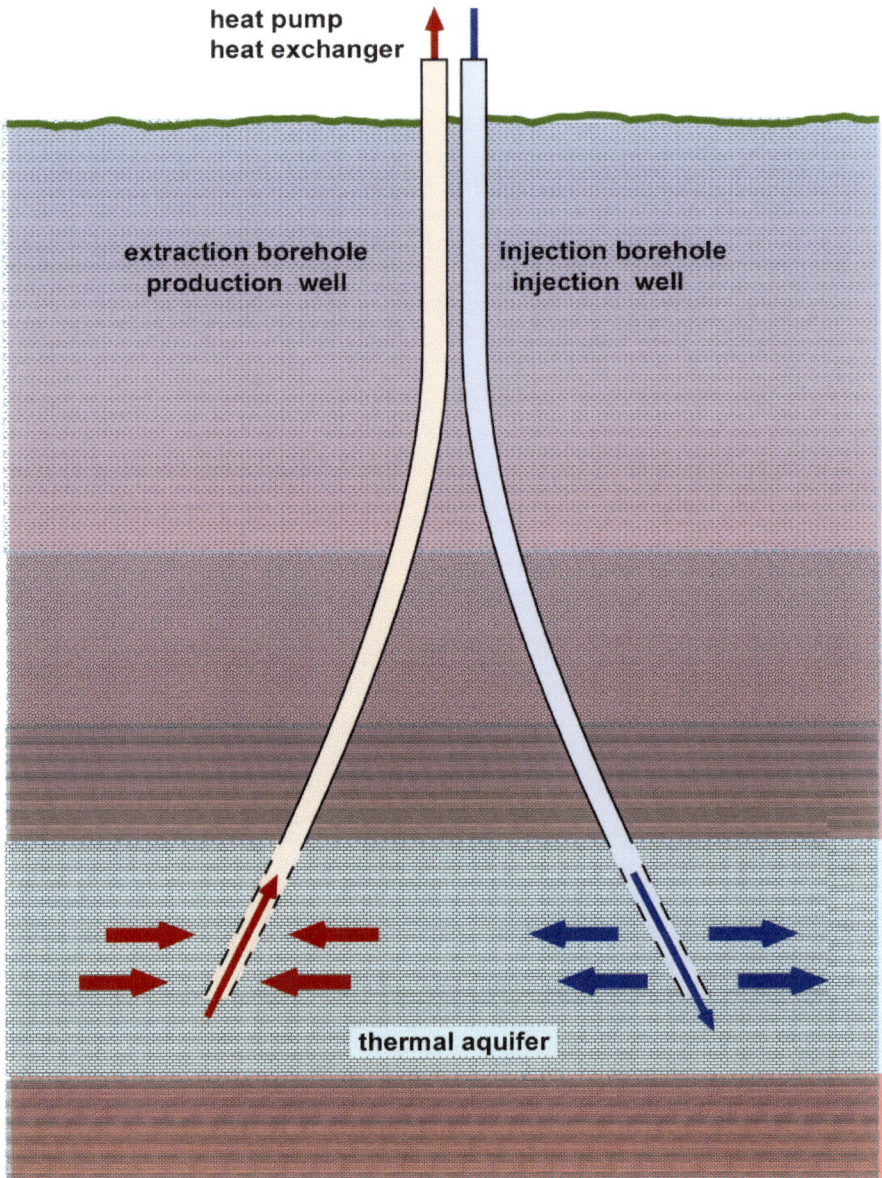

Fig. 4.5 Underground design of a deep geothermal open system installation (doublet, 1 producer, 1 injector)

by numerical modeling of the system. If the wells are too close to each other a thermal short-circuit is at stake. This means that the cooled re-injected water may reach the production well after a relatively short time of plant operation, cooling therefore the produced water. On the other hand, the wells should not be too far

4.2 Deep Geothermal Systems

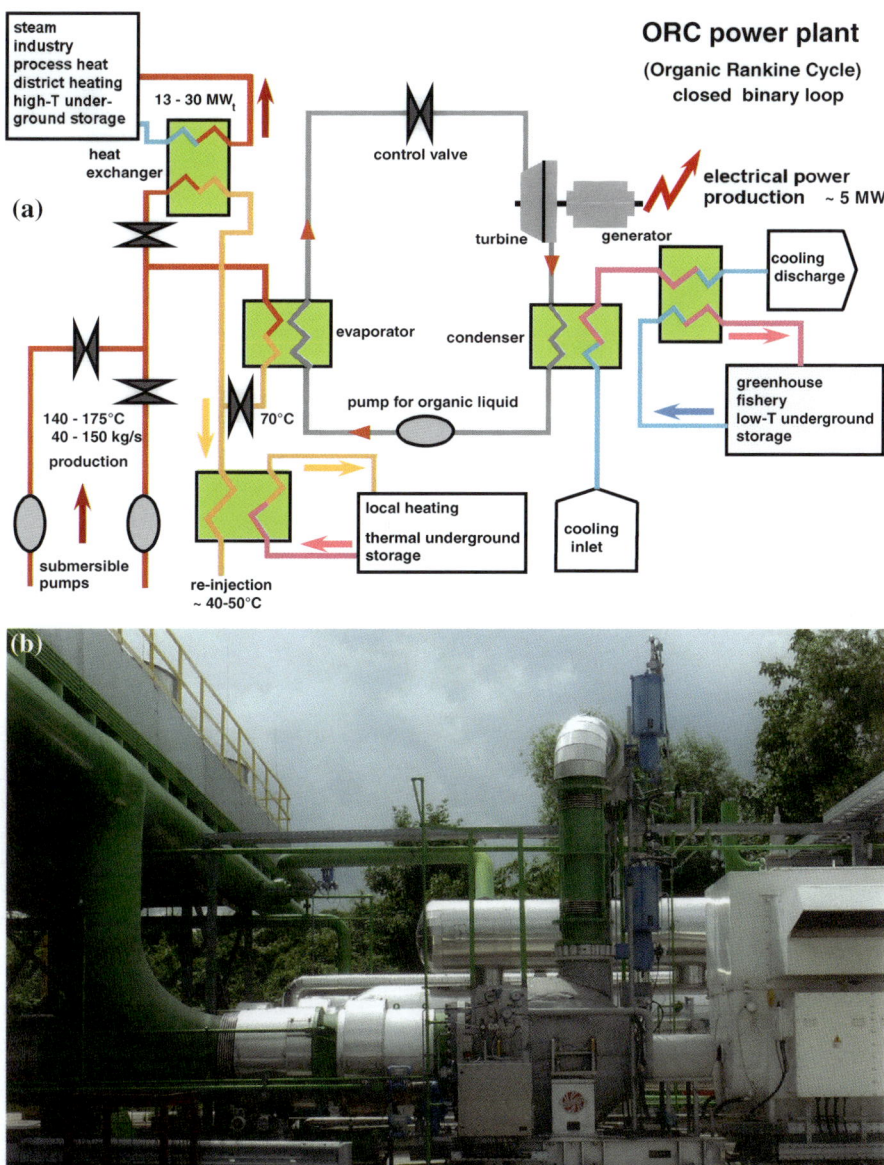

Fig. 4.6 Organic Rankine Cycle power plant: **a** concept and design (modified from Stadtwerke Bad Urach, Germany); **b** ORC cooling system; **c** ORC turbine (photographs **b** and **c** from Soultz-sous-Forets, France)

Fig. 4.6 continued

apart, because in this case the production well does not receive hydraulic support from the injection well. However, production depends on an intermediate time scale on the recharge of the aquifer by cooled water re-injection.

The pumped hot water and, after cooling, reinjected water circulate in a closed system that allows keeping the fluid under a defined pressure. This is necessary to prevent or minimize scales and mineral precipitations from highly mineralized and gas-rich fluids in the installations caused by pressure drop and gas loss. Ca-carbonates (calcite and aragonite) are among the most typical and widespread scales. Degassing of CO_2 from pumped hot water causes carbonates to precipitate in the pipe system even though the carbonates are more soluble in cold water, because CO_2-loss outweighs the temperature effect. In closed pipe systems, the pressure can be adjusted in such a way as to prevent degassing and scale formation. At some sites, small additions of a strong acid (e.g. hydrochloric acid) or other chemicals (organic inhibitors) may be needed to prevent scales (Sect. 14.3). The same applies to EGS systems. The cooled fluid can be recycled to the reservoir by free flow or by pumping depending on the hydraulic properties of the reservoir rocks. Typical reinjection pumps are multistage, single-entry centrifugal pumps in modular design with axial inlet and radial outlet.

Geothermal energy installations typically use two kinds of fluid production pumps: Line shaft pumps (LSP) operated at the surface and electric submersible pumps (ESP) (Fig. 4.8). The pumps for lifting hot fluids to the surface must resist high temperatures, high pressures and chemically aggressive and corrosive

4.2 Deep Geothermal Systems

Fig. 4.7 Wet cooling tower of the Kalina binary cycle power plant at Bruchsal in the upper Rhine river valley (Germany)

fluids thus belonging to the most stressed components of a geothermal power plant (Sect. 14.4). ESP lift the hot fluid with centrifugal force to the surface, where it is directed to a heat exchanger. The extracted thermal energy can then be converted to electrical power or directly fed into a district-heating grid. Combined heat and power improves energy efficiency and reduces emissions. These are particularly environmentally friendly and economical schemes.

Favorable fields for hydro-geothermal plants are above deep aquifers with high natural hydraulic conductivity and high temperatures. If the natural conductivity is too low for extracting hot water from the aquifer at the required rate the hydraulic structure of the aquifer needs to be improved by measures of artificial conductivity enhancement. Improvement measures include well shocking by sudden pumping, acidifying carbonate rocks, stimulation with high water pressures as well as combined stimulation and acidifying by pumping acid solutions with high pressure into the aquifer. Following the knowhow of the oil industry, improved extraction rates can be achieved by side tracking the well.

Utilization of thermal water by means of hydrothermal doublets for heating purposes is for the most part a mature technique. Hydrothermal installations that have been operating for dozens of years are currently in service worldwide.

Fig. 4.8 Fitting an electric submersible pump (*ESP*) into a production well of a hydrogeothermal doublet (2,500 m deep borehole Bruchsal, Upper Rhine River Valley)

Special cases of hydro-geothermal installations are balneological spas utilizing thermal deep waters. In addition to the use of hot water in the bathing pools, the pumped thermal water is also used for the heating of buildings in the local area. After use, the raw sewage is cleaned but not reinjected into the aquifer.

Hydro-geothermal systems include next to thermal aquifers also highly permeable fault and fracture zones in rock masses.

Besides the low-enthalpy hydro-geothermal systems, introduced above, high-enthalpy steam or two-phase systems are used for electrical power and thermal energy production (Sect. 4.4).

The future core systems of deep geothermal energy utilization are petrothermal systems that extract heat from hot rocks characterized by relatively low hydraulic conductivity. The systems are known under a variety of names reflecting the historical development of deep petrothermal techniques. The names include: Hot-Dry-Rock (HDR), Deep-Heat-Mining (DHM), Hot-Wet-Rock (HWR), Hot-Fractured-Rock (HFR) and Stimulated- or Enhanced-Geothermal-Systems (SGS, EGS). The original name HDR reflects the erroneous concept that basement rocks at

4.2 Deep Geothermal Systems

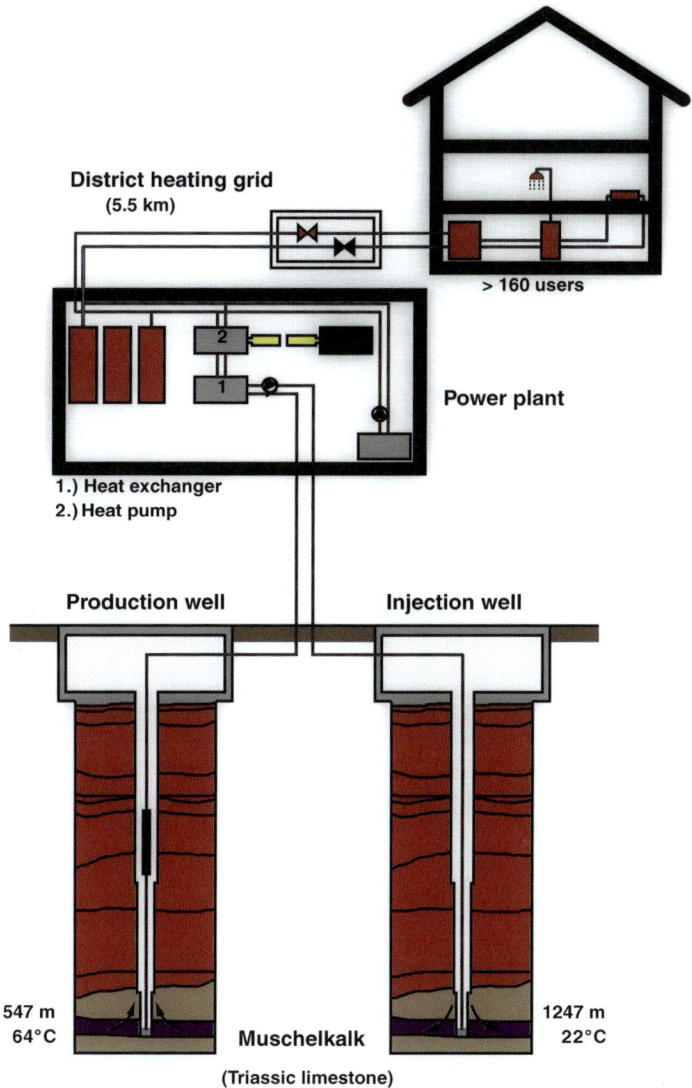

Fig. 4.9 The hydrogeothermal doublet system at Riehen (Basel, Switzerland), redrawn from documents of Gruneko Corp

great depth are dry and devoid of an appreciable permeability. After a large number of deep wells were drilled, it became evident that deep basement rocks (granites and gneisses) at several km depths are generally fractured and that the fracture porosity contains hot and usually salty water. The hydraulic conductivity of the hot rocks at depth is relatively large. In this book we will use the name Enhanced Geothermal Systems (EGS) for these techniques. EGS extract thermal energy stored in the rock mass, in contrast to hydro-geothermal systems that use the thermal energy from water stored in the pore space of rocks. Therefore EGS do not require a heat

50 4 Applications of Geothermal Energy

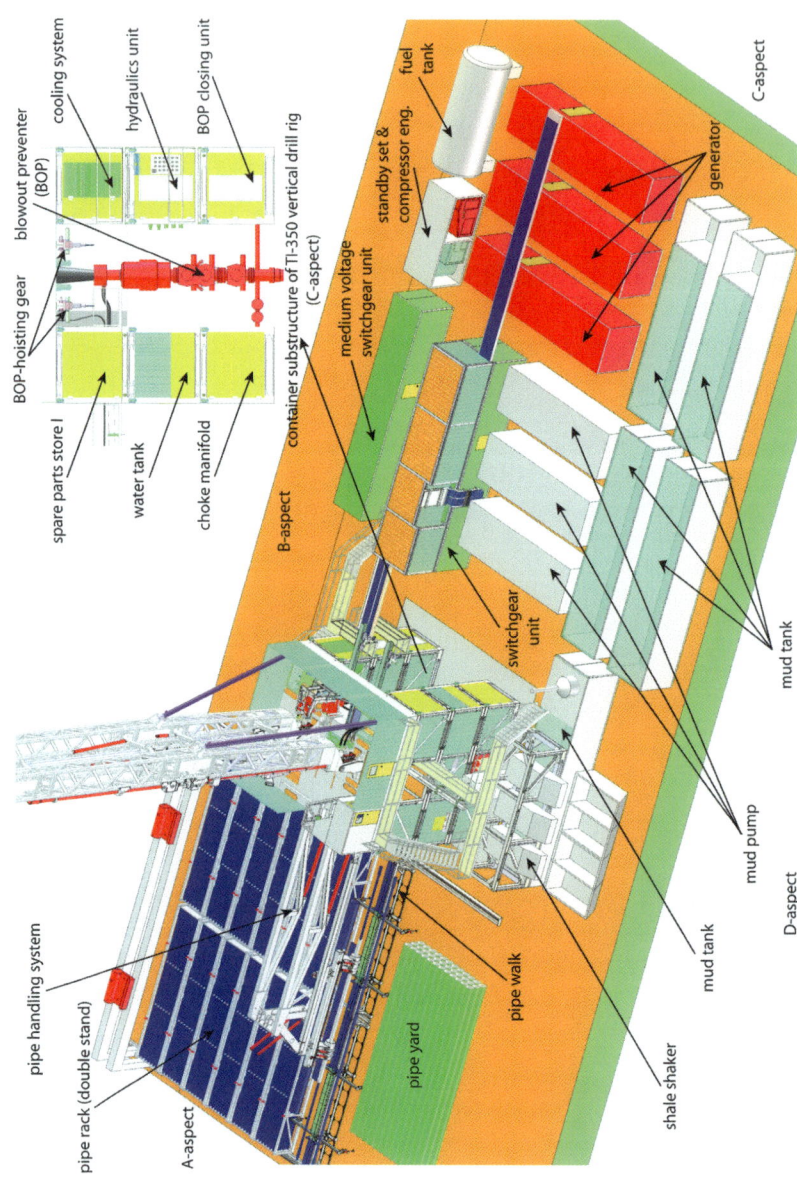

Fig. 4.10 Schematic illustration of a modern drilling site (the figure has been generously and kindly provided by Herrenknecht Vertical)

reservoir with aquifer properties in the hydrogeological sense. EGS primarily have electrical power production in mind. Consequently target temperature is 200 °C and beyond. The hot rocks, usually crystalline basement (granites and gneisses), function as a heat exchanger. Heat transfer to the surface is achieved by natural water present in the fracture pore space of the basement (Stober and Bucher 2007a, b; Bucher and Stober 2010). In crustal sections with average geothermal gradients, 5–7 km deep wellbores are necessary to reach the required rock temperatures (Chap. 9). Ongoing research explores the suitability of dense fractured sedimentary rocks for EGS applications. In the following, the basics of EGS are briefly outlined. A detailed treatment is given in Chap. 9.

The crystalline basement of the continental crust is generally fractured in its upper part. The fractures are the result of failure of stressed rocks in the brittle deformation regime in the uppermost about 12 km thick layer of the Earth. The fractures are flow paths for advective water transport. The hydraulic properties of the fractures depend on fracture aperture, surface roughness of fracture surfaces, connectivity and frequency of fractures and other parameters (Caine and Tomusiak 2003). The hydraulic behavior of the fractured basement corresponds to an infinite homogeneous low-conductivity aquifer (aquitard). High-pressure injection of water into the borehole increases the aperture of natural fractures and unlocks partly sealed fractures therefore improving the hydraulic conductivity.

Injected water passes the fractured rock heat exchanger at depth and scavenges heat from the rock mass. In addition, EGS use water as the heat transfer vehicle. Heat extraction at depth takes place in a nearly closed water cycle. The extracted thermal energy reaches the surface via a production well and can be converted to electrical power or (and) used directly for heating.

The EGS concept uses deep hot fractured rocks with relatively low permeability and does not depend on high-yield aquifers. In principle, an EGS project can be realized anywhere. However, reasonable projects aim for locations with raised geothermal gradients and a suitable tectonic setting. In 2011, only one single EGS plant is operational worldwide. It is located in Soultz-sous-Forêts (France) in the upper Rhine rift valley. The plant is in continuous operation since 2007. Although long-term experience does not exist, EGS will probably play an important role for the electrical power production in the years to come. Their fundamental advantage over other environmentally friendly energy systems is the supply of base load electric power.

Deep geothermal probes are, in principle, also a form of petrothermal systems. Here, thermal energy is extracted from any kind of rock or rock sequence using a closed loop of heat transfer fluid in a deep probe. Deep geothermal probes are used exclusively for heat supply. Because of the relatively low process temperature of the probes, electrical power cannot be generated with presently available technology.

The technology of deep geothermal probes is comparable to the ones of near surface probes. In a deep probe a heat transfer fluid is circulated in a single borehole to depths of up to 3,000 m (Fig. 4.11). The system does not require permeable rocks at depth and thus can be installed wherever. Particularly well suited

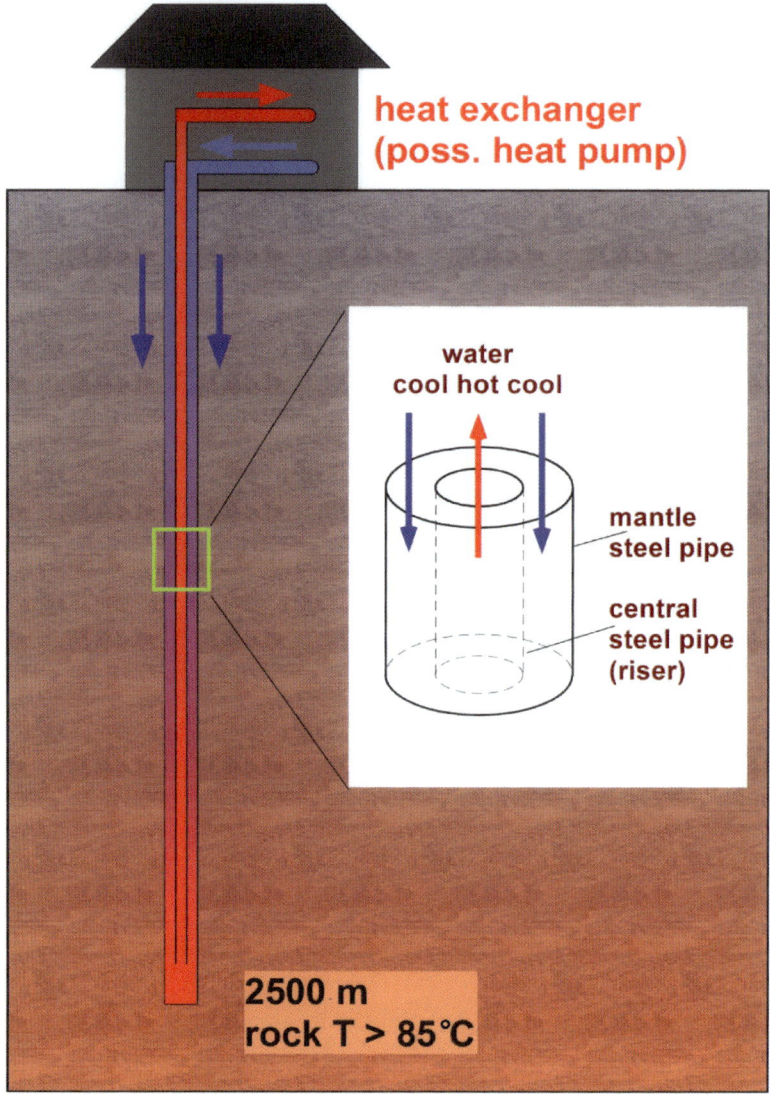

Fig. 4.11 Schematic drawing of a deep geothermal probe. It operates as a single borehole heat exchanger. The probe extracts heat at depth and transfers it to the surface heat exchanger (in combination with an optional heat pump) in a closed loop

for installation of deep geothermal probes are existing old abandoned deep wellbores (e.g. from the oil industry). Because of the closed loop, deep probes do not chemically interact with the deep heat reservoir. The utilization of the deep probes combines other heat producing facilities in an integrated heating central. The heat production of a deep geothermal probe can be in order of 500 kW depending on the local conditions. Examples of deep geothermal probes include the following locations: Prenzlau, Aachen, Arnsberg (Germany); Newcastle (UK).

Heat transfer from the hot rocks occurs by heat conduction through the grouting of the probe and the casing to the advecting fluid. Ammonia is a commonly used heat transfer fluid. The cool fluid slowly flows downward in the annulus of a double-containment pipe system and is gradually heated by the surroundings. The descend velocity is typically in order of 5–65 m/min. In a thermally insulated central pipe thermal energy is lifted to the surface by the heated fluid (Fig. 4.11). At the surface, heat is extracted from the hot fluid in a surface heat exchanger. The cooled fluid (15 °C) is pumped back to the annulus. The heat extraction process cools the underground in the vicinity of the probe.

The amount of effective heat produced by a deep probe depends primarily on the temperature of the ground. Thus areas with a positive thermal anomaly are economically particularly gainful. Further parameters controlling the productivity of a deep probe include the thermal properties of the ground, particularly the thermal conductivity, the total time of operation, the technical layout of the probe, and the thermal properties of the casing and screen materials used. Long and large-diameter probes have evidently a large heat exchange surface.

The structure of thick rock sequences is often characterized by properties that are transitional or mixed between hydro-geothermal and petrothermal systems.

Further fields of use of deep geothermal energy sources include: Heat from deep underground mines, rock caverns and storage of thermal energy in deep geological structures.

4.3 Efficiency of Geothermal Systems

Efficiency characterizes the degree of conversion of primary thermal energy to mechanical and finally electrical energy. Efficiency is the ratio of output to input, or benefit to effort. Because of the second law of thermodynamics, this ratio is always smaller than one. The Carnot efficiency η describes the maximum possible efficiency for any heat engine. It relates the maximum of work that can be produced by the system to the amount of heat put into the system. It is the theoretically possible maximum efficiency for an ideal heat engine. The efficiency of real systems is related to the Carnot efficiency η. The system design aims to reach efficiencies as close as possible to η. The Carnot efficiency η is defined by Eq. (4.1):

$$\text{The Carnot efficiency } \eta = 1 - (T_c/T_h) = W/Q_{th} \qquad (4.1)$$

where T_c corresponds to the temperature of the cold side, the outlet T of the fluid, and T_h the temperature of the hot side, the inflow T of the heat carrier fluid (both in Kelvin). This can also be expressed by the ratio of work done by the system (W) to the thermal energy added to the system (Q_{th}). The Carnot efficiency η of a system with, for example, an inlet $T_h = 100$ °C (373 K) and $T_c = 20$ °C (293 K) has a theoretical maximum of 0.21 (21 %) (Fig. 4.12).

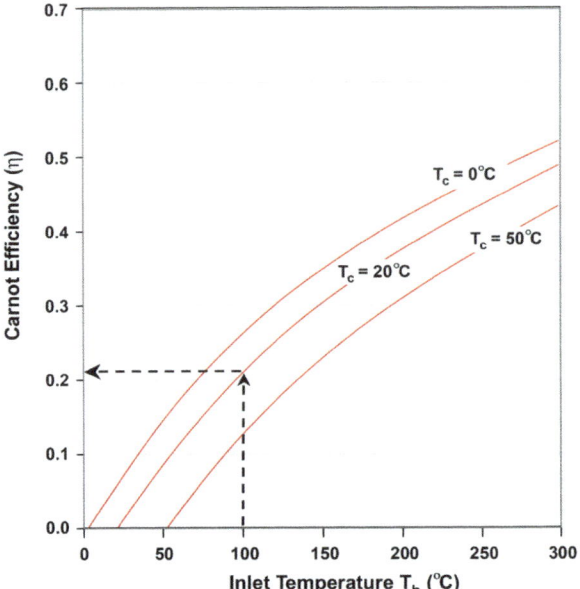

Fig. 4.12 Diagram showing the Carnot efficiency (Eq. 4.1) as a function of the inlet temperature T_i (here in °C) for three different outlet temperatures T_c (also in °C)

The physical upper limit of thermal efficiency for power stations driven by thermal water (hydro-geothermal or EGS plants) of 100–200 °C is about 12–22 %. In this temperature range, electricity production is feasible only with binary loop plants. In 2013, two different systems are available on the market, systems based on the Organic Rankine Cycle (ORC) and systems based on the Kalina Process. ORC systems use organic fluids, typically isobutane, as heat transfer fluid. Kalina systems work with an azeotropic mixture of ammonia and water. Azeotropic mixtures boil over a certain temperature interval called temperature glide. The power output to the grid of air-cooled Kalina systems tends to be higher than that of ORC systems at low input temperatures, whereas ORC plants tend to be better performers at higher input temperatures. Kalina systems withdraw less thermal energy from the thermal water than ORC systems but convert it to electrical power with higher efficiency. In the low temperature range ORC systems suffer from low thermal efficiency that follows from a high auxiliary power requirement of the cooling system, especially when air cooled (Park and Sonntag 1990).

All geothermal power plants produce heat in addition to electrical power (Fig. 4.13). This heat needs to be used in combined heat and power systems. The maximum use of the side product heat determines the economic success of a geothermal power plant. Moreover, production of electrical power only and pointlessly wasting the co-produced heat would be ecologically insensitive and ignorant.

The efficiency of electrical power production is relatively low. Taking the auxiliary power requirement of the production pump and the cooling loop into consideration also, the typical efficiency of the total system is about 5–7 %. Clearly there is room for improvements. However, if the residual heat of the thermal water after

4.3 Efficiency of Geothermal Systems

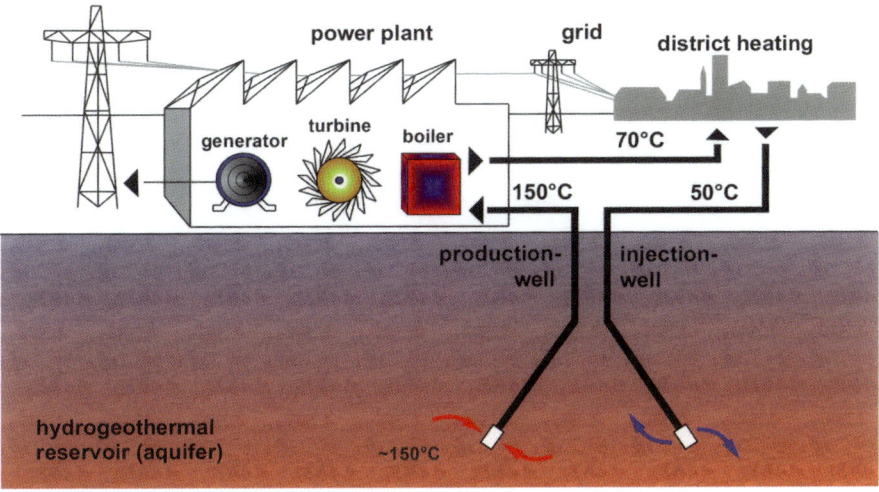

Fig. 4.13 Schematic diagram of a hydrogeothermal system with a binary system power plant converting a part of the produced thermal energy to electricity and utilizing an other part for district heating

electrical power production is used for district heating (Fig. 4.13) and other purposes, the environmental balance of the total system is determined by the quantity of supplied heat. Geothermal energy systems can also be combined with other heat sources including biogas plants and hybrid plants thus improving the environmental balance.

The conversion of thermal energy to mechanical or electrical energy in thermal plants inherently produces waste or process heat that needs to be discarded. If the process heat can be transferred to lake or river water a very low temperature T_c (Eq. 4.1) and a corresponding high efficiency can be attained. However, at many sites potential environmental degradation or lack of cooling water in sufficient quantities requires cooling by means of cooling towers. Wet cooling towers (Fig. 4.7) and dry cooling towers (Fig. 4.6b) transfer process heat to the atmosphere.

4.4 Major Geothermal Fields, High Enthalpy Fields

The annual global production of electrical power from geothermal sources is about 11 GW_{el} (2010), thereof 1.4 GW_{el} in Europe (Bertani 2010; IEA-GIA 2012). Most of the geothermal electricity is produced in high-enthalpy fields that reach high temperatures at shallow depths. The electricity is generated in dry-steam and flash-steam power plants. Examples are the Coso geothermal field at the western edge of the Basin and Range geologic province in eastern California (USA), the Wairakei geothermal field in the Taupo Volcanic district in New Zealand, the Mori geothermal field in Hokaido (Japan), the Hatchobaru geothermal field in central Kyushu (Japan) and many other similar systems worldwide.

These power plants of high-enthalpy fields function as open system geothermal installations. The systems use steam produced by decompressing the thermal heat transfer fluid to drive turbines for electrical power production (Fig. 4.14a–c). The minimum operation temperature in flash-steam plants is 175 °C. The turbine converts geothermal energy into mechanical energy that is converted to electrical energy by a generator. A part of this electrical energy is consumed by pumps and other machinery of the power plant; the net power is fed into the grid.

Electrical energy production from geothermal sources in closed binary-loop low-enthalpy systems such as ORC and Kalina plants (Fig. 4.14c) is a relatively new technology that has been installed at relatively few locations worldwide although suitable locations are far more frequent. There is an enormous potential for future development and expansion of deep low-enthalpy systems. Many projects are in the planning stage (2013). A major disadvantage of high-enthalpy fields is their limited occurrence in volcanic and tectonically active areas along plate boundaries or extensional basins. A major breakthrough for increased geothermal energy utilization must come from petrothermal EGS systems in addition to further development of hydro-geothermal systems.

In Europe, Italy has the highest installed power of about 800 MW_{el} and is well ahead of Iceland with 202 MW_{el} (2005). In Tuscany (Italy) favorable geological settings, very early development and the resulting experience led to a steady growth of the geothermal energy industry. However, the big four in the world are (installed capacity 2005, Bertani 2007): USA 2,564 MW_{el}, Philippines 1,930 MW_{el}, Mexico 953 MW_{el} and Indonesia 797 MW_{el}. Other major producers of electrical power from geothermal resources include (in MW_{el}) Japan (535) and New Zealand (435). Much of the power of the high-enthalpy fields in these countries is produced by dry-steam plants.

Iceland uses mostly geothermal resources from high-enthalpy fields related to the volcanic mid-ocean ridge and the Iceland mantle plume. However, the country also installed some binary loop plants in low-enthalpy fields in recent years. The important Russian high-enthalpy fields and the associated geothermal plants are all situated in Kamchatka and on the Kuril islands. The total installed capacity is 80 MW_{el}. Turkey has a promising geothermal energy potential also, although the installed power is only 20 MW_{el} today (2005). 10 high-enthalpy fields are on the list of identified 170 geothermal heat reservoirs. Some of the boreholes reach 200 °C already at 800 m depth.

The most important and largest high-enthalpy geothermal field in the world is "The Geysers" in California (USA). The field has 888 MW_{el} installed capacity and uses a 300 °C dry steam reservoir at 600–3,000 m depth (deepest well: 3,900 m). The power is produced from 100 km^2 drilled area, 424 production and 43 reinjection wells. Average steam temperature is 235 °C (at 12.4 bar) and the average flowrate per well 5 kg/s. The steam is produced from a sandstone and graywacke reservoir that is heated by a magma chamber at greater depths.

4.4 Major Geothermal Fields, High Enthalpy Fields

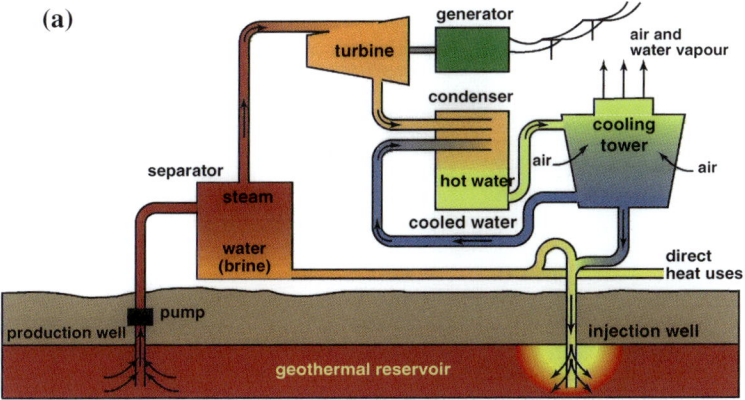

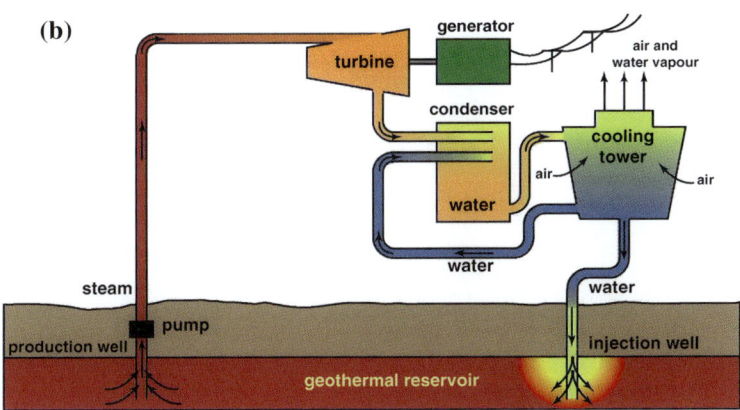

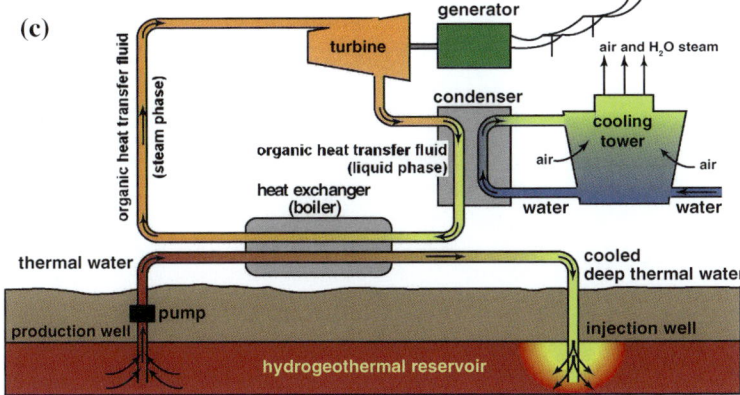

Fig. 4.14 Process diagrams for three common types of geothermal power plants: **a** Flash steam plant. **b** Dry steam plant. **c** Binary cycle power plant

As a result of the power production and the associated reduction of steam pressure the ground became seismically active after about 1975. Seismic events reached magnitude $M_L = 4$ (Sect. 10.1). Seismic tremors correlate with the power production and the related rate of steam extraction from the reservoir even though a part of the condensed and cooled steam is reinjected to the reservoir. The steam pressure in the reservoir is decreasing by about one bar annually since 1966. The increased seismicity relates to reservoir compaction because of reduced pore pressure due to withdrawal of fluid and thermal contraction resulting from cooling (Nicholson and Wesson 1990).

"The Geysers" reached peak production of 1,900 MW in the year 1989. After that, maximum continued steam withdrawal resulted in aging of the reservoir and a decrease of steam pressure. In the last decade additional water injections partly compensate the withdrawals. At present (2011), about 800 kg/s cleaned municipal wastewater from Clear Lake and Santa Rose props up the reservoir. The measures slowed down reservoir degradation and total power output resumed.

The second largest geothermal field is Cerro Prieto, Mexico 720 MW_{el} with 149 production and 9 reinjection wells. The liquid fluid reservoir at 2,800 m depth is in the temperature range of 300–340 °C. The Cerro Prieto Geothermal Power Station is the largest geothermal power station in the world with plans for expansion up to 820 MW_{el} by 2012. The facility is located in south Mexicali, Baja California, in Mexico, and is built in five individual units.

The Malitbog Geothermal Power Station on the Philippines is the largest single geothermal power plant with a capacity 233 MW_{el}.

The history and development of the high-enthalpy field Larderello in Tuscany (Italy) is separately described in Sect. 2.2. Today (2011), the total installed capacity of the Larderello geothermal plants is 545 MW_{el} equivalent to the power of a modern coal-fired power plant. Like with all other plants in high-enthalpy fields, the production costs for the unit of electrical power output is low because there are no fuel costs (coal, oil, fuel rods). Some of the production wells produce up to 350 t/h (100 kg/s) steam at a temperature of 220 °C. The installations at Larderello inject all water not used in the cooling loop back into the reservoir. However, the losses or unbalanced difference between extraction and reinjection caused deterioration of the steam pressure and, as a result, a decline in power production. In the reservoir, the thermal energy is still there but the heat transfer fluid, here steam, is lacking or no longer present in sufficient amounts.

The plant operator ENEL (Ente Nazionale per l'Energia eLettrica) designed a program to revitalizing the high-enthalpy field. The exploited steam reservoirs are replenished with water from neighboring fields. New deep wells replace older shallow wells. This new technique permits to increase the working pressure from presently 4.5 to 5.0 bar to 12 bar. New 60 MW power blocks replace the array of old 20 MW turbines.

Because of the geologic position of Iceland on the Mid-Atlantic Ridge and above the Iceland mantle plume, a multitude of volcanoes is presently active on the island. The geothermal fields associated with the volcanoes are extensively utilized and Iceland is the leading geothermal country (Sect. 2.2). 53 % of the used primary

4.4 Major Geothermal Fields, High Enthalpy Fields

energy is geothermal energy. Five larger geothermal power plants produce 25 % of the island's consumption of electricity and 90 % of the households are supplied with heat. The installed geothermal capacity of the plants on Iceland is about 625 MW_{el}.

The hot water for the capital city of Reykjavik with its 120,000 inhabitants, including the hot water for deicing installations for sidewalks and roads, is supplied by a warm water reservoir, the so-called Perlan at an elevated height above the city making pumps unnecessary. The reservoir consists of five single tanks with 4,000 m^3 capacity of 85 °C hot water each. The hot water is produced from 70 drilled wells in the city.

The hot waters produced from high-enthalpy fields in Iceland, like in any other area, typically contain a large amount of dissolved solids. The hot waters are normally not in chemical equilibrium with the minerals of the reservoir rocks. Therefore, the waters react with the rock matrix in complex hydrothermal reactions. The total mineralization increases with temperature because for many substances and minerals the solubility increases with temperature and since the kinetics of mineral dissolution reactions increases with temperature (Chap. 14). Because of the water–rock interaction, the waters regularly contain high concentrations of dissolved silica. At low temperature, only small amounts of silica can stay in water under equilibrium conditions. Thus, precipitation of silica sinter and silica scale is a common occurrence and problem in high-enthalpy fields. The rate of silica precipitation depends on the temperature and the composition (salinity) of the water, which allows to partly controlling the site of precipitation in the system to some degree. The efficient pressure-controlled separation of steam and liquid (Fig. 4.14a–c) is crucial to avoiding silica scales in surface installations, such as turbines or heat exchangers (Sect. 14.3). The expanded steam from high-enthalpy reservoirs on Iceland contains 5 mg/kg dissolved solids only in contrast to the separated liquid phase that contains 45,000 mg/kg total dissolved solids (Giroud 2008). The major components in most thermal fluids are sodium, potassium and calcium and the associated anion is normally chloride. Dissolved silica is typically in the range of 600–700 mg/kg SiO_2 (this compares to the equilibrium concentration of 6 mg/kg at 25 °C). Boron, fluoride, barium, mercury and other trace elements can be significantly enriched. The high TDS of the produced fluid and solutes that are partly difficult to cope with are a serious challenge to the high-enthalpy geothermal plants. Some of the reservoir fluids also contain high concentrations of dissolved gasses, which are not condensable like CO_2 and H_2S. Degassing of high CO_2 concentrations promote calcite scale formation and CO_2 is corrosive. High H_2S concentrations may cause metallurgical problems, react with metal surfaces, cause corrosion, fatigue and cracking (Sect. 14.3).

The Iceland Deep Drilling Project (IDDP) with several international partners drilled a wellbore into a hot-fluid reservoir containing H_2O in its supercritical state. Reservoir conditions are T > 375 °C at P ~ 225 bar. The critical point of H_2O (CP) is at the coordinates T = 374 °C and P = 221 bar. It is planned to utilize the supercritical fluid for power production. The development and utilization of supercritical fluid reservoirs appears attractive because the system efficiency may be improved by a factor between 5 and 10 in relation to the produced fluid volume.

Wells on Iceland may reach 360 °C at a depth of only 2,200 m in some places, meaning that P–T conditions are close to the critical point of H_2O. The fluids are often toxic and very corrosive. Further challenges are scales that are difficult to control and problematic to remove and dispose without harm to the environment.

The development of very deep high-enthalpy reservoirs for industrial use is presently not workable due to technical reasons. At temperatures of 400 °C and more, the temperature resistance of materials, of drilling mud and geophysical instruments, the materials strength and limited hook load (<500 t) of drilling rigs pose serious hindrances.

Projections predict a total installed world capacity of geothermal electricity production of 140 GW_{el} for the year 2050 (Friedleifsson et al. 2008). This ambitious goal can only be reached if the EGS systems, which are relatively independent on location in contrast to high-enthalpy plants, are being further industrially developed. Additionally, existing geothermal fields must be further developed with increasing numbers of production and reinjection wells. Systems with production wells only are not environment friendly and economically not profitable. Geothermal reservoirs must be backed, that is to say they must be replenished and renewed. Reinjection recycles the high-TDS fluids to the original reservoir, which also helps to prevent disagreeable subsidence formation, reduction of permeability and decrease of production rate. The development of a geothermal field must be inclusive and integrate all potential users from the beginning of the planning stage: In addition to the electricity production side, this includes the utilization of the produced heat in industry, district heating, sports facilities, green houses and other secondary heat users.

Chapter 5
Potential Perspectives of Geothermal Energy Utilization

Drill bits

Geothermal energy is renewable energy in the sense that heat extraction by technical systems is replenished by heat flow from the heat reservoir of the Earth. The latter is virtually inexhaustible at human time scales (Sect. 1.3). Although the ultimate heat reservoir is in effect everlasting, the question of sustainability of geothermal energy utilization must be answered for each individual site, plant and location separately because it depends on the system design and the dimensioning of the installation.

At almost all sites, terrestrial heat flow is too low to balance the heat extracted for use in the power plant. Normally the geothermal system uses heat stored in a reservoir, which consequently cools for a certain period of time before it is replenished by heat flow from depth again. The heat flow density is insufficient in supplying the borehole heat exchangers for heating buildings alone. Because of this, the sustainability of the utilization of near surface geothermal energy is controversial. The removal of heat in shallow ground layers employing geothermal probes may overexploit a limited resource and may lead to continuous temperature decrease of the soil, which would eventually not be economically meaningful. The plausible fears and concerns, however, overlook the significant external heat supply to the ground by solar radiation. The solar contribution to the total of removed ground heat is typically distinctly larger than the amount provided by the heat flow from the interior. All borehole heat exchanger systems gradually approach a steady state where the removed heat is balanced with the recharge heat with its solar and terrestrial components. Along the path to the steady state the soil cools, initially fast then slower and finally approaches a constant steady state temperature. After a few years operating the installations, the annual temperature decrease is minimal (Eugster et al. 1998).

Advective heat transfer by groundwater flow, not considered in the analysis above, may have an additional and often significant effect on the heat budget of a geothermal installation. If borehole heat exchangers intersect groundwater-bearing strata (aquifers) then a large portion of the extracted heat is directly replenished by the advecting groundwater. Advective heat transfer may dramatically increase the efficiency of the ground source heat exchanger (Sect. 6.3.2). Direct use of geothermal energy from groundwater wells draws on groundwater advection (Sect. 7.3).

In deeper ground, outside the reach of solar heat input, the situation is different and the extraction of ground heat mines a heat reservoir that is sluggishly replenished (Sect. 8.3). In open systems, the cooled thermal water must be returned to the reservoir via an injection well for sustainable operation. Depending on local conditions, system design and extraction rates, and the temperature of the produced thermal water may continuously decrease after a certain operation time. If production rate is too high, the distance between production and injection well is too small and/ or the temperature of the re-injected fluid is too low, then the economic efficiency of the geothermal system is at stake. The production could be forced to be reduced or even discontinued until the reservoir temperature recovers. Therefore, deep geothermal systems should not be planned and designed without appropriate numerical modeling of the heat and fluid transfer under various conditions of operation. During operation, modeling must be continued and accompanied with appropriate monitoring programs providing the appropriate data

(Sect. 8.8). The total lifetime of the installation can be estimated from simulation of the reservoir response to plant operation and sound prognosis of system progression may be possible if quality data are available.

Today, the global production of electrical energy from geothermal sources is absolutely dominated by the high-enthalpy fields (Sect. 4.4). The geodynamic setting of the high-enthalpy fields results in high geothermal gradients and correspondingly require short shallow level drillholes to reach the reservoirs with hydrothermal fluids of several hundred degrees centigrade (Sect. 1.4). Depending on the P-T conditions in the reservoir, the systems can be steam or liquid (water) dominated. State-of-the-art also in high-enthalpy fields is the reinjection of the cooled liquid-phase to the reservoir. The condensed steam phase is often nontoxic and it could be discharged into surface waters. However, it is recommended to inject this water also for safeguarding the reservoir (Chap. 10).

Because of the small temperature difference between supply and return flow in low-enthalpy installations, the maximum efficiency of these systems is intrinsically lower than that of high-enthalpy systems. The secondary loops used in low-enthalpy plants (ORC, Kalina) presently consume up to 25 % of the produced electricity for pumps and other equipment (Sect. 4.2). However, low-enthalpy geothermal plants have a great potential and probably an excellent economic future because of their relative insensitivity to the local geological setting.

Direct use of geothermal energy for local and district heating networks is widely used today (2013). We expect that geothermal energy utilization will expand particularly on the heat supply marked. The geothermal heat will save fossil fuels for more valuable products. The development and perspectives of geothermal heat utilization depends strongly on political programs for supporting or subsidizing geothermal energy in the different countries.

The use of near-surface geothermal energy, particularly by borehole heat exchangers and groundwater well systems dramatically increased during the past years. There is a rapidly growing market for installations for heating and cooling of buildings, both for private homes and for business and office buildings. A further rapid development is seen in systems that combine geothermal energy extraction from the ground with solar-thermal systems including storage of excess heat in the ground during summer for retrieval during the cold season. Because these systems operate with electrically driven heat and fluid pumps, energy efficiency and ecological value are determining factors for the economic benefit also.

The continuous increase of demand and price for fossil energy resources in combination with energy political conditions suggest the intensified use of geothermal heat energy from sources that are available everywhere at any time. This is particularly obvious on the background of the fact that about one third of the total energy consumption in areas of middle latitudes goes into heating. Cities and metropolitan areas will revise heat supply concepts and develop efficient district heating systems. Annual and seasonal municipal heat energy management will require new concepts for ground utilization as a heat energy storage reservoir. Geothermal energy utilization techniques will be central and indispensable in the future energy industry.

Chapter 6
Geothermal Probes

Drilling equipment for shallow wells

6.1 Planning Principles

The basic condition of near-surface geothermal energy utilization is the low temperature of the thermal reservoir. The temperature is typically lower than the working temperature of house heating. The heat transfer fluid in house heating systems requires a minimum temperature of about 20–30 °C, whereas ground temperatures are typically in the range of 5–15 °C. Therefore, in order to use the geothermal energy for the heating of buildings the transfer fluid temperature must be increased by means of a heat pump system. The highest reservoir temperatures are accessible to geothermal probes. Depending on the depth of the probe, drillhole heat exchangers may provide fluid temperatures of 10–12 °C depending on the local conditions (Central Europe). The temperature increase needed by the house heating system is then done by the heat pump. Most heat pumps are driven electrically. Electricity is expensive and produced with large losses from fossil energy resources in some countries.

For this reason in the run-up of geothermal energy utilization, a project should make each effort to reduce the heat demand of the planned building or object. This includes thermal insulation measures such as façade and roof insulation, high quality insulating and heat absorbing glass windows.

The economic efficiency of the system and its environmental value depends critically on the required heating temperature. The temperature for underfloor heating systems is typically about 35 °C, active cover or concrete core activation requires about 25 °C only, whereas heating with the classical hot-water radiators craves 45–65 °C flow temperature. These considerations can be easily integrated in the planning of new buildings leading finally to an economically and ecologically optimal heating system. More problematic are restorative measures in existing buildings.

To promote sustainability and long term operation of the heating system it is important to limit the withdrawal of thermal energy from the ground during the annual house heating period to the natural heat influx to the reservoir. Heat extraction must be balanced by natural regeneration.

6.2 Construction of Ground Source Heat Exchangers

Geothermal probes are liquid-filled tubes installed in a borehole. There are different types of geothermal probes including single U-tube probes, double U-tube probes and coaxial probes (Fig. 6.1). Single U-tube probes are closed seamlessly drawn plastic tubes with a U-shaped foot. Double U-tube probes are two independent single U-tubes installed in the borehole. Cool liquid flows downward in the tubes and accumulates heat from the surrounding ground. The warmed liquid turns around in the U-shaped foot at bottom hole and flows back to the heat pump at the surface. The heat pump uses the extracted ground heat to increase the fluid temperature of a secondary cycle so that it can be used for heating purposes.

6.2 Construction of Ground Source Heat Exchangers

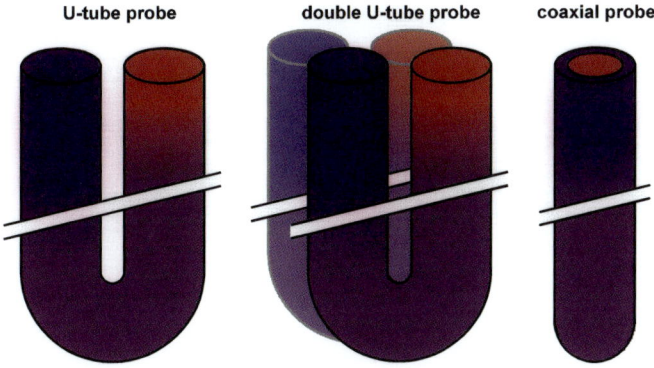

Fig. 6.1 Schematic illustration of types of probe tube design

Coaxial probes contain the return tube to the heat pump in the center of the tube for the descending liquid of the probe.

Double U-tube probes are most commonly used. The advantage is that in case of a tube damage the borehole is not completely lost but can be used as a single U-tube probe with the second tube.

The geothermal probe extracts heat from the ground and cools the vicinity around the probe. The resulting thermal cone is analogous to the drawdown cone in groundwater hydraulics (Fig. 6.2). The geothermal probe receives its heat supply from the surroundings and the efficiency depends on the thermal conductivity of the underground system and its parts.

The length of a probe depends mainly on the system design and the thermal properties of the ground. Essential properties are the thermal conductivity of the individual strata, the temperature distribution in the subsurface and the climatic situation of the area. Also important are the thermal properties of the probe, of the grouting material (backfill) and of the heat transfer fluid.

A special probe design, called heat pipes or thermosyphon, works with phase changes of the heat transfer fluid along the fluid flow path in the probe thus making use of the latent heat of vaporization and condensation. In contrast to the conventional probes heat pipes are typically made of metal (Sect. 6.8.5).

The outside diameter of standard probe U-tubes is normally 32 mm, rarely 40 mm or even 25 mm. Koaxial probes have commonly outside diameters of 63 or 50 mm, occasionally also 40 mm. The central pipe has a smaller diameter of 32, 40 or 25 mm respectively.

The probe pipes are usually made of polyethylene (PE 100). The pipes specification conforms to a nominal pressure of 16 bar (1.6 MPa). This means that probes of more than about 160 m vertical length require special measures and care for correct installation, particularly if the groundwater table is low (Sect. 6.6).

Nearly all probes consist of polyethylene, which is a poor heat conductor with a low thermal conductivity of about 0.4 W/mK. Due to this fact the heat transfer from the ground to the heat transfer fluid is not particularly efficient. New probes

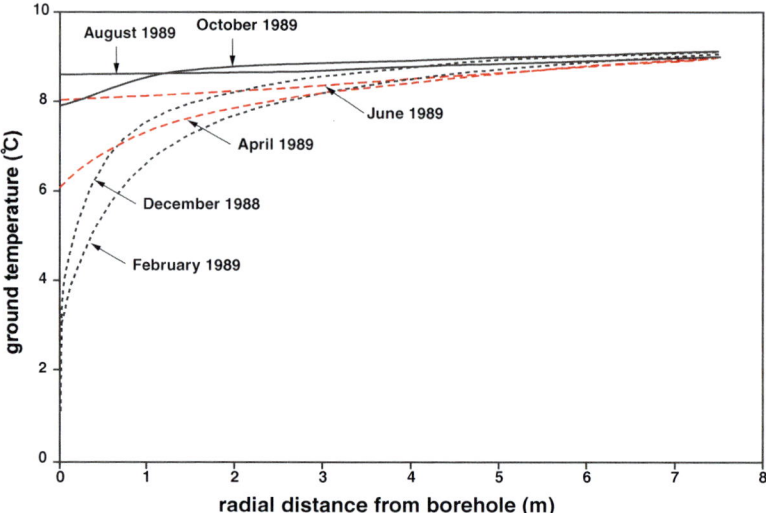

Fig. 6.2 Dynamic temperature decrease around a geothermal probe (measured data and model calculation for a specific installation in Elgg, Switzerland, Eugster 1998). Deepest thermal cone in February 1989, thermal recovery after shutdown of installation until August 1989, renewed development of cone after reactivation until October 1989

have been developed using raw material of increased thermal conductivity of up to 1.0 W/mK and launched on the market recently.

The polyethylene U-tubes come factory-welded, they are not welded at the construction site. Cross-linked polyethylene pipes have a warm-bent probe foot from the manufacturer and do not require welding. Cross-linked polyethylene pipes have a superb resistance to stress cracking and other mechanical damages compared to simple tubes. Probes made from cross-linked polyethylene are also thermally durable and resist long-term exposure to temperatures up to 95 °C. Therefore probes made from this material can be used for transferring heat to the ground for instance in combination with a solar thermal installation. This way excess heat can be transferred to the ground for storage during the warm season. This helps to thermally restore the underground and permits even storage of additional thermal energy. Combining solar thermal and geothermal installations has the further advantage of saving probe length.

The probe foot is mantled with mechanical protection for borehole installation. A heavy weight at the base of the probe (Fig. 6.3) facilitates mounting of the probe into a groundwater-filled borehole. If the borehole is water-filled, it is necessary to fill the probes with heat transfer fluid (or water) prior to inserting them into the borehole to avoid buoyancy and excessive pressure onto the probe tubes. At the construction site the probe comes in the necessary length wound onto a reel ready for installation into the borehole (Fig. 6.4). The probe is wound off the reel and together with a grouting hose inserted into the borehole.

6.2 Construction of Ground Source Heat Exchangers

Fig. 6.3 Geothermal probe with installation weight

Fig. 6.4 Installation of a geothermal probe from the reel into the borehole

Ground source heat probes are closed systems containing a circulating heat transfer fluid. A number of different chemical compounds and mixtures are used as heat transfer fluids. Most common are aqueous solutions of organic antifreeze chemicals that depress the freezing temperature of pure water. Thus, the system can be operated over a larger temperature range than with pure water.

Consequently, the heat pump can extract more heat from the fluid that is eventually returned to the ground with a lower temperature. Accordingly, the lower inflow temperature extends the cone of thermal depression in the ground and increases the temperature gradient towards the probe. However, a certain danger for damages that compromise the durability of grouting and sealing occurs during the first series of freezing thawing cycles. It may have negative effects on the long-term efficiency and may cause conflicts with groundwater protection regulations (Sects. 6.5, 6.7).

Hydraulic and thermal properties of commonly used heat transfer fluids in their typical mixing proportions with water show that pure water has optimal properties. Pure water is the ideal heat transfer fluid (Table 6.1). In addition to the fluids listed in Table 6.1, other substances are used in mixtures with water including potassium carbonate, potassium formiate, bataine, magnesium chloride or sodium chloride. Ethylene glycol—water mixtures is probably the most commonly used heat transfer fluid.

Low dynamic viscosity and low density of the fluid increase the efficiency of a probe because of savings in pump power consumption. High heat capacity means high heat storage and high thermal conductivity implies efficient heat transfer. The higher the product of density and heat capacity is the less fluid must be pumped to transport equal amounts of thermal energy.

From these facts it follows that, as already mentioned, pure water is the ideal heat transfer fluid. However, geothermal probes that are operated with pure water must be precisely dimensioned to avoid freezing conditions during operation. A very positive side effect is that the system avoids damages on the backfill and the adjacent soil caused by freezing (Sects. 6.5 and 6.7).

Hydraulic and thermal properties of different heat transfer fluids (Table 6.1) depend on temperature and vary during circulation of the fluid in the probe. Especially the viscosity of the fluid changes strongly with temperature. The viscosity of ethylene glycol (25 %) at −8 °C is twice as high as at +12 °C, thus operation of the probe at freezing conditions requires considerably more electrical

Table 6.1 Isothermal (25 °C) hydraulic and thermal properties of heat transfer fluids commonly used in geothermal probes (Zapp and Rosinski 2007)

Fluid	Dynamic viscosity μ (kg/m s)	Heat capacity c_p (J/kg K)	Density ρ(kg/m^3)	Heat conductivity λ (W/m K)
Water	**0.0018**	**4,217**	1,000	**0.562**
Ethylene glycol 25 %	0.0052	3,795	1,052	0.480
Ethanol 25 %	0.0046	**4,250**	**960**	*0.440*
Propylene glycol 30 %	*0.0108*	3,735	1,038	0.450
Calcium chloride 20 %	0.0037	*3,050*	*1,195*	0.530
Methanol 25 %	0.0040	4,000	**960**	0.450

Emphasis: bold = best values, italic = worst values for system efficiency

6.2 Construction of Ground Source Heat Exchangers

power for the pump than at temperatures above freezing conditions. Consequently, for economical reasons a geothermal probe should not be operated at freezing conditions. Otherwise the overall efficiency of the entire system is compromised.

Heat transfer fluids also often contain special additives that prevent development of bio films in the probe or to prevent corrosion. Many of these chemicals are toxic and problematic from a legal point of view.

The drilling diameter must be chosen big enough to easily accommodate probe and grouting hose and leaving enough space for the sealing backfill. The total cross-sectional surfaces of the probe pipes and the grouting hose should be less than 35 % of the cross-sectional area of the borehole ($r^2\pi$). This permits a tight backfill with an excellent thermal connection to the ground. A 32 mm diameter double U-probe requires at least a 120 mm drill hole, 40 mm ground probes must have a borehole at least of 150 mm in diameter. The drilling diameter depends additionally also on the planned drilling technique and the geological details of the ground. The recommended drilling diameters above refer to down-the-hole-hammer drilling used typically for hard rock drilling. For rotary drilling (wash drilling) in loose rock the boreholes must have a larger diameter. An overview over commonly used drilling techniques for ground source heat probes is given in Sect. 6.4.

The probe tube is directly unwound from the reel and is carefully placed into the bore together with the grouting hose. For safety reasons motor-driven reels should be used for installing long probes that permit machine-controlled braking. The grouting hose must be attached to the probe before lowering the pipe and hose together into the wellbore. After the installation of the probe pipe it is virtually impossible to put the grouting hose in place separately. In a next step, the grouting material is pumped through the hose to fills the space between the probe pipes and the borehole wall from bottom hole to the well mouth (tremie method, contractor procedure). This method ensures best thermal connection of the probe to the ground and optimal sealing. Before injecting grouting material the ground probe pipes must be liquid-filled and pressurized to avoid damaging the pipes.

Prior to grouting and after binding, complete pressure tests assess if the probes are leak-proof. A flow test verifies the permeability of the ground probe.

Directly touching probe pipes have a significantly lower heat transfer capacity compared to separately guided pipes (Acuña and Palm 2009). The ascending and descending probe pipes at different temperature should be separated in the borehole. This is practically achieved using inner spacers. Outer spacers or centering aids help to install the probe centrically in the borehole in order to reach the best thermal connection to the ground and optimal sealing. Numerous centering aids and spacers in a short distance are necessary to keep the pipes centered in the borehole and separated from each other. The heat conductivity of the material used for these aids is low and increases the thermal resistivity in the borehole. Improved combined spacers are relatively easy to handle and may help to improve the efficiency of the installation (Fig. 6.5). The usefulness of centering aids and spacers is under debate, especially because the current models often slip during installation of the pipes in the borehole. The tools may also obstruct the ascending grouting material, particularly of course when slipped, thus creating blowholes or air-filled cavities.

Fig. 6.5 Combined centering aid and spacer used for geothermal probes (*blue pipes*). Central blue pipe: Grouting pipe

The plastic pipe material tends to be stiff and vulnerable to mechanical damaging in cold weather. During construction work under cold weather conditions, the probe should be stored at a warm place prior to installation or flushed with warm water at the construction site.

6.3 Dimensioning and Design of Geothermal Probes

The design of borehole heat exchangers is made on the basis of the heat demand of the building that should be supplied by the geothermal probe. The heat extraction rate of the probe depends on several factors including the geological and thermal structure of the subsurface at the site, the type of the probe, the heat transfer fluid and the grouting. The hydraulic coupling of the geothermal probe with a heat pump creates an additional dependency. A geothermal probe can be reliably, efficiently and economically operated for long periods of time only, if all relevant parameters have been considered and optimized. For the optimization of the geothermal probe the planning must include the architect and the planner of the building services.

Geothermal installations often have considerable total pipe lengths. Numerous branching, bows and fittings cause an increasing flow resistance with increasing distance from the heat pump and gradually reduce fluid flow and heat transfer. Therefore, it is important to know the flow properties in the system and to keep flow resistance low by optimized dimensioning of the system components. Only a flow-optimized installation has the potential to become an efficient system. The flow resistance of the supply (feeder) lines and in the connecting blocks should be as low as possible. However, in the vertical probe pipes fluid flow should be turbulent for best heat extraction rates. Laminar flow would cause much lower flow losses but the heat transfer from the pipe wall to the heat transfer fluid is much better with turbulent flow. Thus in the pipes of geothermal probes the fluid should flow turbulently.

If probe pipes and lines to the heat pump are of different lengths a hydronic balancing preceding the heat pump improves the heat extraction efficiency.

Vital for the success of heat pump heating systems are the failure-free long-term operation and low electrical power consumption. The technical parameters of the heat pump, the source temperature and the heat requirement of the heating system mutually influence each other. It is difficult to reliably predict the operating behavior and the economic viability of the heating system without computer simulation (modeling).

6.3.1 Heat Pumps

Heat pumps are devices that transfer thermal energy from a source at a low temperature to a reservoir at a high temperature by means of mechanical work. The high-temperature reservoir can then be used, for example, for house heating. These machines make it possible to use a relatively low-temperature heat source such as the near surface ground, soil or groundwater for house heating purposes.

Technical types of heat pumps:

- Compressor heat pumps,
- Sorption heat pumps, subdivided in adsorption- and absorption heat pumps,
- Vuilleumier heat pumps.

Other existing technical solutions for heat pumps may not have potential for being used in house heating and warm water supply in the near future.

Compressor heat pumps are considered state-of-the-art and most widely used. Depending on the type of motor one distinguishes between electric motor and gas motor driven compressor heat pumps. Near-surface geothermal energy installations almost exclusively use electric motor compressor heat pumps. Therefore its technical principle is briefly explained below. In principle, however, compressor heat pumps can be driven also with natural gas, petroleum, or biofuel (biogas, rapeseed oil). In such devices the compressor is driven by a combustion engine. The use of gas-driven compression heat pumps has the advantage that the heat of the exhaust can also be used for heating and thus the primary energy input is used more efficiently than with electric heat pumps.

Sorption is a physical-chemical process where liquids dissolve other liquids or gases (Absorption) or solids such as zeolites capture liquids or gases at their surface (Adsorption). Sorption processes are driven by external physical parameters such as temperature and pressure and can be reversed if the parameters are reset (reversible processes).

The Vuilleumier heat pump works after the principle of a thermally driven regenerative gas cycle process comparable to the Stirling process.

The electric motor compressor heat pump, or simply the heat pump in the following, works like a refrigerator, however, with the difference that heat production of the condenser is the desired power output and not the cooling power of the evaporator (Fig. 6.6).

The fluid of the internal circuit of a heat pump extracts heat from a heat source in a first heat exchanger (the evaporator). The fluid undergoes a phase change from liquid to gaseous. The gaseous fluid reaches the compressor unit driven by an electric motor where the fluid pressure is increased. The resulting associated temperature increase permits the extraction of the heat transferred to the fluid from the original heat source (here the heat from the ground source) in a second heat exchanger, the condenser unit. The high pressure of the gaseous fluid is reduced in an expansion valve cooling the fluid below condensation and it returns to the liquid state. The liquid fluid returns to the evaporator where it is reloaded with thermal energy from the ground source.

Many different heat transfer fluids are used in heat pumps including pure or mixtures of liquids such as partly fluorinated hydrocarbons, pure hydrocarbons (propane, butane) and carbon dioxide. Ammonia is not approved in many countries because of its potential hazards.

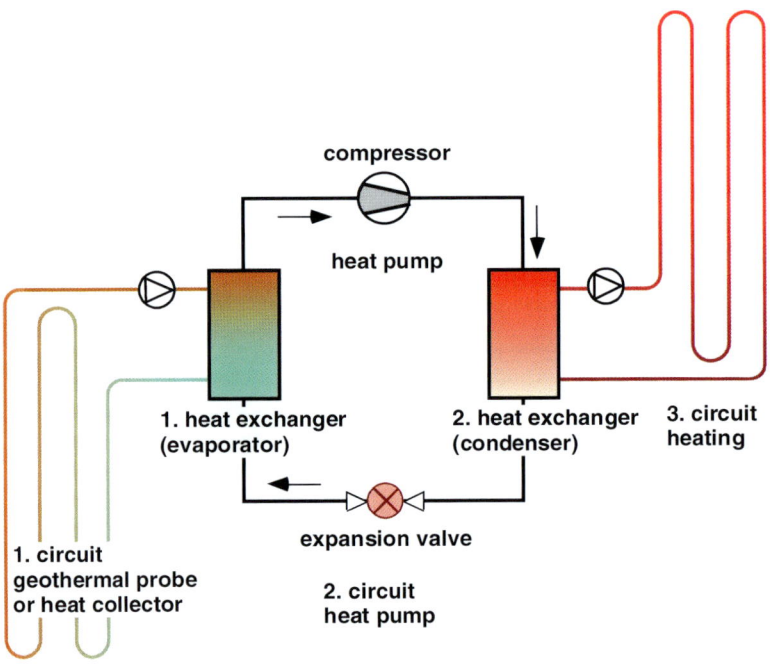

Fig. 6.6 Schematic diagram of a heat pump system utilizing geothermal energy for house heating

6.3 Dimensioning and Design of Geothermal Probes

The primary undisturbed temperature of the ground heat source is given by the thermal and hydraulic properties of the subsurface and the climatic conditions at the location of the probe. During operation of the geothermal probe, heat extraction cools the vicinity of the probe pipe. The effective temperature of the transfer fluid entering the evaporator is still lower than the cooled ground because of unavoidable physical heat transfer losses.

The heat pump needs to produce the required thermal power at the expected lower limit of local conditions. For example, probe systems are typically designed to work at −12 °C surface temperature in Central Europe. They will have problems to heat a house if the temperature falls to −25 °C. For efficiency reasons it is essential to reduce the heat demand of a building by thermal insulation measures prior to planning a geothermal probe heating system. In addition, the supply water temperature of the heating system should be as low as possible. Ideally, the temperature difference between the fluid from the ground source (1. circuit) and the supply water (3. circuit) should be small, saving electrical energy for the heat pump (Fig. 6.6). Typical traditional radiator heating in buildings requires 55 °C warm supply water. Under floor heating can be operated with 35 °C warm water. Large area wall heating requires even lower supply water temperatures.

Invariably, the geothermal probe or the ground source heat exchanger should be generously dimensioned. The source temperature is then high and causes an increased efficiency of the heat pump. The geothermal probe should always be operated at temperatures above zero degrees centigrade.

The dimensioning of the geothermal probe must be adapted to the evaporator capacity of the heat pump. The heat source, specifically the total length of the probe is tailored to the heat demand, the desired extraction rate and operating life. A geothermal probe is not characterized by a defined and constant power at a certain operating point, as it is often specified on heat pumps. A probe may extract high power for a short period of time or lower power for an extended period (Basetti et al. 2006). Therefore, defined extraction profiles (e.g. 1,800 h of thermal performance, monovalent per year) should be verified in advance. The design of a geothermal probe system must also consider the dimensioning of the circulation pump for the heat transfer fluid in the probe (Sect. 6.3.2).

It is generally economically worthwhile using the heat pump for hot water preparation also. However, in this case the system is in operation throughout the year and the time for a certain thermal regeneration of the soil is short. For the hot water preparation the output temperature must be increased to 60 °C, which is higher than for heating. Combining the geothermal probe with a solar thermal installation may give the best economical and energy solution (Sect. 6.8.3). The excess heat produced by the solar thermal installation during the summer can be stored in the ground for supporting house heating during winter with the geothermal probe. During the summer, waste heat from house cooling also can be brought to the subsurface with the geothermal probe (Sanner and Chant 1992).

The so-called coefficient of performance (COP) number facilitates the evaluation of the quality of heat pumps. The COP number is the ratio of electrical power of the compressor plus auxiliary power and the thermal capacity of the condenser (both in kW). COP values increase with increasing efficiency of the heat pumps. COP also increases with decreasing temperature difference between ground source and supply water of the heating system. The COP value, however, does not include the energy demand of the circulation pumps of the geothermal probe and the heating circuit. It is evident that efficient operation of the heating system correlates with high ground source temperature and low required supply water temperature.

For the complete heat pump heating system the seasonal performance factor (SPF) is the most significant and relevant parameter that characterizes the value of the system. SPF is the ratio of the total amount of thermal energy transferred to the heating and hot water production per year and the total amount of electrical energy taken up by the system during the same time (both numbers in Joules, or kWh). $SPF = 4$ means that 1 kWh electricity produced 4 kWh heat output. The energy efficiency of the entire system is the better, the higher the seasonal performance factor is. The SPF and, hence, the performance of the system can only be controlled if an electronic heat meter is installed and measures the thermal energy produced by the heat pump.

COP is a device performance factor that characterizes the heat pump as a machine. It depends on the operating conditions. In contrast, the seasonal performance factor (SPF) is an economical and energy political significant parameter, which depends not only on the installed machine but also on the habits of the user, the climatic situation, operating conditions and other factors.

If the heating system ought to be beneficial from the viewpoint of primary energy consumption, then the seasonal performance factor of an electrically driven system must by clearly higher than the equivalent SPF calculated with the primary energy needed for the production of the electrical energy to run the system. Energy conversion efficiency is different from country to country. In Germany for example, 3 kWh primary energy (from a blend of differently operated power plants and other energy sources) is needed on average for the production of 1 kWh electrical power. Therefore, the legislator requires a minimum SPF of 4 for water-based heat pumps used for house heating. If used also for hot water preparation SPF must be better than 3.8.

In practice, minimizing the temperature difference between evaporation and condensation (heat source and heat output) represents a significant optimization potential for the system engineering of heat pump systems. Each extra Kelvin temperature difference means an excess energy consumption of the compressor of 3.5 %. Thus the hot water part of the building (bath, hot water) requires the highest water supply temperature of the entire installation.

For efficient operation every heating system must be hydraulically balanced. Particularly important is this for heating systems with heat pumps. Hydronic balancing is done on the house/building side of the system right after the heat pump. The installer of the heating system adjusts the flow volume of hot water individually for each radiator of heating circuit of a panel or under floor heating to where the flow volume in each room covers the heat requirement of the room for a given supply water temperature. Thus each room receives exactly the amount of heat needed to reach the desired room temperature. After successful hydronic balancing the heating system can be operated with an optimal system pressure, an optimally low volume of water and the lowest possible supply water temperature, or in other words with the best possible system efficiency.

We recommend considering the following advice for installation of heat pumps:

- Thorough design of the entire system, adjusting the different components (heat source, reservoir, heat sink…) to work as a part of a well orchestrated total system and an integral trade-spanning multidiscipline object specific planning,
- Check and verification of loading strategies of the reservoirs, particularly of combination storages, and monitoring the supply water temperature,
- Careful hydronic balancing and gapless insulation of pipes and components,
- Avoid overly complex hydraulic and reservoir systems,
- Properly designed systems do not require auxiliary electrical heating (heating rod), except perhaps during construction drying.

For high output ranges above about 100 kWth, gas-absorption heat pumps become an interesting alternative to the electric compressor heat pumps. The advantage of the absorption technology is, heat and cold can be used in chorus thus increasing the total efficiency of the system significantly.

6.3.2 Thermal Parameters and Computer Programs for the System Design of Ground Source Heat Pump Systems

A broad-brush dimensioning of a geothermal probe is useful for a cost estimate at best. A ground source heating system must be skillfully planed and professionally dimensioned. An undersized system may cause significant damages to components (Sect. 6.7). If probes have been installed which are too short, the mistake can only be repaired with the installation of a new probe in an additional borehole. We strongly discourage installing electrical heating rods to compensate for the lack of power on economical and ecological reasons. Short-term operation of auxiliary heating rods can possibly be justified in certain emergency situations (Sects. 6.3.1, 6.7).

A rough estimate of the necessary length (l [m]) of the geothermal probe to cover the heating requirement (H [W]) of the object to be heated is normally based on the so-called specific heat extraction rate (E [W/m]). The specific heat extraction rate of a geothermal probe is normally related to the thermal properties of the different ground layers (E_i) that have been intersected by the borehole, specifically to the thermal conductivity of these layers. This is common usage, although the specific heat extraction rate depends on many other factors as well (e.g. grouting material, heat transfer fluid). Strictly speaking, the specific heat extraction rate of a geothermal probe does not exist but merely a potentially recoverable, however, intrinsically variable cooling or heating power of the ground.

In any case, of prime importance is a detailed and precise description of the drilled layers. Laboratory measurements on drilled material give answers to the requested thermal conductivities. These data can then be used to verify the predicted profile based on assumed thermal conductivities of the layers.

The thermal conductivity laboratory data often fail to precisely characterize the real situation in the subsurface for various reasons. The used samples may not representatively reflect all the heterogeneity including fractures and local variability of a certain layer. Laboratory data also do not characterize the effects of stagnant or flowing groundwater in the pore space or the thermal consequences of air-filled pore space in the vadose zone. Therefore, it is important and recommended to acquire data by direct in situ measurements using so-called Thermal Response Tests (Sect. 6.3.2).

The thermal properties of a particular rock (thermal conductivity, heat capacity) may vary widely depending on its detailed local structure and water-content (Table 6.2). To be on the save side, it is recommended to work with the pessimistic lower parameter values if any ambiguity on the true values exist during the planning stage of the project. The listed values for the potential extraction rates (Table 6.2) should be considered as rough estimates for orientation. However, it follows from the compilation that the potential extraction rate varies by a factor of three depending on the material present underground.

Also note that the extraction rate for a specific geological unit may vary from place to place and cannot be safely predicted using Table 6.2, or any other fixed compilation. The parameter cannot be transferred from site to site. Remember that the true achievable extraction rate also depends on local climatic factors, the installed system, the heating habits of the residents and other factors. It does not exclusively depend on the properties of soil and rock. Here it is used as a parameter for rough

Table 6.2 Compilation of thermal conductivity λ, heat capacity s and formal heat extraction rate data for various types of rocks (from VDI 2001)

Ground, rock	Thermal conductivity (W/mK)	Heat capacity (MJ/(Km^3))	Extraction rate (W/m)
Gravel, sand dry	0.4	1.4 to 1.6	20 to 30
Gravel, sand moist	0.6 to 2.2	1.2 to 2.2	30 to 50
Gravel, sand wet[a]	1.8 to 2.4	2.3 to 3.0	55 to 70
Moraine	1.7 to 2.4	1.5 to 2.5	40 to 55
Clay, loam moist	0.9 to 2.2	1.6 to 3.4	30 to 50
Limestone dense	1.7 to 3.4	2.0 to 2.6	45 to 65
Marl	1.3 to 3.5	3.0	40 to 60
Sandstone	1.3 to 5.1	1.6 to 2.8	40 to 70
Conglomerate	1.4 to 3.7	2.1	40 to 65
Granite	2.1 to 4.1	2.1 to 3.0	50 to 70
Basalte	1.3 to 2.3	2.3 to 2.6	35 to 55
Andesite	1.7 to 2.2	2.4	45 to 50
Quartzite	3.6 to 6.0	2.1	65 to 92
Breccia	2.2 to 4.1	2.1	50 to 70
Schist	1.5 to 2.6	2.2 to 2.5	40 to 55
Gneiss	1.9 to 4.0	1.8 to 2.4	50 to 70

[a]Water saturated

The extraction rate data refer to a single geothermal probe and for an operating time of 1,800 h per year. The values are prone to the reservations made in the text verification of the system design. The parameter should never be used as the only tool for the dimensioning of the system without any further critical reflection.

In practice, for a specific heat demand of an object (H[W]) the necessary length of the geothermal probe is commonly estimated from the known stratigraphy and thicknesses of the individual layers (h_i[m]) by summing the contributions of the layers to the specific extraction rate:

Single layer case:

Probe length h (m) = heat requirement of the object H(W)/specific extraction rate E (W/m)

Multilayer case:

$$\sum (E_i \cdot h_i) = H \quad [W] \qquad (6.1a)$$

$$\text{Probe length h [m]} = \sum h_i \qquad (6.1b)$$

The heat extraction rate of geothermal probes varies between 20–100 W per meter probe length, with outliers in both directions. The indicated variation is large and may be even larger under unfavourable circumstances.

The definition of one heat extraction rate that characterizes a geothermal probe is only sensible if it does not fluctuate much in the building/object (Sect. 6.3.1). Large-scale plants but also small units (<20 kW) with high power variations need to be designed with dedicated expert tools. The same holds true for bivalent systems, systems that include hot water production or swimming pools, for integrated geothermal fields and combined systems for heating and cooling. Furthermore, geothermal probe projects in areas where the average annual surface temperature is below about 10 °C should be planned with special dimensioning tools. At such low surface temperatures, which is equivalent to average temperature of the near surface soil layer, the extraction rate of the probe rapidly decreases.

The extraction rate of a geothermal probe, as described above, depends on many factors including (Kohl and Hopkirk 1995; Signorelli 2004):

- The conductive and convective heat transfer power of the ground and of individual layers (thermal conductivity, flow rate, …),
- The temperature distribution in the subsurface, the climatic situation at the building site,
- The duration of heat extraction from the ground (annual operating hours),
- The diameter of the drill hole,
- The thermal properties of the grouting (backfill) material,
- The type of heat transfer fluid used, the type of the probe, material of the probe, position of the pipes in the borehole (Sect. 6.2),
- Design and geometry of the probe field: Distance of the probes from each other, number and arrangement of the probes.

It follows from the diversity of the listed factors how complex the assessment of the true extraction rate of a geothermal probe really is (Sect. 6.3.1). Therefore, the planning engineer relies on computer codes and tools for the optimal dimensioning of a system. Some of the more popular tools are:

- The application EWS by Huber (2008) (http://hetag.ch/)
- The application EED (Earth Energy Designer) by Sanner and Hellström (1996) and Hellström and Sanner (2000), (http://www.buildingphysics.com/index-filer/Page1380.htm)
- The application TRNSYS by The University of Wisconsin Madison (http://sel.me.wisc.edu/trnsys/)
- The code PILESIM (Version 2, 2007) by Pahud (1998).

Such applications compute and model the expected temperature evolution in the heat transfer fluid at the entry point to the heat pump as a function of time. The programs are of variable complexity and some do not consider all listed factors that may influence the heat extraction rate. Thus under detrimental circumstances the programs may produce erroneous results. Some of the programs also handle simultaneous computation of multiple probes or of entire ground source heat exchanger fields. The number of geothermal probes, their arrangement and length can be varied and selected to where the target settings for the temperature course can be achieved with the lower and upper limits. Further programs on the marked such as EWS, EED or PILESIM also consider the heating requirements of the building, that is the monthly energy heating and cooling requirement and the heating and cooling load. An example of a modeling program for heat pump heating systems is WP-OPT© (www.wp-opt.de). It permits planning and optimization of heating systems using heat pumps (Sect. 6.3.1).

Because of variable pumping temperatures from site to site and from application to application it is evident that a standardized geothermal probe does not exist. Thus the professional dimensioning of the geothermal probe is of vital importance for its later successful operation. Dimensioning of geothermal probes with rule-of-thumb values such as 45 W per running meter of probe that do not consider specific particularities of site and building is unfortunately widespread practice. Such an unprofessional approach may lead to irreversible damages to the system (Sect. 6.7).

Fundamental research on complex systems requires three-dimensional models such as the 3-D finite element program FRACTure (Kohl and Hopkirk 1995). Using FRACTure Signorelli (2004) has demonstrated for example that the temperature of the top soil layer may have a larger effect on the dimensioning of geothermal probes than the thermal conductivity of the underground.

Numerical modeling using different programs consistently suggest that if two geothermal probes are placed at a distance of 10 m and more no significant mutual interference (>1 °C) will occur even after many years of operation. Most commonly a distance of about 7 m is sufficient. But again, these specific distances should not be generalized as they vary with the number of ground source heat exchangers to be installed, their arrangement in space, their depth, if the probes are placed in flowing ground water or if the ground consists predominantly of clayey sediments with low thermal conductivity.

The thermal range of geothermal probes used for heating is principally larger if the ground is predominantly clayey and silty compared to aquifers consisting

6.3 Dimensioning and Design of Geothermal Probes

of sand and gravel. This is because of the low hydraulic conductivity of clay and silt, which results in extremely low ground water flow velocities even if hydraulic gradients are high. Numerical experiments have shown that the thermal range, defined as effects greater than 1 °C after year-long operation, may exceed 10 m for soils with low hydraulic conductivity. The findings have been confirmed by observations in the test-field Elgg near Zurich (Eugster 1998).

Guidelines for the design distance of borehole heat exchangers are based on the assumption that the boreholes are accurately vertical. In practice, it is not always feasible to drill perfectly vertical for justifiable expense. Placing a standpipe in the upper meters of the borehole helps to accomplish relatively vertical drilling. In addition, also the plastic probe pipes do not run strictly vertical but are coiled in the drill hole. Taking all this into account it is wise to observe a recommended minimum distance for the probes of 10 meters.

Thermal Response Tests (TRT) are well established tools for the in situ measurement of the thermal properties of the ground (Morgensen 1983; Gehlin and Nordell 1997; Gehlin 2002; Sanner et al. 2000). The tests inject a defined amount of heat energy into one end of the loop of the probe over a period of several days and measure the outflow temperature at the other end of the loop. During a test the principal parameters measured are the temperatures of incoming and outgoing heat transfer fluid, the flow rate of the fluid, and the heat energy input. The particulars of the temperature increase provide information on the thermal properties and structure of the underground in the vicinity of the borehole analogous to the hydraulic properties with a pumping test (Sect. 13.2) (Fig. 6.7). Instead of hydraulic conductivity and the specific storage coefficient, which are measured with

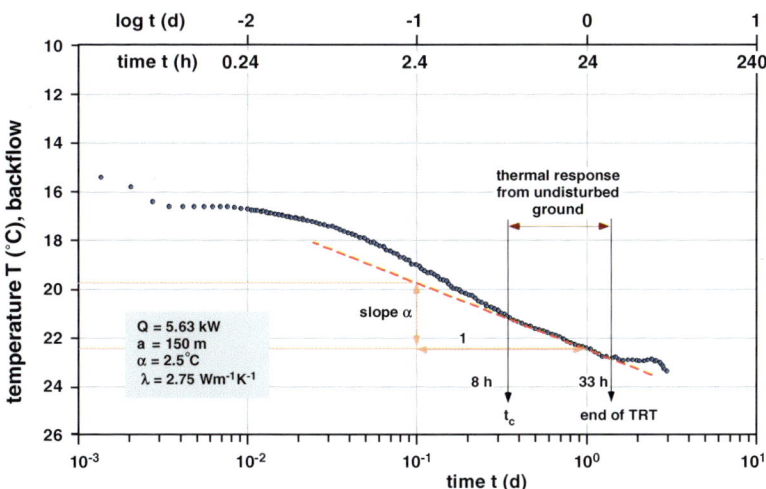

Fig. 6.7 Interpretation of Thermal Response Test data: Temperature of the heat transfer fluid versus logarithm of time. The slope of the straight line through the thermal response of the infinite homogeneous matrix provides the average effective thermal conductivity

hydraulic pumping tests, Thermal Response Tests provide thermal conductivity and the specific heat capacity. The hydraulic effects of the wellbore, skin and wellbore storage (Sect. 13.2) have their equivalents in thermal tests and are lumped to effects of the "inner zone". This structurally damaged area around the borehole shows a complex sequence of thermal resistances between circulating groundwater and intact ground, heat transfer to the probe pipe, the grouting material and other structural elements. The "inner zone" thermal structure depends also on the type of probe (single U-pipe, double U-pipe, coaxial pipe) and the diameter of the borehole. The thermal borehole resistance R_b [K m/W] is the sum of all effects of the "inner zone".

The comprehensive hydraulic computation tools that have been mostly developed by the oil and gas industry during the last decades are ultimately based on analytical solutions of the line source problem (Theis 1935). Originally the mathematical solution for the line source stems from the subject matter of heat conduction (Carslaw and Jaeger 1959) and has been adapted by Theis (1935) to the evaluation of pumping tests. Today, highly developed and sophisticated hydraulic evaluation tools can be modified for thermal evaluation methods such as the Thermal Response Test (Sect. 13.2).

Numerical models for the interpretation of Thermal Response Test data have been presented for example by Wagner and Clauser (2005) and by Gustafsson (2006). The advantage of using numerical models is that they allow considering simultaneously any kind of boundary conditions and spatially heterogeneous ground properties (e.g. Diersch 1994). In practice, however, algorithms based on analytical solutions still dominate the daily working routine.

In order to access the undisturbed original rock and its thermal properties and the inhomogeneity of these properties beyond the near field of the borehole ("inner zone") with a Thermal Response Test, the test must be of adequate duration by analogy to a pumping test. Otherwise the TRT is useless and the obtained pseudo-data should not be used as basis for the system design. As a general rule, a successful TRT typically lasts for several 10 s of hours. On-line recording of the data stream and continuous evaluation of the data allow for stopping the test as soon as the data can be implicitly modeled. The interpretation of TRT data is usually based on the analytical solution of the line source equation with the asymptotic solution for sufficiently long periods in analogy to hydraulic pumping tests (Cooper and Jacob 1946). The temperature T(°C) at distance r(m) from a thermal line source with a constant heat output Q (W) in an infinite homogeneous and isotropic ground with a thermal conductivity λ [Wm^{-1}K^{-1}] and a volumetric heat capacity s (Wsm^{-3}K^{-1}) can be computed from Eq. (6.2):

$$T(r, t) = T_0 + Q/(4\pi\lambda H) \times [\ln(4\lambda t/sr^2) - 0.5772] \qquad (6.2)$$

where H [m] denotes the length of the geothermal probe (test length), T_0 is the original undisturbed ground temperature. Equation 6.2 does not consider the thermal borehole resistance.

The effect of the thermal borehole resistance R_b can be approximated by adding the following term to Eq. 6.2 (Hellström 1998):

6.3 Dimensioning and Design of Geothermal Probes

$$\Delta T_{iz} = QR_b/H \tag{6.3}$$

This approximate description is sufficiently accurate for $t > 4sr^2/\lambda$, that is for long periods of time. Temperature changes tend to become proportional to the logarithm of time ln (t) at large t. This correlation is used to describe the effective thermal conductivity of the volume of ground around the borehole and includes effects of groundwater flow. The thermal conductivity of the "inner zone" is expressed as an average parameter value for the total length of the geothermal probe H [m]. The effective thermal conductivity λ_{eff} follows from the slope α [Ks^{-1}] of the straight line through the data points representing the undisturbed ground in Fig. 6.7 after conversion of the logarithms from ln to log (log x = ln x/ln 10):

$$\lambda_{eff} = 2.303Q/(4\pi H\alpha) \tag{6.4}$$

A set of TRT data is shown as an example in Fig. 6.7. In this example, the average effective thermal conductivity of the drilled formations is $\lambda_{eff} = 2.75$ Wm^{-1}K^{-1} using the parameters given in Fig. 6.7. The graph clearly shows the thermal regime during the first period of about 8 h of the experiment is strongly influenced by processes other than heat transfer through undisturbed rock matrix. During this initial experimental phase effects of the "inner zone" that correspond to wellbore storage in hydraulics and other effects prevail and the criterion for the asymptotic approximation ($t_c = 4sr^2/\lambda$) not yet applies. Beyond that critical time t_c the thermal response of the undisturbed ground dominates. It can also be seen in Fig. 6.7, that a meaningful interpretation of TRT data requires a test lasting for about 33 h (1.5 days) in this particular example.

After determination of the thermal conductivity λ, the thermal resistance of the borehole R_b is obtained by rearranging Eq. 6.2:

$$R_b = H/Q\left(T(r,t) - T_0\right) - 1/(4\pi\lambda) \cdot [\ln(4\lambda t/sr^2) - 0.5772] \tag{6.5}$$

The heat capacity s is very similar for different kind of rock materials and thus has not much effect on R_b. For the example made here, $R_b = 0.15$ KmW^{-1} using the same parameters as above and an assumed heat capacity of $s = 2.7 \times 10^6$ Jm^{-3}K^{-1}.

The range of temperature changes in the ground is estimated from:

$$r = \sqrt{2.25\lambda t/s} \tag{6.6}$$

It follows from Eq. 6.6 that the range of temperature changes is independent of the heat input. Using the example parameters given above, Eq. 6.6 predicts that after one year of operation the zero influence line has reached a distance of 8.5 m from the borehole. A noticeable temperature effect at this distance, however, may not be expected before 10–30 years of operation.

The results of the described interpretation practice for TRT data are average values for the total test interval. It means that all derived thermal properties of the subsurface and the thermal resistance of the borehole integrated averaged values across all drilled layers for the total depths of the geothermal probe.

It is possible to record a vertically resolved temperature profile with an improved TRT technique. Special tools can measure temperature profiles inside the borehole. But also external fiber-optical temperature measurements from fixed devices outside the geothermal probe produce depths-resolved temperature data. These data can be used to identify thermal effects of groundwater flow. Hydraulically active aquifers can be discerned. If data from multiple temperature measurements has been collected, thermal conductivities of individual layers can be derived and separated from the thermal resistance. For this extended type of TRT fiber-optical temperature measurements proved to be beneficial. The depth-connected thermal parameters of the ground gained with this method can be correlated with the local stratigraphy known from drilling. If additional advective heat transport resulting from groundwater flow occurs at the site then the derived thermal properties for that layer are effective system properties not representing the rock only.

TRT is a relatively costly analytical method and is typically used for larger projects such as borehole heat exchanger fields or for larger development projects where it is planned to drill many geothermal probes. The dimensioning of a geothermal probe for heating a single-family home is done in practice on the basis of a predicted stratigraphy and the climatic situation at the site with the methods described above. Typical thermal conductivities are attached to the rock layers underground, the heat extraction rates and the dimensioning of the geothermal probe modeled with programs described above resulting in an estimate for the required length of the probe to be installed. Therefore, it is important that the drilled geology is determined and documented carefully, and the predicted or assumed stratigraphy is verified.

6.4 Drilling Methods for Borehole Heat Exchangers

Drilling methods are established routine and typically the client may choose from many competing service providers (Figs. 6.8a, b). Drilling the borehole for a geothermal probe should go on fast and at low cost. However, fast and cheap should not compromise the long-term durability of the system. The probes installed in the borehole are the foundation for the yearlong sustaining functioning of the heating system.

The boreholes must be drilled with the required diameter, the correct caliber, and straight vertical to the needed depths (Sect. 6.3). The required diameter of the borehole depends on the size of probe pipes, the type of ground to be drilled and the drilling method used. The borehole must be sufficiently wide to lower the probe pipes and the grouting hose into the borehole without being damaged. Certain geological conditions may require the use of packers and several separate grouting hoses. The chosen borehole diameter also needs to consider spacers or centering aids (Sect. 6.2). A forceful insertion of the probes into the borehole may result in severe damage.

6.4 Drilling Methods for Borehole Heat Exchangers

Fig. 6.8 **a** Drilling equipment for borehole heat exchangers. **b** Drillmaster at the operator panel of a drilling rig

For budget reasons the small-diameter boreholes for geothermal probes are drilled by direct mud drilling technique rarely by dry drilling or indirect mud drilling methods. In mud drilling the cuttings are continuously hauled by the mudflow, in dry drilling intermittently with a tool. Indirect mud drilling pumps the cuttings through the drill string to the surface whereas direct mud drilling lifts the cuttings through the annular gap between the drill pipe and wall of the borehole. An installed casing ensures the stability of the borehole in dry drilling; a density-adjusted drilling mud stabilizes the borehole in mud drilling methods. Commonly, dry drilling methods are used for installing the standpipe for the mud rotary drilling subsequently used to complete the borehole (Fig. 6.9).

Two different types of direct wash-drilling methods can be distinguished depending on the drive used. Rotary mud drilling is mostly used in soil and unconsolidated sediments, the borehole being stabilized by the drilling mud. Rotary percussive drilling (down-the-hole-hammer) is used in hard rock. Because an air stream transports the cuttings, the borehole wall must be stable. Therefore, the borehole is cased to the depths where solid hard rock is reached (Fig. 6.10).

A flushing flow of water and, if necessary, additives remove cuttings in rotary drilling methods. Rotary percussive drilling (pneumatic drifter drilling) produces the cuttings and drives the top hammer with compressed air.

An interesting variant of a rotary percussive method is the so-called Geothermal Radial Drilling. The method permits to drill a series of inclined boreholes radially from a fixed central drilling site. The resulting installation is a hybrid of ground heat collectors and vertical geothermal probes and is of advantage on smaller properties for utilizing all available land.

Still, the optimal layout of a geothermal probe depends on a detailed and correct description of the drilled geological layers independent of the drilling method used.

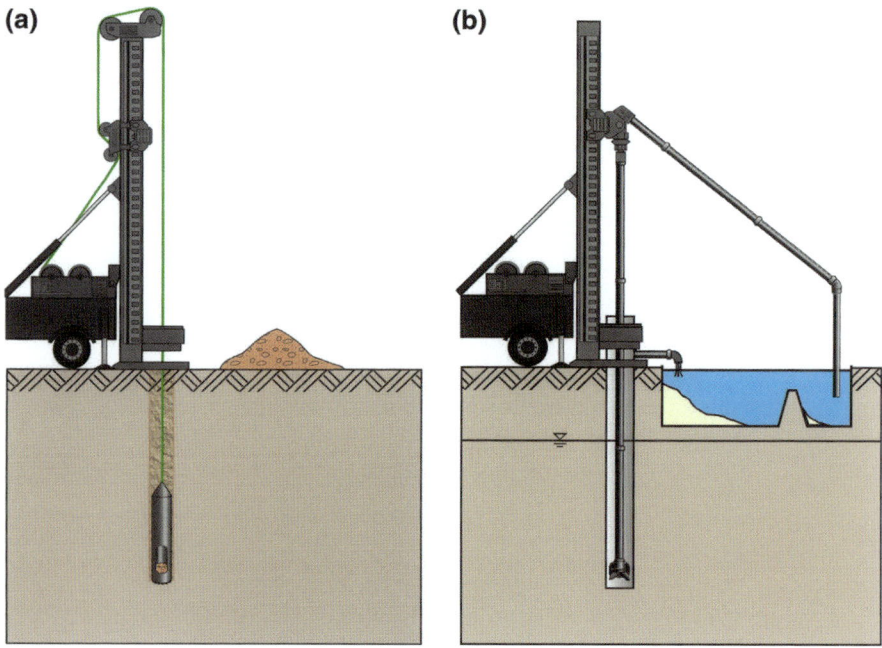

Fig. 6.9 a Cable drilling. b Washing fluid drilling

Fig. 6.10 The picture from a drilling site shows the standpipe with the drill string and the down-the-hole hammer in place

6.4.1 Rotary Drilling

Rotary drilling uses a top drive and the upwelling wash-fluid brings the cuttings through the annular space to the surface. The pumped fluid reaches an adequate flushing pool or settlement tanks or sedimentation ponds (Fig. 6.11). Samples of the cuttings for geological analysis need to be taken immediately from the washing-fluid before it reaches the tanks/pools. In the tanks, the cuttings settle from the drilling fluid. After sedimentation, the clean fluid is pumped through the drill string back to the drilling tool where it picks up new cuttings. The pumping pressure needs to be high enough to overcome friction losses in the drillstring. Direct-wash rotary drilling uses typically piston or centrifugal pumps. Centrifugal pumps can generate large flow rates but depend on the delivery height, in contrast to piston and displacement pumps.

The stability of the borehole depends critically on the overpressure of the drilling fluid, which corresponds to the difference between the groundwater table and the drilling fluid table. Consequently, the density of the drilling fluid must be adequately adjusted.

The flow velocity of the drilling fluid in the annular space needs to be about 0.5–1.0 m/s to produce the cuttings. Velocity depends on the density of the fluid and the size of the cuttings. The downward flow velocity of the drilling fluid in the

Fig. 6.11 Example of a simple settling tank for drilling fluid from a geothermal probe installation site. Behind the tank: Geothermal probe with mounting aid ready for installation

drill string is much higher and the flushing jet forcefully hits bottom hole, where it loosens the cuttings and washes them into the annular space. The flow velocity of the drilling fluid in the drill string depends on the inner diameter of the string. A small diameter results in a high flow velocity, however, friction losses increase considerably also. The upward fluid flow velocity in the annular space controls the maximum size of the cuttings that can be transported. Typical fluid ascending speeds of 0.5 m/s can transport cuttings not larger then about 8 mm.

Wash boring is a boring system to drill larger size holes in soft formations. Casing with a casing crown attached is rotated into the ground and water is used to flush out the drilled formation.

The drilling fluid has many functions in wash boring techniques. It stabilizes the wellbore, it produces the cuttings, it supports the drilling process and a clean bottom hole, it cools and lubricates drill bit and drill string and, in some techniques it drives the drilling tool (down-the-hole hammer, drilling turbine). Furthermore, a suitable fluid composition may help, to some degree, to control spontaneous variations of formation pressures.

Typical additives to the washing fluid in drilling boreholes for geothermal probes include: Bentonite clay powder, carboxymethyl cellulose (CMC) products and loading agents. Bentonite clay powder increases the viscosity of the drilling fluid and thus promotes the discharge of the cuttings, particularly if the fluid speed is low. Bentonite-bearing washing fluids tend to produce filter cakes in the wellbore. CMC products are polymeric additives that produce a thin filter cake grouting the borehole wall. The sealed wall facilitates the removal of the clay fraction from the borehole. The filter cake prevents loss of drilling fluid to aquifers and overburden. The seal stops infiltration of washing fluid into drilled clay layers thereby precluding clays from swelling. Because the additive bentonite cannot swell in a CMS fluid, CMC's must be added later to a bentonite fluid. Increasing the density of the drilling fluid with loading agents such as calcite or baryte powder is valuable in drilling artesian groundwater aquifers.

6.4 Drilling Methods for Borehole Heat Exchangers

With this increased density of the drilling fluid many artesian aquifers or aquifers with slight hydraulic overpressure can be controlled. If, for example, an artesian aquifer with an overpressure of 0.3 bar is drilled 60 m below surface, adjusting the density of the drilling fluid to $\rho = 1.1 \times 10^3$ kg/m³ will produce the necessary pressure of the drilling fluid:

Pressure of drilling fluid: 60 m × 1.1 × 10³ kg/m³ $\cong$ 6.6 bar
Pressure of artesian aquifer: (60 m + 3 m) × 1.0×10³ kg/m³ $\cong$ 6.3 bar
Overpressure of drilling fluid: 6.6 bar – 6.3 bar = 0.3 bar

The feed regulation of the drilling fluid depends on the ground to be drilled, on the planned drilling method, the power of the pump and the ascent rate of the fluid. The drilled solid materials need to settle sufficiently in the sedimentation tanks, for not to alter the density of the drilling fluid. Available standard methods can easily measure the density of the drilling fluid with hydrometers, the viscosity with a so-called Marsh funnel and the water release rate (water binding capacity) with a Ring apparatus.

In addition to the cuttings, several drilling parameters such as drilling progress, drilling (wash) fluid pressure, rotational speed and drill bit pressure give valuable information about geological changes at bottom hole during drilling. These parameters are in sight of the operator at the control panel of the drilling rig (Fig. 6.8b). The parameters can also been continuously digitally recorded and may serve as important legal proves if necessary and for the geological interpretation of the drilled stratigraphy.

6.4.2 Down-the-Hole Hammer Methods

Rotary percussive drilling is mostly done with down-the-hole hammer (Fig. 6.12). The cuttings are brought continuously to the surface through the annular space by a powerful air stream. A top drive rotates via the drill pipe the down-the-hole hammer at speeds of several tens of revolutions per minute. At the same time, a

Fig. 6.12 Down-the-hole hammer used for geothermal probes

compressor pumps air at 15–35 bar through the drill pipe to the down-the-hole hammer. The compressed air drives a plunger that causes the bit of the hammer to strike the bottom of the borehole at speeds up to 3,000 times per minute. The compressed air leaving the hammerhead cleans the bottom hole and transports the cuttings trough the annular space to the surface.

Drilling with down-the-hole hammer is favorable in hard rock and hard cohesive soil. It is of limited use in loose sand and gravel. The great advantage of this method is that water ingresses to the wellbore from water conducting layers and structures can be recognized immediately during drilling.

Specially designed down-the-hole hammers can be operated with water instead of air or with a water–air mixture.

Double head drilling is a further popular method for drilling boreholes for geothermal probes. Two separate top drives for the inner drill pipe and the outer casing work in tandem until the casing is put in place. After that, drilling is continued with the top drive for the inner drill pipe.

6.4.3 Concluding Remarks, Technical Drilling Risks

Drilling boreholes for geothermal probes is relatively rapid and short-lived, however, it suffers from cramped and limited space. Therefore compact and mobile drilling equipment is ideal. Space limitations make it essential to carefully plan and organize the drilling site. Before starting to drill, the exact location of existing pipework (water, gas), cables and other obstacles on the property must be cleared up.

We strongly recommend keeping details of the drilling progress with an appropriate drilling data recorder and documenting the data. Mechanical loggers record drilling progress, depth, drill bit pressure and drilling fluid pressure. These data give valuable information on the geological properties and structure of the ground. For example, drilling data from digital loggers can be combined with data from a gamma ray logger used to probe the borehole to considerably ease the geological interpretation of the drilled stratigraphy.

If several boreholes are to be drilled, minimum distances between the wells must be strictly observed (Sect. 6.3). If the individual geothermal probes that ought to be connected to the heat pump have different connection lengths or reach to different depths, then hydronic balancing is also necessary on the geothermal probe side.

If the borehole is exposed to water influx or even artesian water, natural gas etc., special measures are required to fight these threats. The procedures include protection by casing, increase of the density of the drilling fluid, installation of a packer or, in extreme situations, tight sealing of the borehole.

The most common *technical drilling risks* are briefly described in the following:

Swelling clays in clay-rich strata may represent a drilling risk. Drilling such strata requires a special drilling fluid that prevents swelling of clays (Sect. 6.4.1). After installation of the probe, the borehole must be professionally and permanently sealed excluding any water access to the clays in the future, otherwise swelling pressures may build up to where the installed probe may be destroyed or in the worst case may cause damage to nearby buildings.

Any geological formations that contain the mineral anhydrite ($CaSO_4$) must be drilled highly alert and attentive. Anhydrite reacts in contact with water to gypsum ($CaSO_4 \cdot 2H_2O$). The mineral transformation causes a volume increase of more than 60 %. Extreme caution is required if aquifers are drilled in the footwall or hanging wall of the anhydrite rocks. Drilling should be aborted and the borehole should be firmly sealed when artesian aquifers are drilled in the footwall or aquifers with negative pressure in the hanging wall of an anhydrite layer (Fig. 6.13). Similar effects and challenges can be caused by any kind of other mineral in geological strata that react with water. Therefore, it is highly recommended to be aware of the problem and study available geological maps and other local geological stratigraphic information before drilling.

Drilling quicksand layers require casing of the respective sections to prevent sand loss to the borehole and potential erosion of the hanging wall and danger of subsidence of the surface. Wellbores without casing trough strongly karstified

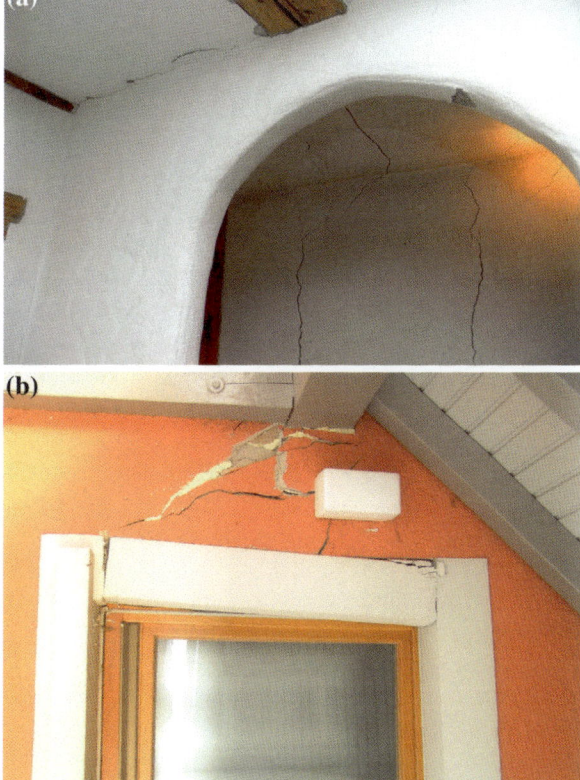

Fig. 6.13 Drilling risks: Structural damage of buildings caused by swelling layers in the ground. Here the differential surface movements have been caused by the reaction of anhydrite with water from an aquifer to produce gypsum accompanied by a 60 % volume increase (City of Staufen, SW Germany). **a** Damage pattern inside a building, **b** Complex structure of damage around a window

ground beneath unconsolidated clastic sediments tend to draw material from the hanging wall causing subsidence and sinkholes at the surface. Thus such drillholes should also be equipped with a standing pipe or a casing to the solid hard rock. Surface subsidence also may result from leaching and dissolution of soluble minerals in geologic strata such as halite and other salts by abundant water infiltration along hydraulic conduits produced by drilling but also other water conducting structures in the ground.

Serious drilling problems may occur if larger cavities or caves are drilled in susceptible rocks including karstified limestone, dolomite, salt- and gypsum-bearing strata, fault zones and coarse banked fractured hard rock aquifers. The drill pipe may break through and massive loss of drilling fluid may occur. Secondary problems resulting from this geological situation include difficulties with the backfill of the annular space. Special backfill materials and packers may be required. In the worst case the borehole can be lost and must be backfilled and firmly sealed.

During drilling through fault zones and other tectonically damaged ground the drill pipe may jam. The borehole crossing faults may close after its completion to where the geothermal probes cannot be installed. Commonly the only remedy is to over drill or drilling a new borehole.

Sections of wellbore can be hydraulically separated using geothermal probe packers. Packers are also used to control weak artesian aquifers or separate aquifers at different hydraulic pressures. Specially designed hose-like geothermal probe packers (Fig. 6.14a) consist of two packer sealing elements, a fabric hose and mountings. Also the packer elements are made of textile fibers. The fabric hose is put over the geothermal probe and placed at the appropriate position. Both ends of the textile hose are then fixed to the probe by rubber sleeves and sealed (Fig. 6.14b). A grouting pipe is placed through the two rubber sleeves so that the section below the packer can be sealed. A second grouting pipe runs through the upper rubber sleeve and is used to fill the packer. A third grouting pipe is used to backfill the section above the packer. The assembly of geothermal probe, mounted packer and installed grouting pipes are lowered into the borehole and fixed at the desired position of the packer. After grouting the section below the packer, the packer itself can be inflated with backfill material in this way sealing the lower from the upper section of the borehole (Fig. 6.14). Then follows backfilling of the upper section. In principle it is possible to install two packers for the isolation of a certain section of the borehole. In practice, however, placing the entire assembly of probes, packers and grouting tubes into a borehole against hydraulic overpressure high forces and a large borehole diameter are needed.

Drilling for a geothermal probe in certain regions may possibly expose layers with pressurized natural gas or artesian groundwater. Drilling into an artesian aquifer may cause large volumes of water to flow from the borehole causing additionally washout of fine-grained solid material. The occurrence may severely damage the borehole to where sealing is hindered later or made impossible. Moreover the ground around the borehole may subside or even collapse. Therefore it is necessary to plan with precaution when drilling in regions where artesian aquifers are known to exist. When drilling at sites where artesian aquifers

6.4 Drilling Methods for Borehole Heat Exchangers

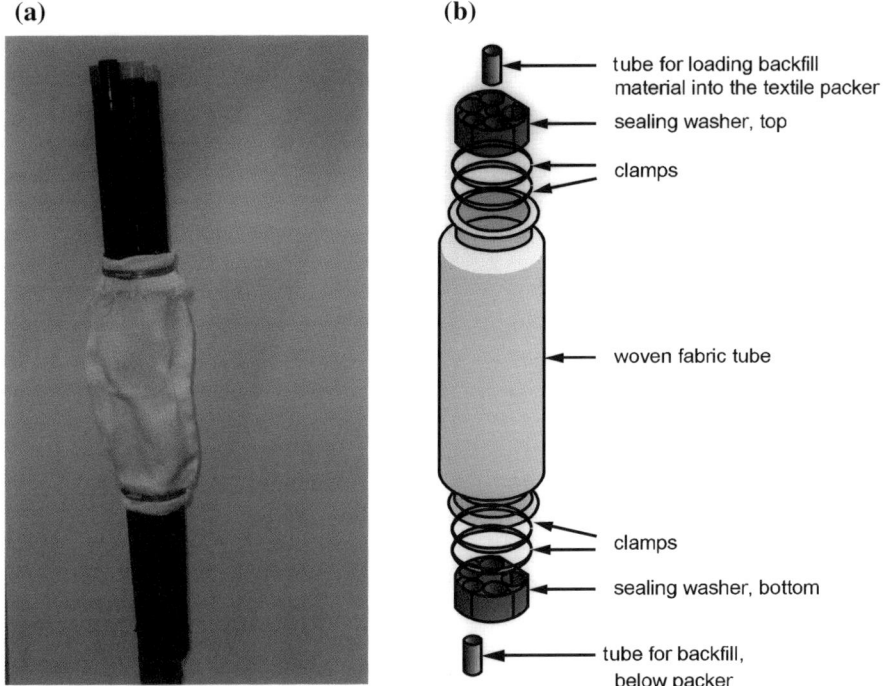

Fig. 6.14 a Geothermal probe packer (textile fabric packer). b Schematic assembly of a geothermal probe packer

pose a potential hazard the casing must be firmly anchored in the solid ground of the hanging wall (Fig. 6.10). Because loaded high-density drilling fluid will become necessary if artesian aquifers are drilled, appropriate amounts and types of loading agents must be kept ready. Also a suitable packer system for reliably blocking the water ingress must be hold in readiness. A weakly confined aquifer with low overpressure can perchance be sealed with a geothermal probe packer (Fig. 6.14). We recommend sealing the lower section or the entire borehole using a "lost" packer if drilling into a stronger artesian aquifer or the borehole must be abandoned.

Natural gas may stream from the borehole under high pressure. Also possible are diffuse slowly developing gas seeps. If a gas reservoir is drilled, similar measures to the ones controlling an artesian aquifer are recommended. Additional precautions are of need, however, because gas may burn or explode (methane), it can be highly toxic (e.g. hydrogen sulfide) or it may cause a suffocation hazard (e.g. carbon dioxide, nitrogen). Therefore, the gas must be analyzed and identified before any decisions are made concerning the further procedure. If the gas is hazardous the geothermal probe cannot be installed and the borehole must be gastightly sealed and may not be overbuilt later.

Methane seeps are known particularly from coal-bearing strata (Carboniferous), from clays rich in organic material (e.g. middle Jurassic Opalinus Clay in central Europe) but also in other strata. 5–14 volume % CH_4 in air is explosive, higher concentrations can cause difficult to control fires.

In the Swiss village of (canton Obwalden), natural gas has been drilled at 125 depth which then streamed into the borehole at an overpressure of 3 bar (Wyss 2001). The judicious and rapid reaction of the drillmaster controlled the discharge. The gas was burned. Later on the borehole was sealed and refilled. A geothermal probe has not been installed.

Groundwater may contain high amounts of dissolved gasses. The commonly used probe pipe material, PE pipes is permeable for gases. Carbon dioxide, for example, can easily diffuse through the walls of the PE pipes, due to the size and structure of the CO_2 molecules. Restart of the circulation pump of the primary circuit after extended downtime brings the gas-rich water to the surface where it degasses possibly with massive foam generation. Conventional air separators in the inlet line may get overcharged and cannot separate all of the large amount of gas exsolved from the pumped water. The foam may reach the evaporator of the heat pump and may considerably reduce the heat extraction power of the system and may eventually shutdown the heat pump. Considerable corrosion hazard can be associated with CO_2-rich salty waters, which may damage the heat exchanger of the evaporator of the heat pump. Geothermal probe projects in regions with known free CO_2 gas in soil, strata or hard ground should use gas-proof probe material that is impervious to gas diffusion.

6.5 Backfill and Grouting of Geothermal Probes

Backfill and grouting of a geothermal probe is an essential component for an efficient, ecologically meaningful operation, for the durability of the probe and for observing groundwater protection regulations. The backfill must be permanently impervious. It must efficiently physically and chemically stably connect the probe with the surrounding ground. Inefficient grouting is very hard to repair and remediate. It must be kept in mind that the land owner and client is liable for any damage of any kind resulting from a unprofessional and incompetent construction of the ground source heat exchanger system. Nevertheless, in some countries backfill is not an authoritative requirement.

Immediately after drilling of the borehole, the probe pipes are taken from the reel and mounted in the borehole. Grouting must be started without delay. Backfill of the probe seals the different drilled layers of the ground against each other and thus is important for protecting groundwater horizons, preventing hydraulic short cuts along the borehole and restoring the sealing properties of aquitards/aquicludes (Fig. 6.15). A proper backfill represents also an additional barrier for leaking heat transfer fluid from damaged probe pipes to the groundwater. An appropriate backfill thermally connects the geothermal probe optimal to the surrounding ground. The backfill stabilizes the borehole, must fill it permanently and without settling.

6.5 Backfill and Grouting of Geothermal Probes

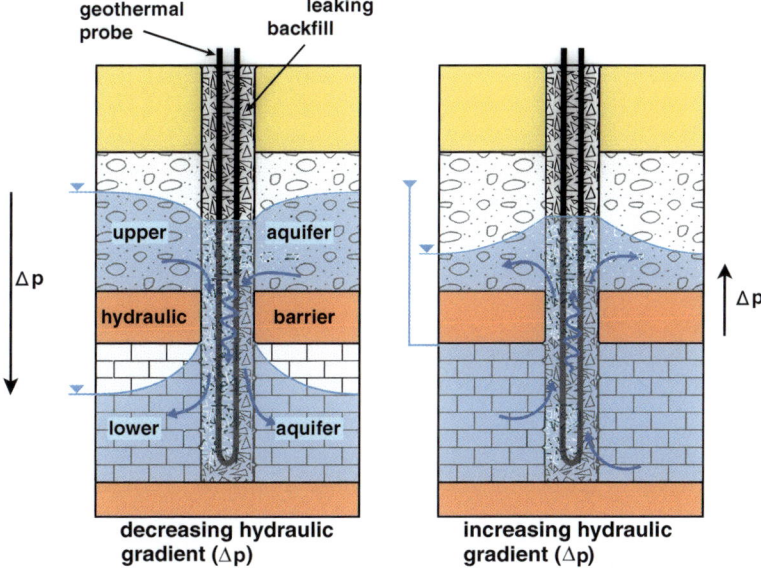

Fig. 6.15 Requirements of backfill regarding prove of layers

From the described function and requirements, optimal backfill material has the following properties:

- Low permeability ($k \leq 10^{-9}$ m/s), long-lasting tightness,
- Approved for use in aquifers, water-hygienic unproblematic, not hazardous for water,
- Easy to use and secure workability at the construction site, easy to pump,
- Sedimentation-stable, volume constant and low-shrinkage setting behavior,
- Resistant to chemical attack (concrete-corrosive waters, sulphate-bearing rocks),
- Thermal and mechanical stability,
- Very good flow properties,
- Void-poor structure of the set material,
- Excellent thermal conductivity.

Many different products are used as backfill material. The components of backfill, normally cement, bentonite, clay or quartz sand, are mixed with water to a suspension. The different components have different functions. Cement provides compressive strength and sealing properties. The swelling properties of clay are responsible for volumetric stability of the suspension. The features of bentonite are similar to clay. It has a particularly high swelling capacity and a distinctly thixotropic behavior. Admixing quartz sand or quartz powder increases the thermal conductivity of the backfill. Cement makes operation of the geothermal probe at freezing temperatures possible. However, cement should be added sparingly to keep the backfill slightly plastic. Flexible backfill assists accommodation of the thermal expansion of the probe and averts water-conducting structures along the probe pipes to form.

Typical backfill materials have a high thermal conductivity in the range of 0.6–1.0 W/mK. Improved products may reach 1.6–2.2 W/mK and enhance transport of thermal energy to the heat transfer fluid in the probe pipes. Presently new pipe materials with high thermal conductivity are being tested (Sect. 6.2).

Factory-ready backfill materials should be used only. The backfill suspension needs to be mixed to the density specifications given by the provider within the limits of the accuracy of the measuring instrument (about ± 0.05 g/cm^3). Deviations can only be accepted for defensible exceptional cases. Otherwise the desired backfill material properties cannot be warranted. The components must be mixed at the construction site and the sealing slurry must be handled instantly (Fig. 6.16). Special backfill and grouting mixers have been designed for geothermal probe installation (batch mixer, colloidal mixers, continuous mixers). The density of the suspension should be within 1.3 and 1.9 t/m^3 (Fig. 6.17). The Marsh funnel time, a measure for the viscosity of the suspension should vary between 40 and 100 s/l (28 s/l for pure water). The backfill should achieve minimal settlements in the range of 0.5–1.0 % and the strength should be ≥ 1 N/mm^2. The temperature during setting of the backfill materials must stay below the danger of heat damage to the probe pipe material.

In order to achieve a dense, void-poor backfill it is necessary to grout from bottom to top (Tremie process). Density and viscosity of the backfill material should be high enough to push remaining drilling fluid and water from the borehole during ascent. The backfilling work should not be interrupted and carried out without delay after completion of the borehole. The suspension-filled grouting line should remain in the borehole. If the grouting pipe is to be removed from the borehole,

Fig. 6.16 Use of a colloidal mixer at the construction site

6.5 Backfill and Grouting of Geothermal Probes

Fig. 6.17 Use of a density balance at the construction site for measuring the density of the backfill slurry

this should be done carefully and slowly and with re-injecting backfill material. The actual volume of the backfill material must be bigger than the difference between volumes of the borehole and the geothermal probe. Backfilling is not finished before the density of the suspension discharging from the top borehole has not reached the density of the original backfill slurry.

In practice, however, very often too "thin" low-density low-viscosity suspensions are being grouted because they are easier to pump. Such suspensions have a high water/solid ratio and thus produce a incomplete and short-lived backfill of the probe, which results in a reduced heat transfer capacity from the ground to the heat transfer fluid in the probe pipes (Fig. 6.17). Centrifugal pumps build up relatively little pressure and thus can only be used for backfilling of geothermal probes with shallow depths. Positive-displacement pumps (worm pumps) are suitable for backfilling of deeper probes.

One of the important functions of an appropriate backfill is to prevent a permanent hydraulic connection of drilled aquifers at different levels. If aquifers at different levels have been connected by a borehole, several quantitative and qualitative effects may be triggered, particularly if the aquifers that were originally separated by groundwater barriers are at different hydraulic pressures. Firstly, hydrochemical alteration and exchange of groundwater contaminants between aquifers become possible. For example, seepage of highly mineralized deep groundwater may deteriorate shallow fresh-water aquifers. Anthropogenic contaminants from shallow groundwater levels may gain access to deeper aquifers and nitrate, pesticides, bacteria and organic solvents can be transferred to unpolluted groundwater. Normally groundwater in different soil and ground layers is of distinctive chemical composition. Migration of water from one layer to another because of a hydraulic short-cut along an improperly grouted borehole may cause a multitude of chemical reactions between the water and the solid aquifer rock including dissolution of rock thus creating cavities and precipitation of minerals reducing the hydraulic conductivity of the aquifer. Furthermore, a short-cut of aquifers at different hydraulic potentials (water table) can cause hydraulic disturbance such as changes of the water tables including drying-up of springs. For this reason, if aquifers at different levels are drilled, water tables must be observed and regularly measured and if the hydraulic potentials are significantly different appropriate precautions need to be taken to isolate the aquifers.

Drilling a borehole for a geothermal probe may, like any other drilling, cause turbidity and microbial contamination of groundwater. This possible danger is particularly relevant for neighboring groundwater wells, mineral water or thermal water developments. Experienced drilling companies should be aware of these vulnerabilities and with all associated legal aspects and responsibilities. They should be capable of performing the necessary tasks without affecting other uses of the ground at specified distances from the geothermal probe borehole.

6.6 Construction of Deep Geothermal Probes

Geothermal probes that shall be installed to depths significantly below 150 m must consist of special probe materials resisting the higher pressure and the excessive stresses associated with conventional installation. Special installation techniques for the overly long probe may be necessary for avoiding damages to the probe. Because this special probe pipe material has much ticker walls the geothermal probe has a strongly increased thermal borehole resistance (Sect. 6.3) so reducing the efficiency of the system. Therefore, these pressure-resistant materials should be avoided whenever possible. Installing probes with standard wall thicknesses in deep boreholes requires a more elaborate installation procedure for not compromising the pressure-stability of the probe pipes. More than 300 m long probes should not be built with conventional pipes, however.

Probe pipes for overly long probes in boreholes deeper than 150 m should be made of polyethylene resistant to crack (PE-RC) pipes or cross-linked polyethylene (PEX) pipes. Both materials are more resistant to crack formation than e.g. PE 100. The pipes withstand pressures of 15–16 bar during a 50-year lifetime.

The relevant pressurization of the probe pipe corresponds to the pressure difference between external pressure (fluid pressure of the saturated zone) at depth and the internal pressure in the pipe (hydrostatic pressure of the heat transfer fluid + system pressure of the geothermal probe loop). The pressure conditions need to be carefully examined for geothermal probes deeper than 150 m. Typical system pressures of the probe loops vary between 1.0 and 2.5 bar during operation.

The backfill material used for geothermal probes has a density (ρ_V) between 1.4 and 1.9×10^3 kg/m^3; the density of the heat transfer fluid (ρ_w) is much lower around 1.0×10^3 kg/m^3. The difference between hydrostatic pressures of the fluid-filled probe and the grouting suspension determines the maximal possible depth of the probe.

For example: With a density of the backfill suspension $\rho_V = 1.8 \times 10^3$ kg/m^3 and a water-filled probe pipe the pressure difference reaches 16 bar ($= 16 \times 10^5$ Pa) at the probe foot in 200 m depth.

This follows from:

$$(\rho_V - \rho_w) \cdot 9.81 \, \text{m s}^{-2} \cdot 200 \, \text{m} = 16 \times 10^5 \, \text{Pa}$$
$$\text{Dimension}: \ 1 \, \text{Pa} = 10^5 \text{bar} = 1 \, \text{kg m}^{-1} \, \text{s}^2 \tag{6.7}$$

6.6 Construction of Deep Geothermal Probes

The resulting pressure difference implies that for the densities used in the example, construction site practice advises that a quality-assured installation of a probe deeper than 200 m is not possible.

A reliable estimate for the maximum possible installation depth (D_{max}) of a geothermal probe follows from the relationship:

$$D_{max} = 15 \times 10^5 \text{ Pa} / [(\rho_V - \rho_w) \cdot 9.81 \text{ m s}^{-2}] \qquad (6.8)$$

With the parameters used in the example above the resulting $D_{max} = 191$ m. The underlying assumption here is that the pipe material tolerates not more that 15 bar pressure difference.

The mounting and installation procedure for very deep geothermal probes depends on the planned depth and the water table in the borehole.

If the borehole is water-filled nearly to the top, the probe pipe must be filled with water already on the mounting reel and inserted to the borehole this way. The density of the anticipated backfill material must be in harmony with the planned depth of the probe (Eq. 6.8). For a 400 m deep geothermal probe the maximum tolerable density of the backfill is $\rho_V = 1.375 \times 10^3$ kg/m^3. These considerations need to be coordinated with all other requirements for the backfill (Sect. 6.6).

If the water table in the borehole is more than 150 m below surface, the installed geothermal probe should not exceed 300 m depth. In principle, deeper probes are possible but conventional conditions at typical construction sites may not warrant quality assured installation, which is considerably more intricate than at high water tables. Quality assured installation of deep probes into completely dry boreholes is likewise difficult and convoluted.

Moreover, the pressure losses for circulating heat transfer fluids in standard 32 mm diameter probe pipes gradually increase with length (depth) and become exceedingly high at lengths higher than 130 m, which seriously compromises the profitability of the installation.

6.7 Operating Geothermal Probes: Potential Risks, Malfunctions and Damages

Ground source geothermal probes are established and mature systems for heating and cooling of buildings and other structures. These installations require a professional dimensioning, design and construction. Inappropriate construction and operation normally results in a case of damage. For this reason, official authorities in Switzerland released a damage catalogue addressed to all parties involved in planning, installation and operation of geothermal probes with the intension to reduce damage cases (Bassetti et al. 2006). In Switzerland and SW-Germany an institution called "heat-pump-doctor" assists in especially difficult instances. Several damage cases have been documented (Wyss 2001). Special risks related specifically to drilling have been discussed in Sect. 6.4.3 above.

A common problem results from excessive heat extraction from the geothermal probe causing the surroundings of the probe to freeze; resulting normally in reduced efficiency of the system and may finally lead to a complete failure of the system.

Cases of damage occasionally result from to long operation times for the heat pump caused by a flawed control system for the heat pump—geothermal probe system and (or) by inadequate coupling to the heat distribution system of the building.

Commonly mistakes that lead to later damages on the geothermal system are already made in the planning phase of the project. Typical errors include wrong determination of the heat requirement of the building, ignoring the necessary continuous hot water demand, using inadequate thermal parameters and properties of the drilled ground, under dimension of the geothermal probe (too short probe) and not correctly considering the thermal effects of other nearby geothermal probes. Damage cases occur also after installation and commissioning caused by incorrect adjustment of the heat pump or the hydraulic system. It is also problematic if the client adds further buildings, annexes, or installations to be heated by the geothermal probe for which it has not been designed for.

During construction it is important to insist on a correctly emplaced backfill (Sect. 6.5). Is the backfill lacking or incomplete, or has excavation or other improper material has been used as backfill then the system imposes a threat to groundwater safety. Moreover the stability of the geothermal probe is compromised resulting in poor extraction rates combined with operation at freezing conditions. Operating a geothermal probe under freezing conditions (negative °C) may result in damages to the probe, the backfill and the ground itself because ice may form at the outer probe pipe surfaces and in the ground surrounding the probe. Related to the freezing-thawing cycles increasing numbers of fractures form together with ice in the backfill and the enclosing ground is concentrated at material boundaries between probe and backfill and backfill and ground. The fractures gradually increase in size. The ice cover steadily displaces the backfill around the probe pipes with the result that the seal for groundwater protection becomes broken. Ice formation may cause ground swell in the vicinity of the probe during the cold season. After thawing, the ground surface may subside and even collapse locally (sinkholes, cones around the probe, ground settling above the supply lines). Further damage in the probe environment slowly develops. Additionally, the electrical power requirement of the entire system gradually increases. However, not all damages become manifest at the surface (swelling, subsidence). Damage to the structure of the system grows to be visible first in performance deterioration of the installation. In extreme cases the entire system may break down.

If the thermal properties of the drilled ground layers or of deep groundwater tables have been considered for the design of the system incorrectly or insufficiently the danger exists that the probe may be too short and thus operated later under freezing condition with all the negative consequences for the entire system as previously explained.

Unprofessional and improper drilling of boreholes combined with unmindful supervision may lead to very serious or even catastrophic damage to structures and buildings of a large area (Sect. 6.4.3) as the example of Staufen (SW Germany) shows where 250 buildings have been damaged and the total cost of the reconstruction exceeds 200 millions of dollars. These drilling risks, however, jeopardize any drilling and are not a risk exclusive to building geothermal probes.

The probe pipes can be defiled by forceful installation into the borehole, especially when the diameter of the borehole is too small or the ground unstable.

Installation of the probe in the cold season without prior warming of the pipes is difficult because of the stiff and inflexible probe plastic. The probes often develop damages during installation and consequently later the system suffers from leakages.

In case the geothermal probe is intended to be used for heating and cooling or it is planned to use the probe for storing excess heat in the warm season in the ground (e.g. heat from a solar-thermal installation) one must bear in mind that conventional probe material does not tolerate temperatures above 30–40 °C. Such applications require high-pressure cross-linked polyethylene pipes (PEX pipes).

Geothermal probes that are much longer than 150 m call for special stress and pressure resistant probe materials or a special installation protocol so that the probe is not damaged (Sect. 6.6).

6.8 Special Systems and Further Developments

If a geothermal probe in combination with a heat pump is used to heat a house this classic scheme is called a monovalent heating system. Only one energy source is used, therefore ground source heat pump, ground source heat exchanger, ground source loop and similar equivalent expressions are used. If many geothermal probes are installed in many boreholes to cover the larger heat requirements of a bigger building one speaks of a geothermal probe field.

If several geothermal probes are needed to cover the heat requirement of a building the mutual interferences need to be carefully evaluated. The usable heat reservoir per probe decreases with the number of geothermal probes installed in the available volume of ground. Geothermal probe fields require well-considered management plans to avoid the risk of continuously decreasing temperatures in the ground resulting from overexploitation. Dimensioning and planning of probe fields must utilize adequate modeling tools.

Commonly, geothermal probe fields are also used for cooling in addition to heating. Combining geothermal probes with solar-thermal systems is becoming increasingly popular. These bivalent heating systems consequently use two sources of thermal energy.

Further new developments in the field of geothermal probes are systems that utilize the thermal effects associated with a phase change of a refrigerant (Sect. 6.8.5).

With further developments and increasing popularity of geothermal probe systems there is a growing demand for controlling the professional installation of the geothermal probe and to quantitatively measure its performance during operation (Sect. 6.8.4).

Geothermal systems should not excessively heat or cool drilled aquifers, if present, which are kept for further utilization because this may lastingly alter the chemical and microbiological properties of the groundwater.

6.8.1 Geothermal Probe Fields

Many geothermal probes are typically required to cover the heat requirement of larger buildings. In contrast to single family homes with one or two installed geothermal probes and monovalent systems, polyvalent systems are of interest for larger objects or buildings where air conditioning, cooling, utilization of process heat, utilization of other regenerative energies including solar and/or biomass, in combination may be of interest (Sects. 6.8.2, 6.8.3). Combined heating and power systems may be integrated into complex systems where a geothermal probe field is an integrated part. To meet the diverse requirements and all the individual profiles of the components and to deploy all the multifarious plant technologies an overall energy concept is utterly required. One must keep in mind that geothermal probes respond very slowly to changing conditions therefore they are not the right instruments for short-term storage of thermal energy.

The term geothermal probe field is used for groups of more than five closely arranged geothermal probes. Geothermal probe fields that also function as seasonal heat storage devices for solar-thermal installations (Sect. 6.8.3) or for cooling of larger buildings (Sect. 6.8.2.) may involve ordered arrays of 100 or more geothermal probes. The geothermal probes transfer the solar thermal energy collected during the warm season to the ground reservoir. During the cold season the stored heat is mobilized and used for heating with the assistance of a heat pump.

Particular requirements must be complied for construction, installation and operation of geothermal probe fields. The design of large geothermal probe fields as heat reservoirs (sinks and sources) presupposes detailed knowledge of the thermal and hydrogeological structure and property of the ground. This necessitates the drilling of one or several test case geothermal probes at the planned site and collecting thermal and hydrogeological parameters of the ground. Thorough and detailed geological documentation of the drilled ground is compulsory. Water table, hydraulic conductivity and important hydrochemical parameters (e.g. concrete corrosive groundwater) must be measured if aquifers have been drilled. Thermal-Response-Tests (Sect. 6.3.2) provide the project management the necessary data on the thermal properties of the ground. Local climatic conditions and variations must be known and implemented into the project planning.

Geothermal probe fields used for heating and cooling, particularly if sited in aquifers may create considerable thermal effects in the ground with remarkable cold or warm thermal plumes. The resulting temperature changes may cause local chemical or biological alterations in the groundwater. Increasing groundwater temperature, for example, can promote microbial activity in the ground that may cause chemical reactions between groundwater and the minerals of the solid ground thus changing groundwater composition. These changes may be inoffensive or harmful and toxic (mobilizing arsenic, cadmium, uranium for example). Stinted dimensioning of probe fields may generate the feared ice barriers in groundwater.

Heat storage in the ground changes the chemical composition of groundwater with the consequence that some minerals may dissolve and other minerals may precipitate from the water as the mineral-water equilibria adjust to the changed temperatures. Particularly critical is the supply of additional dissolved oxygen via the borehole to the aquifers, which typically results in precipitation of iron oxides and oxyhydrates and production of concrete corrosive sulphate. Even seemingly small temperature reductions from 10 to 2 °C cause feldspars to dissolve and clays and silica to form in sandy soils. These silicate-involving reactions are generally slow. However, reaction progress of dissolution-precipitation reactions rapidly increases with temperature. Various geochemical modeling tools are in use for analyzing groundwater composition data and for predicting the response to changing temperature, matrix minerals and chemical composition of the groundwater. The most popular of these tools is the code PHREEQC (Parkhurst and Appelo 1999) published and maintained by the USGS (Boulder, Co) (Sect. 14.3).

Geochemical and physical processes in the reservoir also rearrange the conditions for microbial prosperity. Population density and amount of biomass may dramatically change. High-temperature heat reservoirs may develop zoned microbial populations with thermophile microorganisms in the core region and mesophile microorganisms in an outer zone. The original population may be killed off. The adjustments to the biomass seem to be of local not regional extent (Ruck et al. 1990), however, the feature needs to be observed.

Several countries have issued strict legal constraints for minimizing the potential environmental impact on the ground or on groundwater. Such constraints may require that the temperature increase in the aquifer may not exceed 5 °C at 10 m distance from the nearest geothermal probe and must be less than 2 °C at 50 m from the probe field.

6.8.2 Cooling with Geothermal Probes

Air conditioning and cooling are typical planning issues for larger buildings. However, with climate change these subjects may also receive increasing attention for residential developments and family homes. Geothermal probes in 10–12 °C temperate ground are well suited for cooling tasks. In combination with a heating system interesting synergy effects may develop.

Geothermal probes can be used to transport excess thermal energy of the warm season to the ground where it can be stored for later use. The excess thermal energy may result from high room temperatures in the summer but also from technical facilities and production processes such as heat from IT equipment. Storage of excess heat in the ground accelerates recovery of the ground heat reservoir that has been exploited in the cold season for heating.

The operation of a geothermal heating system with a geothermal probe extracts thermal energy from the ground during the cold season. The ground cooled by some Kelvin at the end of the heating period. In the beginning of the warm season the building can be air-conditioned by pumping cold heat transfer fluid from the cooled ground alone. The heat transfer fluid warmed by the excess heat of the building is piped back to the ground where it is cooled again (building heat source rather than ground heat source = reversed geothermal probe). The cool ground slowly warms and regenerates. Later in the warm season it will be often necessary to operate the heat pump of the loop as refrigerator to transfer the excess heat of the building to the ground. The ground continuously warms and recovers as a heat reservoir. The reversible heat pumps are needed for such kind of operational mode that reversibly heats and cools ground and building. Note, however, the portion of geothermal energy is a small fraction of the total energy budget in this mode of operation. The ground has predominantly the function of a heat storage device and not that of a heat source.

A geothermal probe can also be exclusively used for cooling purposes. Sustainable operation of such a cooling system requires that the stored thermal energy can be removed from the reservoir during the downtime e.g. in the cold season. In aquifers the stored thermal energy is dispersed by groundwater flow but also by heat conduction of the ground. Problems may arise if the probe is placed in clay or cohesive material with very low hydraulic conductivity.

6.8.3 Combined Solar Thermal: Geothermal Systems

The advantage of combining solar thermal with geothermal systems is easy to understand. In the warm season plenty of solar thermal energy can be harvested but there is no use for it in house or building heating. Thus solar thermal installations are primarily used for preparing hot water. If the solar installation is correctly planned then the furnace can be switched off during the warm season because the solar device produces sufficient heat for preparing hot water. If the solar heat source shall be used for house heating, a long-term heat reservoir is necessary. It makes it possible to store the solar summer heat and extract it when needed in the winter for heating. Geothermal probes can efficiently transfer the thermal energy to a ground reservoir. If the geology of the ground is favorable and if the environmental regulations can be respected, the combination of geothermal probes and solar devices can make very efficient bivalent heating systems.

As explained above, the bivalent system uses two different sources of renewable energy. The solar thermal energy that is transferred to the ground in the warm season improves regeneration of the heat reservoir. During the heating period thermal energy can be extracted from the ground and the lost heat is restocked in the warm season. The solar thermal system efficiently refills the used heat in the season when more solar energy is collected than is needed for hot water production. During this crucial period regeneration of the ground heat reservoir is efficiently enhanced and improved. It makes sense to elevate the reservoir temperature above its natural level prior to the heating period using the ground around the probe as a heat reservoir. At the beginning of the heating period, the ground source heat pump finds improved starting conditions. This results in a significantly improved annual energy efficiency of the heat pump operation, giving rise to a distinctly increased seasonal performance factor. This is an important economic aspect for all systems with electricity-driven heat pumps.

Remember that normal PE-100 probe pipes don't have a sufficient long-term stability if operated and temperatures above about 30 °C (Sect. 6.7).

The authorities of many countries require that near-surface aquifers may not be permanently heated to temperatures higher than 20 °C because of microbiological concerns.

A further emerging application is the seasonal heat storage in high-temperature ground reservoirs. The reservoirs operate with hundreds of borehole heat exchangers. Large solar thermal systems collect thermal energy in the warm season that is continuously transferred to the ground reservoir via the installed geothermal probe field. The reservoir may be heated up to 90 °C. The thermal energy supports heating of buildings during the cold season that are connected to a local heating network. About 50 % of the heat requirement of the connected buildings can be covered by renewable energy on a long-term average (Schmidt et al. 2003). For geothermal probe heat reservoirs, large numbers of closely spaced probes (1.5–4 m) must be installed and padded with thermal insulations at the surface. The optimal relation between surface and volume of the reservoir is for the probes to be arranged in a circular array and they should not reach very deep. For loading the reservoir the heat transfer fluid flows through the central probes first and then through the peripheral probes in order to optimize the temperature distribution in the ground. For the heat extraction the flow direction is being reversed.

6.8.4 Geothermal Probe: Performance and Quality Control

Regardless of the large number of installed geothermal probes, some issues remain unsettled. Several important aspects of performance and quality control have not received the attention they deserve. Particularly the subjects related to the quality and durability of the backfill in the borehole and to its efficiency as hydraulic seal, the control of the thermal performance of the probe and also the actual precise position of the probe pipes at depths.

Related to these issues, measurement methods providing in situ data from geothermal probes that allow for conclusions regarding the quality of the backfill would be highly desirable and welcomed. Technical problems include the small diameter of the boreholes and the coiled course of the probe pipes. Most probe pipes have an outer diameter of 32 mm. Therefore standard tools for geophysical borehole logging typically used for groundwater monitoring and production wells, do not work in geothermal probe pipes because of this size limitation. Recently, measuring equipment fitting narrow probe pipes has been developed.

At present, the quality of the backfill is very difficult to examine. In addition to the space problem with the small pipe diameters, it is almost impossible to differentiate between the signals from inadequate and incomplete backfills and the nearby probe pipes next to the tested pipe. Also the grouting hose is normally next to the probe pipes and adds additional complexity to the geometry and signal interpretation. The precise location and orientation of probe pipes and grouting hose may vary from central in the borehole to peripheral at the borehole wall over short vertical distances compounding the situation further. The problem is related to the utility and layout of spacers and centering aids. A rigorous interpretation of borehole data would require precise knowledge of the spatial position of all probe pipes. Data must be collected in all probe pipes.

At present, a cable-free small-size data logger, NIMO-T, that fits into 32-mm U-tube probes measures temperature and pressure in geothermal probes and records temperature as a function of depth (pressure) (Fig. 6.18). The temperature logs are measured in completed and installed probes that have not been operated yet. The data logger is 23 mm in diameter and 219 mm in length. The internal diameter of a 32 mm U-tube is 26 mm thus it is a bit fiddly to insert the logger into the probe pipe. The logger sinks in the probe pipe through its controlled and adjustable weight with a velocity of about 0.1 m/s to the probe foot and registers P and T. The sinking velocity can be adjusted to the needs at the specific site by changing the weight of the device. The measured pressure relates to the depths. The logger is recovered by flushing the device to the surface from the other side of the U-tube. The recuperation process disturbs the temperature profile. Because of this, the undisturbed temperature profile cannot be measured before a certain delay for thermal equilibration. The data can be downloaded and processed at the construction site immediately after recuperation of the logger. The logger operates to 350 m depth and the temperature resolution is 0.0015 °C (Forrer et al. 2008).

Fiber-optical measurements also produce temperature data with vertical resolution. The fiber-optic cables can be attached to the probe before installation or may be placed into the probe pipe after installation. Temperature is being measured every 0.5 m and interpolated in between. Optical fibers serve as temperature sensors because of their specific physical properties. The fibers produce simultaneous and synchronous temperature data with high temperature and depth resolution along the entire length of a geothermal probe. The simultaneous T measurement with distributed sensor technology is a significant advancement over measuring T with temperature sensors at discrete points along the vertical probe (Hurtig et al. 1997). Because the temperature sensor cable can be permanently installed with the probe pipe, the vertical temperature profile can be retrieved any time or even continuously. Fiber-optical temperature measurement can be ideally combined with thermal response tests providing detailed thermal conductivity data of a layered ground, which in turn helps to locate faulty backfill sections. The fiber sensor also enables monitoring the backfill implementation of the geothermal probe borehole during and immediately after the termination of the grouting work.

Fig. 6.18 Cable-free mini data logger for temperature and pressure down-pipe measurements (23 × 219 mm)

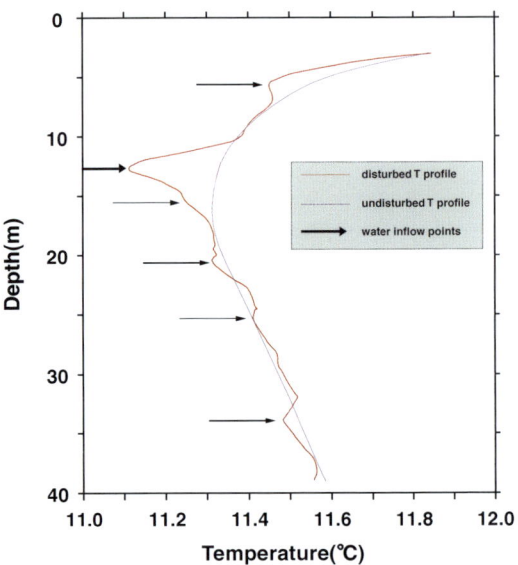

Fig. 6.19 Example of a disturbed temperature profile with indications of strong and weak water inflow (*arrows*)

A new cable-connected temperature sensor with 18 mm diameter can be lowered and hauled up in the probe pipe (Fig. 6.20a). The sensor can be used to control the setting of the backfill.

Deviations from the mean vertical temperature profile (Fig. 6.19) may indicate leaks or water inflows, hence an incomplete backfill. In contrast to fiber optical temperature measurements, the temperature sensor registers upwelling or descending waters, generally water of a markedly different temperature than the solid material at the same depth. Vertical groundwater flow results from different hydraulic potentials in separate aquifers. Leakages without significant vertical flow may be detected with the temperature sensor only in exceptional cases.

The cable-connected temperature sensor can be extended and used as gamma ray logger also. The method of measuring naturally occurring gamma radiation is used to characterize the geological strata in a borehole (Fig. 6.20b). The individual elements of the logger for temperature and gamma ray can be connected with flexible connectors so that the sensor chain can follow the coiled probe pipes. If the backfill has an added radiating tracer, the gamma ray logger can be used to control the backfill, in analogy to the classical borehole gamma logging in wells. Labeled grouting materials are presently being tested. As already mentioned, the presence of (three) additional probe pipes in the borehole contributes to the measured data and complicates the analysis and interpretation of the data. The variable spatial orientation of the probe pipes complicates the situation further. For the use of labeled backfill legal requirements may exist.

The flexible geophysical multi-tool (Fig. 6.20a) can be expanded by a magnetic inclinometer for precise measurement of the 3D-orientation of the probe pipes. A three-axis sensor measures inclination and orientation of the pipes. Separate flexible inclinometer tools measure the dip and orientation of the geothermal probe

6.8 Special Systems and Further Developments

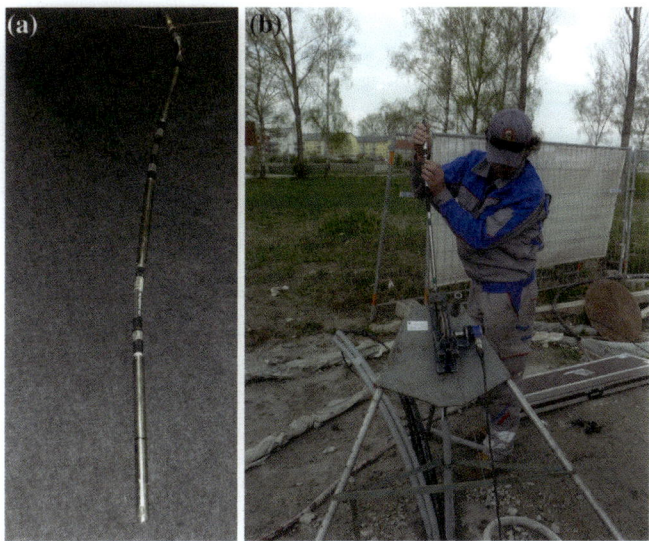

Fig. 6.20 **a** Flexible tool string for geophysical measurements in geothermal probes. **b** Inserting a gamma ray logger into a geothermal probe

pipes. Maximum length of these tools is presently about 100 m. With a diameter of 27 mm, the inclinometer tool fits into a standard probe pipe. Measurement precision is about 0.001 Kelvin. Testing the tool has shown that the pipes are too strongly twisted so that the inclinometer chain every so often got jammed and could not be lowered in the geothermal probe completely.

A newly developed gamma–gamma logger for geothermal probes measures density of the surrounding material and thus gives information about the completeness and quality of the backfill. The gamma–gamma method is based on nuclear physical interaction of gamma rays from a source in the sensor and the atoms that make up the rocks and backfill. The interaction processes absorb gamma ray energy when passing through the ground (rock, backfill). The detected remaining back-scattered gamma irradiation is thus inversely proportional to the density of the ground. The method measures density of the formation and detects incomplete backfill and other cavities in the wellbore. The device has a length of 80 cm and a diameter of 15 mm. Also with this method, the presence of other water-filled probe pipes in the borehole and their complicated spatial position complicate the processing and interpretation of the data. Measuring with a gamma–gamma logger may not be inoffensive at all sites and buildings.

Also with a new ultrasonic tool being presently developed it is hoped to detect irregularities regarding density, homogeneity and structure of the local environment of the geothermal probe and thus also control the completeness of the backfill. The ultrasonic probe uses a pulse-echo technique. A further tool under development is the kappa probe measuring magnetizability of rocks and ground, indirectly permitting conclusions on the quality of the backfill.

More tools and methods are currently being developed and tested. Many of these methods have originally been developed and used in the oil industry, then adapted by the groundwater well industry and now being modified and tailored for the needs of the geothermal energy industry.

The future development is evident: For the approval and acceptance of geothermal probes and for the quality control of these systems such quantitative data and measurement based investigation procedures will become standard and an integral part of the construction project.

6.8.5 Geothermal Probes Operating with Phase Changes

Geothermal probes may integrate the thermal effect of a phase change which is heat from a reaction associated with a phase change such as boiling or condensation into the geothermal system. Other names for such systems include thermosyphon and heat pipe. The principle is as follows: Instead of using a conventional heat transfer fluid for extracting heat from the ground, the thermosyphon uses a liquid with a low boiling temperature and directly evaporates the liquid in the geothermal probe by the geothermal heat added by the ground. The ground provides the reaction enthalpy of the liquid—steam phase change, the latent heat of evaporation. At the surface the steam is condensed to the liquid phase and the reaction enthalpy gained. This latent heat of condensation, can be used for house heating or other purposes. Like in conventional geothermal probes also thermosyphons work in cased boreholes. Thus the phase changes are independent of the geological structure and properties of the ground. At present heat transfer fluids used for thermosyphons include propane, ammonia and carbon dioxide. Some important physical properties of these transfer fluids are listed on Table 6.3.

The refrigerant ammonia is toxic, difficult to handle and could be hazardous to the groundwater after a leakage. Propane is flammable and, if unprofessionally handled, explosive. Thus there are legal requirements for using both these fluids. In some countries their use in geothermal probes is not permitted at all. For these reasons, geothermosyphons operating with carbon dioxide, so-called CO_2-probes have been further developed and improved.

The liquid CO_2 flows along the walls of the probe pipe downwards, evaporates because of the heat uptake from the ground. The warmed CO_2 gas flows in the

Table 6.3 Properties of heat transfer fluids in geothermosyphons

	Ammonia	Carbon dioxide	Propane
Boiling temperature (°C) at 1 bar	−33	−78	−42
Density (10^{-3} kg/m^3) at boiling T	0.682	1.032	0.58
Vapor pressure (bar) at 0°C	4.82	34.91	4.76
Enthalpy of phase change (kJ/mol)	21.4	23.2	19.0
Critical temperature (°C)		31.1	
Critical pressure (bar)		73.8	

6.8 Special Systems and Further Developments

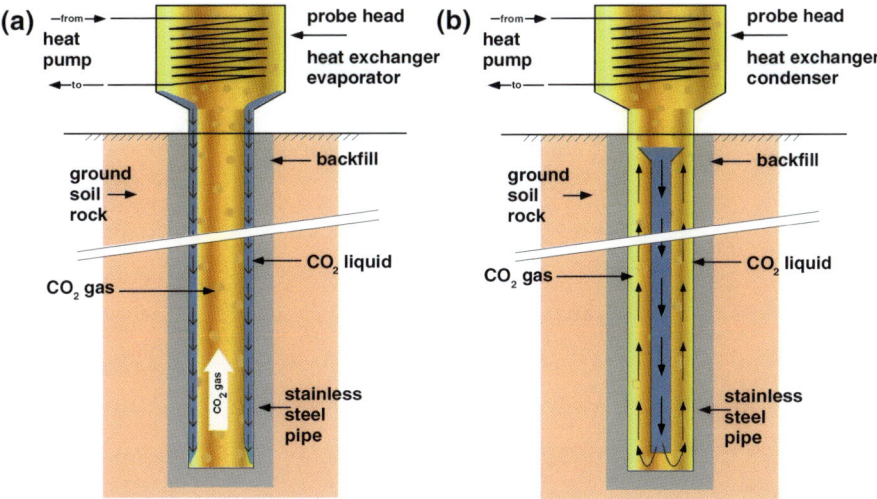

Fig. 6.21 Schematic diagram of a geothermosyphon (heat pipe). **a** Single pipe probe; **b** Two pipe probe

interior of the probe to the surface and supplies heat to the heating loop by condensation (Fig. 6.21).

The great advantage of thermosyphons is that they do not require a pump for circulating of the heat transfer fluid in contrast to conventional geothermal probes. There is no heat loss like with single and double-U probes where the heat transfer between both probe sides has to occur compulsory.

Standard PE pipes used for geothermal probes are not CO_2 proof. Therefore, CO_2 based heat pipes are made from a flexible, pressure-resistant, helically corrugated stainless steel or aluminum. The deployed pipes have diameters in the range of 40–60.3 mm. The thin film of liquid CO_2 flows downward in spirals along the corrugated pipe protected by the coil. After evaporation the gas rises in the free center of the pipe without obstructing the liquid film to the heat exchanger where it condenses. The heat exchanger disposes of copper coil tubing bundles in a pressure-tight strong housing (Fig. 6.21). The heat transfer fluid of the heat pump circulates to the probe head of the CO_2 heat pipe where it evaporates in the coil bundles (Fig. 6.21). The latent heat of evaporation is taken from the warm CO_2 gas. The thermosyphon is presently in the state of market launch. The technique certainly has a great potential for house heating applications also.

In addition to the described so-called single-pipe probe (Fig. 6.21a) CO_2 thermosyphons with two pipes have recently been developed (Fig. 6.21b). The two-pipe CO_2 probe separates the liquid and the gas phase by a coaxial pipe. The CO_2 gas streams upwards in the outer pipe to the heat exchanger where it condenses to liquid CO_2 that flows downward in the inner pipe. The design prevents obstruction of the CO_2 gas stream in small-diameter pipes by the downward trickling liquid CO_2.

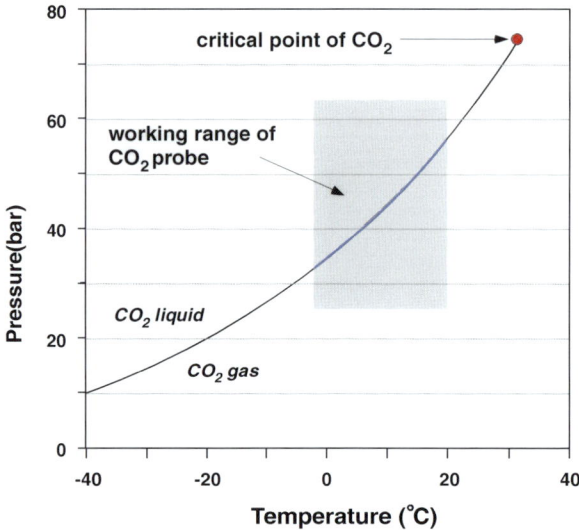

Fig. 6.22 Phase diagram of CO_2, blue section of the phase boundary is the typical working range of a CO_2 geothermosyphon (CO_2 data from Weast and Selby 1967)

Fig. 6.23 Heat pipes used to stabilize permafrost soil along the Trans Alaska pipeline

Two-pipe CO_2 thermosyphons can be used for cooling also, but this mode requires a pump for the liquid CO_2.

The boiling curve of CO_2 (Fig. 6.22) separates the pressure and temperature conditions where CO_2 is in a liquid phase from those where CO_2 is in a gas phase.

6.8 Special Systems and Further Developments

The curve ends at the critical point above where CO_2 is in a supercritical state. In order to operate the thermosyphon in the temperature range relevant for near surface geothermal applications (about -2 to $+20$ °C), the probe needs to be run in a pressure range of 35–55 bar. It is fully possible to operate a CO_2 thermosyphon above freezing conditions of water by selecting an appropriate working pressure, thereby avoiding all undesired perils and troubles associated with freezing of the probe and the surrounding ground (Sect. 6.7). However, also CO_2 probes may suffer from heat losses in the cold season because the probe is exposed to the low ambient temperatures in the uppermost meters and may give off some of the collected heat to the near surface ground. Generally, however, the seasonal performance factor of a CO_2 geothermal probe is considerably higher than for convectional geothermal probes.

Ammonia-driven heat pipes have a well-established range of applications and have been in use for years. With ammonia as refrigerant they are used for example to stabilize permafrost ground along the Trans Alaska pipeline (Fig. 6.23). Inside the vertical support pillars two heat pipes filled with ammonia take up the heat from the soil, this causes the fluid to boil and in the gas phase it condenses at the top of the heat pipe giving of heat of condensation through the finned radiators to the air. The condensate returns to the bottom of the heat pipe as thin liquid film along the pipe wall. Heat pipes are also used for snow thawing on sidewalks and pedestrian zones and deicing of railroad track switches (Narayanan 2004).

Chapter 7
Geothermal Well Systems

Operator panel of a drilling rig

Geothermal well systems utilize the thermal energy of clean groundwater of hydraulically highly conductive aquifers with water tables close to the surface (Fig. 7.1). The thermal energy of water produced from the well can be extracted by means of heat pumps (Sect. 4.1). Such systems are also called two-well-systems, water–water-heat-pump-systems, or groundwater heat pump. Geothermal well systems are a form of direct-use systems of near surface groundwater. The use of geothermal energy from groundwater can be particularly energy efficient in these systems. The direct-use of groundwater as a heat transfer fluid minimizes energy losses in heat exchanger systems. The relatively constant temperature of the groundwater flow is ideal for heat extraction by heat pumps. The advective heat transfer by groundwater flow has clear advantages regarding efficiency and economy compared with the conductive heat transfer utilized in geothermal probes. Limitations of direct use of groundwater are imposed by the availability of sufficient volumes of groundwater and appropriate aquifer properties, technical feasibility of well system development and the admissible thermal impact of the well system on groundwater and aquifer.

7.1 Building Geothermal Well Systems

The geothermal utilization of shallow groundwater requires a production and an injection well on the property for small and medium-sized systems (homes, smaller buildings). Larger buildings entail a system of several two-well units (two-well gallery). The geometrical arrangement of the two-well units and the power of each unit must be evaluated with a numerical model. Prerequisite for a successful model is an expert knowledge of the subsurface. Numerical simulation of heat and cold storage in the ground permits an optimized design of the well system for long-term operation and reliable prediction of the effects on the surroundings.

Engineering of production and injection well is similar to normal standard groundwater wells or groundwater measuring points with full-section pipes and screens, with an adequate gravel bed and appropriate sealing in sections with aquitards and in the near-surface area. However, geothermal well systems have relatively small diameters because of small production rates compared to ordinary groundwater wells. Because of different flow patterns of production and injection wells (Fig. 7.1), the production well should be screened at greater depth than the injection well. The pump must be installed above the screened section because of the increased flow velocity close to the pump inlet. The screened section in the injection well should begin higher up to prevent overflow of the well particularly during periods with a high groundwater table. This also helps to retard ageing processes. Experience shows that injection wells age faster than production wells. To prevent early ageing, the return pipe must be placed and connected to the injection well deep below the undisturbed water table (Fig. 7.1) and the screens must be in the groundwater under all operation conditions.

7.1 Building Geothermal Well Systems

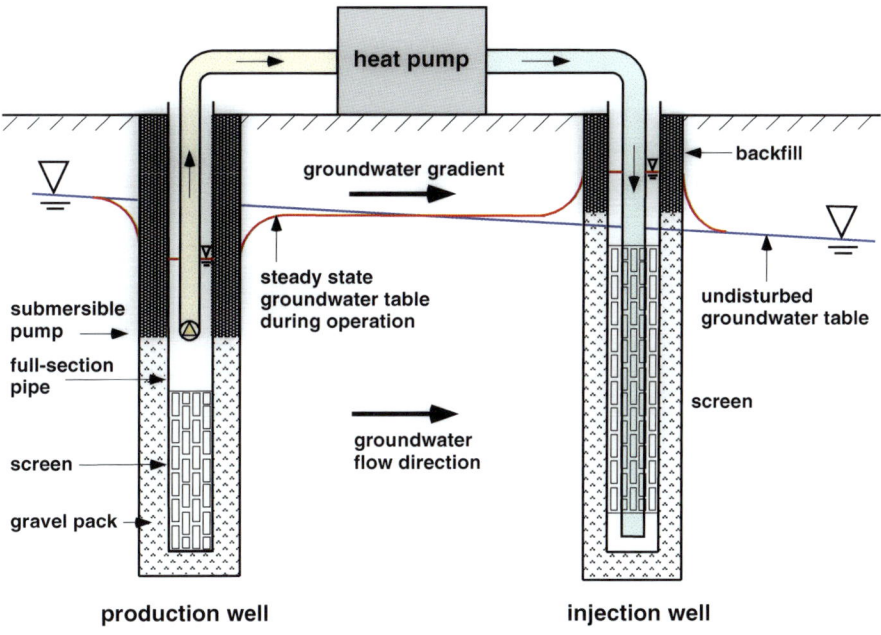

Fig. 7.1 Geothermal well system with production and injection well

The submersible pump (Fig. 7.1) produces groundwater of about 10 °C (Central Europe) to the surface; a heat pump extracts thermal energy from the groundwater and cools it to about 5 °C. The cool water is returned to the aquifer passing through the injection well. Reinjection of the thermally depleted groundwater assures a quantitative mass balance and conserves the resource groundwater.

The two wells may not interfere thermally with each other. The re-injection of the cooled water should not be upstream of the production well, of course. Ideally, if space permits, the injection well is placed normal to the hydraulic gradient (normal to the groundwater flow direction) or, second best geometry, downstream from the production well. Furthermore, some waters tend to become critically oversaturated with respect to certain minerals when cooled. The chemical effects of cooling should be carefully considered and chemical composition data must be available for modeling mineral scaling processes during project planning.

Before starting the operation, the yield of the two wells needs to be evaluated by means of appropriate pumping tests to ensure a sustainable utilization (Chap. 13). If measured yield is insufficient for meeting the planned heat extraction needs, then the wells can be deepened or the well diameter increased. Well depths for two-well direct-use systems are typically in the range of 5–15 m. Two-well systems work best if groundwater-surface distances are small and the hydraulic conductivity of the aquifer is high. Two-well systems for single-family homes are operated with flow rates of less than 1 l/s.

The supply lines for the wells must by laid frost-proof. The supply pipes must have a gradual slope to the wells so that they can be emptied easily if necessary. It must be ensured that the line position is below the groundwater table at all times.

Temperature variations of groundwater are relatively small. The mean annual temperature of shallow groundwater is related to the annual mean air temperature and typically ranges from 7 to 12 °C in moderate climates. The constant temperature enables very efficient operation of the heat pump. Heat extraction should not cool the groundwater by more than about 6 °C. Monovalent operation is usually trouble-free. The seasonal performance factor (Sect. 6.3.1) of geothermal two-well systems should be near 5. The German legislator, for instance, prescribes seasonal performance factors of 4 or better for water–water heat pumps used for house heating to ensure the energy efficiency of the system. If the heat pump is also used for hot water preparation a minimum annual performance factor of 3.8 must be achieved.

Heat output of 7 and 10 kW respectively require groundwater abstraction in the order of 2 and 3 m^3/h corresponding to 0.6 and 0.9 l/s flow rate respectively. The necessary flow rates are low so that 3″ or 4″ low-capacity submersible pumps are sufficient. The finished diameter of the production well should not be chosen to small despite the relatively small required extraction rates particularly in the upper section where the pump is installed. A generous well diameter minimizes hydraulic resistivity during pumping avoiding energy losses and uncontrollable power consumption during operation. The production well must be designed so that the pump can be installed deep enough for unproblematic operation also if groundwater table is low. Increasing the length of the screened section and the well diameter will improve, within limits, well performance. It is advised not to scrimp on the well diameter. Large well diameters generate long-term performance reserves for balancing later aging and degradation.

Attention must be paid to a sufficient distance between production and injection well (Eq. 7.1). This is important and avoids unwanted thermal interferences at the production well during operation. Typically, the distance between the two wells is some tens of meters, but shorter distances are possible depending on the conditions at the specific location. Possible long-range thermal effects on the groundwater by the operation of a two-well system must also be kept in mind and should be avoided. Specifically, the measured aquifer parameters from a qualified pumping test (Sect. 13.2) in the first drilled well can be used to determine the reach of the depression and injection cones. The model cones and the deduced minimum distance between the wells can then be verified by a properly devised pumping test in the second drilled well. If necessary, a third well must be drilled.

The necessary minimal distance (d) in meters between production and injection well can be estimated from Eq. 7.1:

$$d = 0.6 \, Q / \left(i \, k_f \, H \right). \tag{7.1}$$

assuming the aquifer has constant thickness (H) [m] and hydraulic conductivity (k_f) [m/s] and equally long filter sections and a continuous operation of the heat pump. Q [m³/s] represents the production and infiltration rate respectively and i is the dimension-less hydraulic gradient. The equation is valid if the wells are positioned normal to the direction of groundwater flow.

It is important that the injection well has sufficient conductivity for trouble-free take up of the pumped and cooled water from the heat pump. Water injection produces a positive cone around the injection well, which may become problematic if the difference between surface and groundwater table is small. The dimensioning of the injection well should consider the typically highest groundwater tables of the year thus avoiding overflow from the well. Here also, skimping on well depth, well diameter and length of the filter section of the injection well may not be economically advisable.

The power reserves created by generous dimensioning of well depth, well diameter and length of the filter section at the two wells invariably increase the durability and lifetime of the wells significantly. They also delay potentially necessary regeneration work. An increased filter section or a larger final diameter of the well reduces water flow velocity in the production well substantially and slows down well ageing markedly. Like in any other well, the filter section in the production well of a geothermal well system must begin considerably below the depression cone also during periods of low groundwater table and during maximum extraction rates. Otherwise atmospheric oxygen enters the well past the filter section producing sinter and mineral scales (sedimentation of iron ochre).

7.2 Chemical Aspects of Two-Well Systems

Ageing of wells is a cumulative effect of a number of different processes including iron-manganese ochre formation, scaling, siltation, corrosion, and formation of biofilm (slime). Geothermal well systems typically have contact to atmospheric oxygen. This may support increased microbiological activity and resulting biofilms of bacteria and algae. Access of oxygen also promotes oxidation of iron and manganese to insoluble Fe- and Mn-oxydes and oxyhydrates. The processes may also impair the aquifer itself. If the systems are also used for cooling, microbiological deterioration is to be expected.

Corrosion of casing, screen and other components of the system is mainly a consequence of access of atmospheric oxygen to the pumped water. The resulting oxidation of sulfide minerals in the soil and rock causes increased sulphate and low pH. These parameters are a secondary indication of oxygen access and ongoing corrosion. Also increased chloride and carbon dioxide support corrosion (Sect. 14.3). If hydrogen gas can be detected in the produced water then the system has a severe corrosion problem that needs to be fixed without delay.

Geothermal well systems should never be installed within the chemical plume of a waste deposit, a landfill, legacy assets and other groundwater damage areas. In aquifers with high natural sulphate concentrations sulphate-resistant materials must be used for constructing the well. Access of oxygen is typically the result of a drawdown cone in the production well reaching into the filter section.

Groundwater quality has a strong influence on the operation and lifetime of the geothermal well system. Biofilms, iron hydroxide scales and other deposits, particularly on heat transfer components reduce the efficiency of the system rapidly. The efficiency of the system is also compromised by scale deposits in the filter section, because of the required increase of pump pressure to maintain the flow rate.

The scaling and corrosion potential can be evaluated and modeled from chemical composition data and the temperature of the groundwater by using computer codes such as PHREEQC (Parkhurst and Appelo 1999). If conditions are very unfavorable, the project must be abandoned.

7.3 Thermal Range of Influence, Numerical Models

Computation of the thermal energy utilization must first distinguish between monovalent use for heating or cooling or for a combination of heating and cooling. After that, the energy need of the building or the facility determines the necessary amount of groundwater to cover these needs. The maximum annual groundwater demand controls the number of wells and their engineering details. These settings can then be used to model the thermal effects of operating the well system. These modeled thermal effects must be appraised and, if necessary, the array and spacing of wells must be adjusted and optimized as to minimize the effects on the aquifer.

Numerical models of geothermal utilization of near-surface groundwater have been computed already in the 1980s of the last century. Successful approximation solutions have been developed for small and medium size systems that are still valid and widely used. The layout of two-well systems, particularly the distance between the two wells should be assessed using approximation methods (e.g. Eq. 7.1).

During operation of the geothermal two-well system a thermal anomaly relative to the undisturbed temperature distribution gradually develops around the injection well. The temperature anomaly decays nearly exponentially along the groundwater flow direction. The outer limit of the created temperature anomaly may be defined by the boundary where the temperature difference to the undisturbed temperature field is smaller than one Kelvin. For stationary conditions in the one-dimensional case and by ignoring longitudinal dispersion the cooling distance L can be approximated by Eq. 7.2a (Kobus 1992):

$$L = \ln 10 \, \rho_W \, c_W \, n_d \, H \, H_D \, u / \lambda_D \quad (7.2a)$$

where L is the cooling length or length of the anomaly in meters, ρ_w stands for the density of water [kg/m^3], c_w for the heat capacity of water [J/(kg K)], n_d is the flow porosity of the ground (−), H the thickness of the aquifer [m], H_D represents the thickness of the cover layer [m]. λ_D the thermal conductivity of the cover layer [W/(m K)] and u the effective groundwater flow velocity [m/s].

7.3 Thermal Range of Influence, Numerical Models

The cooling distance L (m) for the case of radial flow caused by the injection rate Q (m³/s) can be estimated from Eq. 7.2b (Kobus 1992):

$$L = \sqrt{(0.733 \, \rho_W \, c_W \, Q \, H_D / \lambda_D)} \tag{7.2b}$$

where ρ_W stands for the density of water [kg/m³], c_W for the heat capacity of water [J/(kg K)], H_D represents the thickness of the cover layer [m]. λ_D is the thermal conductivity of the cover layer [W/(m K)].

Equation 7.2b is valid for very small hydraulic gradients corresponding to nearly stagnant groundwater. The cooling distance L characterizes the symmetrical circular anomaly around the injection after very long times of operation.

For appreciable hydraulic gradients and significant groundwater flow, the thermal anomaly at the injection well becomes oval shaped. The maximum length of the anomaly from the well to the <1 K boundary in flow direction can be estimated using an iterative procedure developed by Ingerle (1988).

The maximum lateral extent B_T of the temperature anomaly normal to the flow direction can be approximated by Eq. 7.3 for the hydraulic width B_H [m]:

$$B_H = Q / (i \, k_f \, H) \tag{7.3}$$

with Q being the flow rate in m³/s, i is the dimensionless hydraulic gradient [m/m], k_f the hydraulic conductivity [m/s] and H stands for aquifer thickness [m].

In many areas, the groundwater flow direction changes with periodic or irregular fluctuations of the groundwater table. From other areas data on dispersion coefficients of the aquifers have been determined experimentally. If such parameters change and the variability of aquifer properties are known and significant they can be considered in Eq. 7.4. The lateral extent of the thermal anomaly influenced by the seasonal variation of groundwater flow direction and dispersion effects can be considered by the propagation angle α. From experience the propagation angle α varies from about 5° for the pure dispersion case and 15° for dispersion and strong seasonal variations of the flow direction. The total width of the thermal plume B_T can be estimated as a function of the downstream distance x (m) from the injection well:

$$B_T = B_H + 2 \, x \tan \alpha \tag{7.4}$$

where B_H (m) denotes the hydraulic width defined by Eq. 7.3.

Another approach for modeling the temperature changes in the aquifer around the injection well has been described by Kobus and Mehlhorn (1980). They consider four specific points on an isotherm around the injection well and compute the temperature changes using Eqs. 7.5 and 7.6. The flow trajectory through the injection well is defined as x-axis. The intersection of the isotherm ΔT representing a considered temperature difference relative to the undisturbed temperature field with the x-axis is given by:

$$x_0 = (4 \pi \, \alpha_T)^{-1} \, (Q \, \Delta T_E / n_d \, u \, H \, \Delta T)^2 \tag{7.5}$$

where α_T stands for the transvers dispersivity [m], u for the effective flow velocity [m/s], n_d the dimensionless flow effective porosity, H the aquifer thickness (m), Q the flow rate [m³/s] and ΔT_E the temperature difference of the water between the two wells (K). The intersection of the isotherm ΔT with the y-axis that is normal to the flowline through the injection well is given by Eq. 7.6 for $x \leq x_0$:

$$y = \pm \left[4\alpha_T\, x\, \ln \left\{ Q\, \Delta T_E\, /\, n_d\, u\, H\, \Delta T\, \left(4\pi\, \alpha_T\, x\right)^{0.5} \right\} \right]^{0.5} \quad (7.6)$$

The simple computing approaches presented above are supplemented by a large number of more or less user-friendly software for modeling geothermal two-well systems for home heating and cooling. The software ranges from relatively simple spreadsheets to sophisticated (and expensive) packages. Two examples: GED (Poppei et al. 2006); EGON (Rauch 2009). This kind of software does not provide the user with a general solution for the immense variety of possible well configurations, well operation conditions, variable groundwater and heat flow conditions, and parameters of technical equipment such as heat pumps. Computed model solutions are valid for simplified conditions and special settings such as for single two-well systems in ideal isotropic aquifers and special boundary conditions. The software Groundwater Energy Designer (GED) by Poppei et al. (2006) successfully computes, for example, the groundwater flow field and heat transfer in homogeneous isotropic aquifers for several well systems decoupled. It is, however, not designed to model transient conditions and variable and complex geological structure of the ground. The code EGON (Rauch 2009) models the heat plume in a vertical section along the flow trajectory through the injection well. The computation is done for a single well and solves for decoupled groundwater flow and heat transfer using analytical, numeric and empirical methods. The program handles transient flow and heat conditions.

Geothermal well projects for larger buildings with several wells used for heating and cooling definitively require groundwater flow models for planning the engineering details. Coupled groundwater flow and heat transfer can be modeled using numerical finite difference or finite element codes. For the direct vicinity of the well a 3-D model is needed. The results of decades of research and plentiful of experience is available for general simulation of heat- and mass-transfer. The basic physics of flow mechanics and of heat transfer has been outlined in classic books of e.g. Bear (1979) and (Carslaw and Jaeger 1959). Heat transport in groundwater by combined conduction, convection and dispersion has been comprehensively treated by Sauty (1980). The models implemented in the codes cover a large variety of relevant processes. Well known program packages include FEFLOW by WASY (Diersch 1994), TOUGH2 of Lawrence Berkeley National Laboratory (Pruess 1987) and HST3D of the US Geological Survey (Kipp 1997).

Meaningful system planning and decisions on the number of required wells, well dimensioning and layout requires a sound knowledge of hydrogeological parameters including hydraulic conductivity, storage coefficient, aquifer thickness, aquifer structure and hydraulic gradient and its temporal variations. Furthermore, serious models are based also on injection temperature, the pre-operation underground temperature and flow fields, hydraulic gradients, specific heat capacity and thermal conductivity of all rocks present in all strata, just to name the most important parameters. These derived data are input for the programs and can be used to compute the time-dependent temperature field, the temperature evolution in space and time related to the reinjection of the thermally depleted water, the range of the thermal effect during operation and, if applicable, a possible thermal breakthrough.

The relatively large specific area of highly porous or intensely fractured near-surface rocks and soil promote rapid heat exchange between ground and re-injected water primarily by heat conduction. The resulting decrease of the temperature difference between rock and water reduces the thermal plume in space and time. The process has formal similarities to the process of mixing and distribution of a sorptive tracer. Therefore, indirect solutions for heat transfer can be gained from pure groundwater flow models such as MODFLOW (Harbaugh 2005) used for modeling contaminant transport and associated sorption processes.

The presented computational concepts, programs and models are also valid for deep aquifers and their hydrothermal systems (Chap. 8).

Chapter 8
Hydrothermal Systems, Geothermal Doublets

Production test at a geothermal power plant

Hydrothermal systems use the thermal energy of an aqueous fluid at greater depths. Depending on the heat content of the fluid, systems with high enthalpy can be distinguished from low enthalpy systems. High enthalpy systems produce electrical power directly from hot steam or from a high-temperature two-phase fluid (Sect. 4.2). Low-enthalpy systems use the warm or hot water directly or via a heat exchanger to feed local or district heating systems, for industrial or agricultural utilization or for balneological purposes. Profitable electrical power production is possible at fluid temperatures above 120 °C. The thermal water is produced from deep groundwater reservoirs (aquifers). In principle, thermal water may also be retrieved from water conducting faults and fault zones, however, hydrothermal systems typically connect to aquifers.

High-enthalpy hydrothermal systems are related to regions with extreme geothermal gradients and very high ground temperature at shallow depth typically found in volcanic active areas, young rift systems and similar geological conditions. Low-enthalpy systems can be developed in any region with average or slightly elevated geothermal gradient. Therefore, the potential for low-enthalpy systems is evident because they can be installed in normal continental crust. The present day situation is, however, strongly focused on high-enthalpy systems and most of the worldwide installed electrical power capacity from geothermal sources relates to high-enthalpy systems (Sects. 1.3 and 3.4).

Deep aquifers that permit the installation of geothermal doublets may also be used as seasonal heat storage systems. This can be attractive if for example the seasonal excess heat collected by a photovoltaic system can be transferred to the deep ground for use in the cold season. An aquifer heat storage system uses, in contrast to a geothermal probe storage system (Sect. 6.8), the heat capacity of water and rock of a natural aquifer that is hydraulically sealed at the bottom and the top of the conductive layer. The aquifer heat storage system is developed by means of a production and an injection well, similar to geothermal doublets. For charging the system, water is produced from one well, heated in a heat exchanger and injected into the aquifer through the second well. The process is reversed for discharging the system (Hasnaina 1998a, b).

8.1 Geologic and Tectonic Structure of the Underground

Hydrothermal systems use natural deep groundwater residing in geological reservoirs with high hydraulic conductivity. The reservoirs are embedded in other geological units with different properties. Therefore detailed and thorough knowledge of the geological structure of the underground is absolutely compulsory for exploring and constructing hydrothermal systems. The prime exploration target is to prove the existence of suitable geothermal reservoirs and determine the depth and thickness of the thermal aquifers. The exploration process involves a long series of data accumulation before the first exploration borehole is drilled. The pre-drilling exploration first collects all geological and hydrogeological data that are already

8.1 Geologic and Tectonic Structure of the Underground

known about an area. The general structure of the underground of a potentially interesting area is then explored with geophysical tools, primarily with seismic soundings, supported by gravimetrical, geomagnetic and aeromagnetic measurements if necessary (Sect. 12.1). Collected seismic field raw data must be processed with complex mathematical tools and algorithms. The correctness of the modeled structure of the underground depends much on the sound assumptions about rock properties that have been used in converting time to depth data. Seismic data may image the geological structures along vertical sections (2D seismics) or in three-dimensional space (3D seismics). Remember that an exploration drillhole provides 1D information about the structure of the subsurface.

The deep and consequently hot aquifer used by a hydrothermal system should have a high hydraulic conductivity. The decisive aquifer parameters are temperature and yield. The attainable yield or flow rate results from an economically and technically manageable drawdown in the production well. The productivity index (PI) is defined as the ratio of flow rate and drawdown (Sects. 8.2 and 8.6). This crucial parameter for hydrothermal systems cannot be derived from geophysical pre-drilling data from the surface. However, the productivity index can be determined, like other hydraulic parameters (Sect. 8.2), from hydraulic test data from well tests in a borehole.

Nevertheless, geophysical prospection may find indirect clues for elevated hydraulic conductivity such as indications for faults and other major brittle deformation structures or gradual facies changes in sedimentary units. Seismic data also give hints on the stress regime of an area, if the region is under compression or extension. Even stress changes with time may be revealed. Generally, the chances for finding zones with elevated hydraulic conductivity increase if fractured units and fault zones are being drilled, although brittle deformation structures may be sealed by secondary minerals and the structures may not conduct hot fluids. Similarly, increased hydraulic conductivity may be associated with extensional stress regimes whereas water-conducting structures in compressional regimes could be closed. The pre-drilling prediction of the structure of the underground must finally be verified by an expensive drillhole. Because drilling is the most expensive part of a hydrothermal energy project, pre-drilling exploration is important and should be taken serious.

Figure 8.1 shows an example of a geologically interpreted seismic cross section through the upper Rhine river valley south of Strasbourg. The valley is a Tertiary rift structure with Paleozoic basement exposed east (Black Forest) and west (Vosges) of the graben structure. The existing seismic 2D section has been calibrated with data from several existing deep boreholes in the area. The borehole data have been projected onto the plane of the section. The seismic data show the depths and the thicknesses of the potentially valuable hydrothermal aquifers. The section also shows a general half-graben structure and the presence of an inverted flower structure in the western part of the section. Both structural features are evidence for an extensional regime. The section exhibits several faults with associated vertical displacements of strata (reflectors) in the deeper parts of the section. The faults are lacking in the upper part of the section, they are absent in the younger

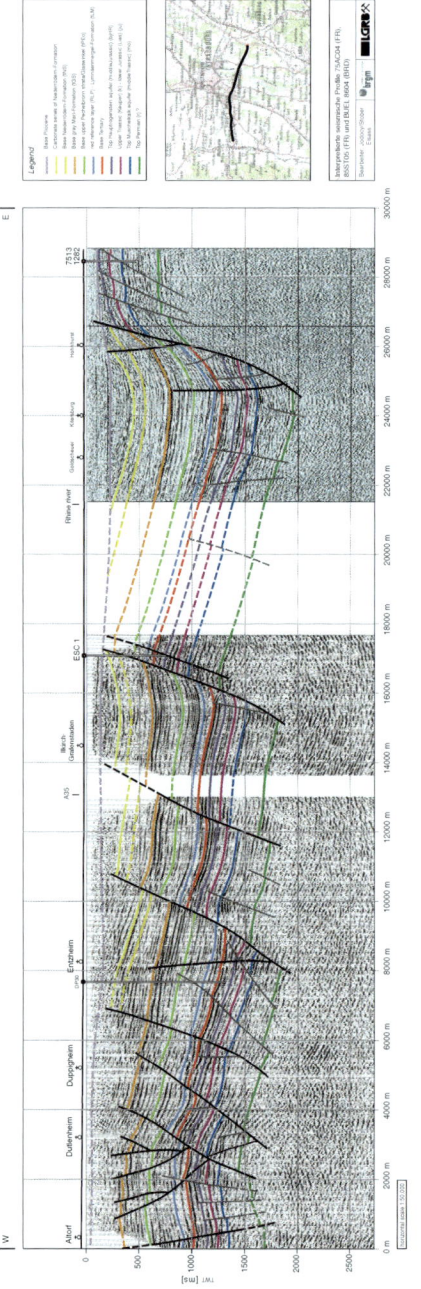

Fig. 8.1 Seismic section across the upper Rhine graben with geological interpretation (Jodocy and Stober 2008)

deposits. The structures are thus older faults that are no longer active. These brittle structures may have been sealed by deposition of secondary minerals and may not be associated with high hydraulic conductivity. Thus these structures may not be a prime exploration target.

Already existing seismic data and data from deep boreholes need to be searched for and carefully studied during the early stages of pre-drilling exploration. Reprocessing of old seismic data may be possible and has proven to be very helpful in many cases. Existing deep boreholes greatly facilitate geological interpretation of seismic data. Furthermore, seismic travel time data can be calibrated using borehole data (Sect. 12.1). If boreholes are not available within a suitable distance from the investigated new site, then seismic travel time data need to be converted to depth using uncertain model assumptions. The sound interpretation of seismic data establishes the stratigraphy, depths and thickness and in some cases the sedimentary facies of the geological strata underground. A further important goal of the seismic data interpretation is the survey of deformation patterns and fault systems. Fault mapping may show that large fault displacement disrupts hydrothermal target aquifers. Hydraulic connection of the displaced aquifer may be indirect through the fault. Surveying fault patterns is generally more difficult, if not impossible, in crystalline basement compared to sedimentary cover sequences. Identified and mapped faults in the cover can be extrapolated to the crystalline basement in some favorable cases. The research results from existing seismic exploration data and from related boreholes enable a sensible decision on the necessity of further seismic investigations.

8.2 Thermal and Hydraulic Properties of the Target Aquifer

The most important thermal parameters in the context of hydrothermal system development are thermal conductivity λ [W m^{-1}K^{-1}] and the specific heat capacity c [J kg^{-1} K^{-1}] (Sects. 1.4 and 1.5) of the geological material and the fluid phase(s) contained in it.

The thermal conductivity refers to the ability of material to transport thermal energy; the heat capacity represents the ability to store thermal energy. The heat capacity of material is a particularly important parameter for characterizing the effects of time dependent systems and transient states.

The heat flow density q [W m^{-2}] is a further important parameter in the framework of hydrothermal system development. It quantifies the heat flow per surface area and contains the parameter time in the dimension Watt [J/s], energy per time. The heat flow density q relates to the product of thermal conductivity λ and the temperature gradient grad T [K m^{-1}] and is defined by the Fourier equation of conductive heat transfer (Sect. 1.4; Eq. 8.1):

$$q = \lambda \,\text{grad}\, T \qquad (8.1)$$

The Fourier equation (8.1) states that the driving force of a non-equilibrium temperature distribution {grad T} causes the flow q of an extensive parameter, here thermal energy per time, to reduce the imposed driving force of the process. The magnitude also depends on material properties, here the heat conductivity λ. If λ is high, high heat flow results from the imposed driving force, if λ is low the same driving force decays over a much longer period of time. These simple relationships have significant consequences for hydrothermal projects.

Thermal conductivity λ varies between 2 and 6 W m^{-1} K^{-1} in hard rocks. Water has a very low λ of only 0.6 W m^{-1} K^{-1} at 20 °C. Highly porous aquifers have a lower thermal conductivity than impervious rock units with a low porosity. The specific heat capacity c of hard rocks varies within the very narrow range of 0.75–0.85 kJ kg^{-1} K^{-1}. However, the specific heat capacity of liquid water of 4.187 kJ kg^{-1} K^{-1} is five-times greater than that of rocks. Water poorly conducts thermal energy but it is highly capable to store heat.

Density ρ [kg m^{-3}] of rocks ranges from 2,000 to 3,000 kg m^{-3}. Some rocks, peridotite and eclogites, may be as heavy as 3,300 kg m^{-3}, other rocks, for example coal, may weigh less than 2,000 kg m^{-3}. Typical continental basement rocks, granites and gneisses, weigh about 2,700–2,800 kg m^{-3}. Liquid water has a density of about 1,000 kg m^{-3} at the temperature of +4 °C and surface pressure of 1 bar (density anomaly of water). The density of rocks and water depends on the temperature and pressure. The p–T dependence of rock density can be normally ignored in planning hydrothermal facilities. The density of water as a function of temperature and pressure is shown in Fig. 8.2a. Density decreases with temperature regulated by the thermal expansibility of water and it increases with pressure controlled by the compressibility of water. Water under hydrostatic pressure at 7 km depth and 80 °C has a density of ~1,000 kg m^{-3}, similar to surface water. However, in regions with normal geothermal gradients, the temperature effect slightly outweighs the pressure effect and, consequently, the density decreases slightly with depth. Decreasing hydraulic conductivity with depth and increasing mineralization and salinity of the deep water efficiently inhibits upwelling of deep hot water. The density of water depends also on the total amount of dissolved solids (TDS). At a given pressure and temperature the density increases with TDS. Deep water is commonly highly mineralized. Several hundred grams of dissolved solids per kg liquid is not uncommon. The density decrease of pure water with depth in thermally normal areas is more than compensated by the concurrent density increase caused by the growing TDS with depth. The net-effect is a slight density-increase with depth under normal conditions.

The thermal conductivity λ of water increases with temperature to a maximum at about 140–150 °C depending on pressure and then slightly decreases with further T-increase (Fig. 8.2b). λ also increases with pressure.

The dynamic viscosity μ [Pa s] of a fluid describes the internal resistance to flow; it is a measure of fluid friction. It strongly depends on temperature (Fig. 8.2c) and decreases from 0.2 at 0 °C to 1.75 × 10^{-3} Pa s at 200 °C. The three orders of magnitude decrease of μ is in sharp contrast to the small variation of density with T. Viscosity largely regulates the flow properties of thermal groundwater. Kinematic viscosity ν [m^2 s^{-1}] is defined as the ratio of dynamic viscosity and density ($\nu = \mu/\rho$).

The compressibility c_F [Pa^{-1}] of a fluid describes the volume change with pressure at constant temperature normalized to the volume at a reference pressure (Eq. 8.2).

$$c_F = 1/V \cdot \Delta V/\Delta p \qquad (8.2)$$

8.2 Thermal and Hydraulic Properties of the Target Aquifer

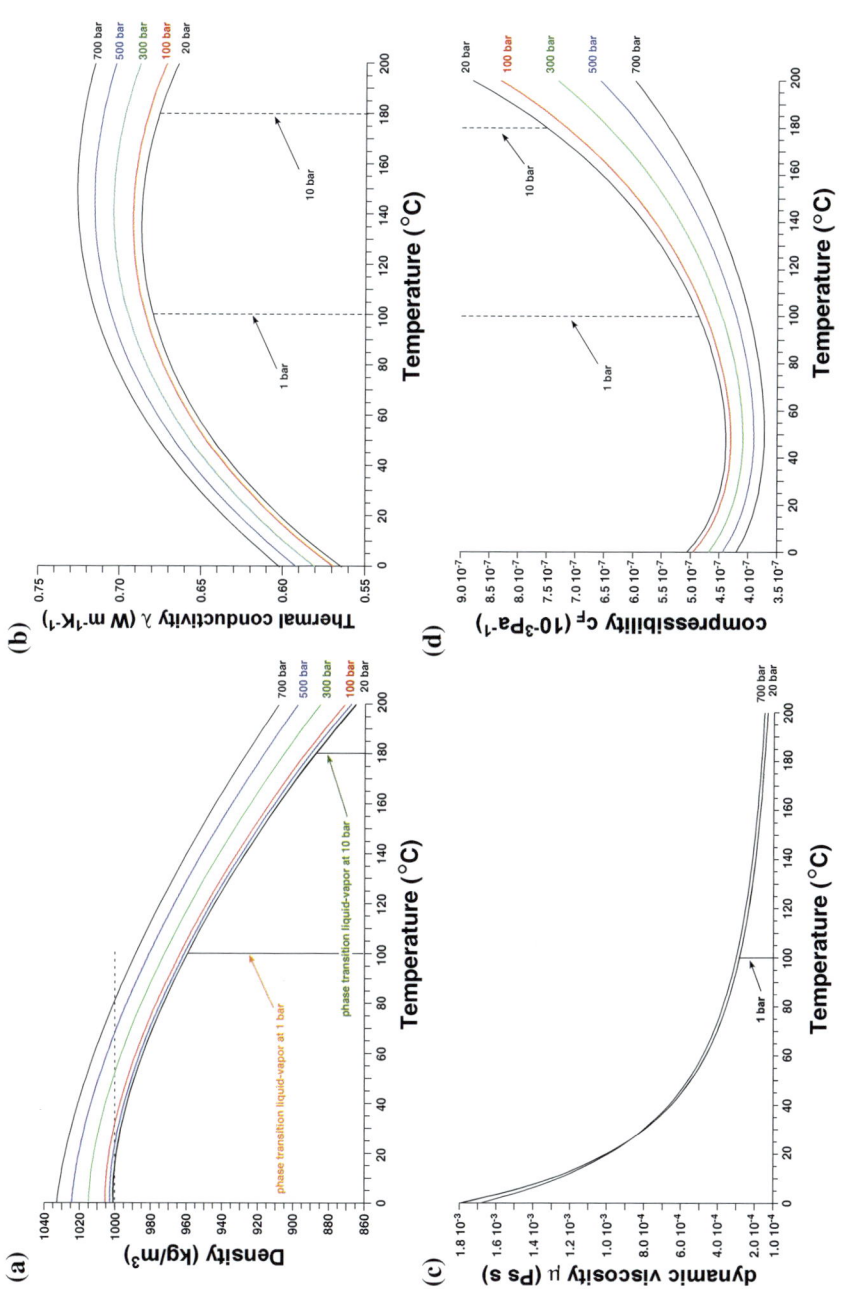

Fig. 8.2 Pressure and temperature dependence of important properties of water (Wagner and Kretschmar 2008): **a** Density. **b** Thermal conductivity. **c** Dynamic viscosity. **d** Compressibility

The compressibility of H_2O is inversely proportional to the pressure. At temperatures above 50 °C c_F increases with temperature, whereas c_F decreases with temperature below 50 °C (Fig. 8.2d). The compressibility generally varies from about 4.0×10^{-10} to 5.5×10^{-10} Pa^{-1}.

Permeability and hydraulic conductivity describe the ability of a system (rock) to let a viscous fluid pass through its pore space. The permeability characterizes the conductive properties of the rock matrix (soil, hard ground). Hydraulic conductivity characterizes the conductive properties of the system including both the porous solid and the flow properties of the fluid flowing through the pore space including fractures (fracture pore space). A rock with a given permeability has a low conductivity for "sticky" fluids and a high conductivity for "thin" fluids. Permeability and hydraulic conductivity should not be confused. Fluid flow in rock with a given permeability structure and for an identical flow force (hydraulic gradient) may differ considerably for aqueous fluids of different temperature, salinity, gas content, total of dissolved solids etc. Hydraulic conductivity k_f [m s^{-1}] represents the material related proportionality factor in the Darcy flow law, which relates fluid flow Q [m^3 s^{-1}] through the area A [m^2] to a hydraulic gradient i [dimensionless], representing the driving force for fluid flow.

$$Q/A = k_f \, i \qquad (8.3a)$$

$$k_f = Q / (i \, A) \qquad (8.3b)$$

Equation 8.3a represents the phenomenological Darcy flow equation of the form $J = L \, X$, where J is the flow of a quantity per unit area depending on the material property L and driven by the force X. It has the same structure like the Fourier Eq. 8.1. Equation 8.3b defines the hydraulic conductivity.

Permeability κ [m^2] and hydraulic conductivity k_f are related by the Eq. 8.3c, which considers the flow-relevant properties of the fluid (viscosity μ, density ρ)

$$k_f = \kappa (\rho g / \mu) \qquad (8.3c)$$

where g is the acceleration due to gravity. It is evident from Eq. 8.3c that the hydraulic conductivity of the ground increases with temperature, because the viscosity of water strongly decreases with temperature (Fig. 8.2c).

The pressure and temperature dependence of physical and thermal properties of H_2O (Fig. 8.2) has consequences for developing and operating hydrothermal systems. The borehole drilled after termination of pre-drilling exploration needs to be extensively tested. A well test in a thermal aquifer first pumps the relatively cool water standing in the borehole, later during operation the temperature of the produced water gradually increases and, consequently, the density of the fluid decreases. The density of the water is typically higher thus the water table is lower at the early stages of a pumping test. Density is lower and the water table higher at later stages of the well test. This density effect is particularly prominent in highly conductive aquifers (Fig. 8.3).

The dependence of the hydraulic conductivity on the physical properties of water means in practice that the conductivity of an aquifer at 70 °C is about three

8.2 Thermal and Hydraulic Properties of the Target Aquifer

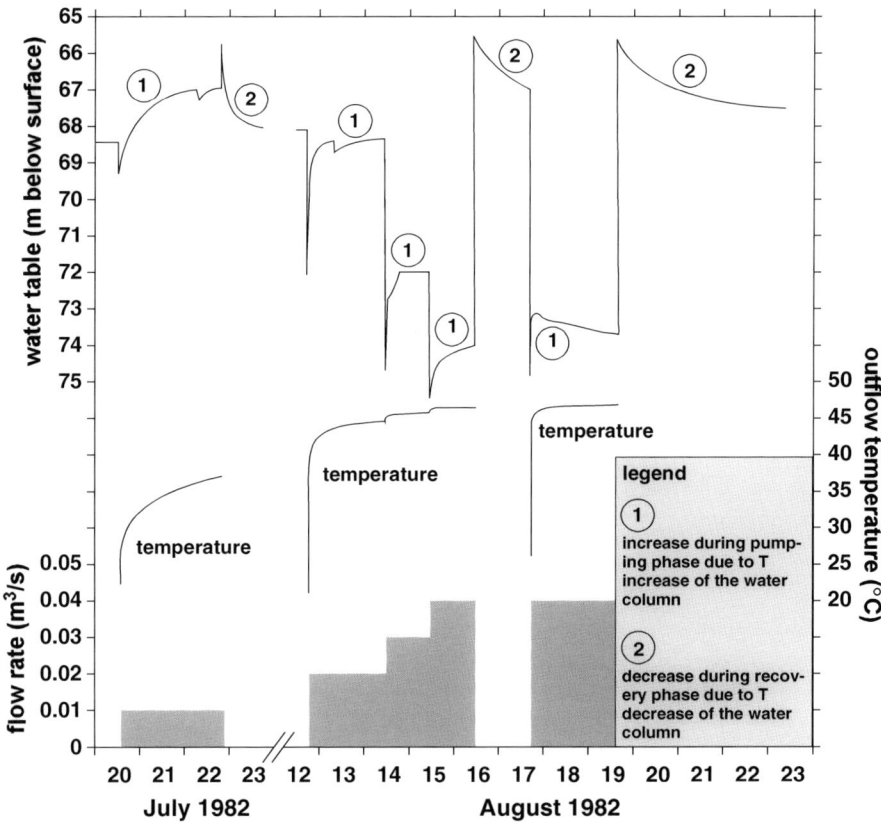

Fig. 8.3 Apparently paradox behavior of the water table during a pumping test in a thermal water well (Stober 1986). The figure shows the daily measurements of (from *top* to *bottom*): water table, outflow temperature, and the pumping rate

times higher than at 10 °C, all other aquifer properties being the same (Fig. 8.4). The dynamic viscosity dominates the temperature variation of the hydraulic conductivity (Fig. 8.2c). The effect of P–T variations of density on the hydraulic conductivity is negligible (Fig. 8.2a). Dynamic viscosity varies many times over the density in the temperature range from 0 to 200 °C. It is, therefore, of prime importance for the flow behavior of thermal groundwater.

The dependence of the hydraulic conductivity on the physical properties of water has direct consequences for the design of hydrothermal doublets. Extracting heat from the pumped deep water at the surface results in cooled, heat depleted water that needs to be reinjected into the aquifer. Consequently, the hydraulic conductivity and the water uptake capability of the injection well decrease dramatically. This, in turn, has the consequence that the injection cone is much more prominent than the cone of depression at the production well (Fig. 8.5). These interrelations should be considered in the planning phase of a hydrothermal project. If ever possible the well with the highest conductivity should be used as injection well.

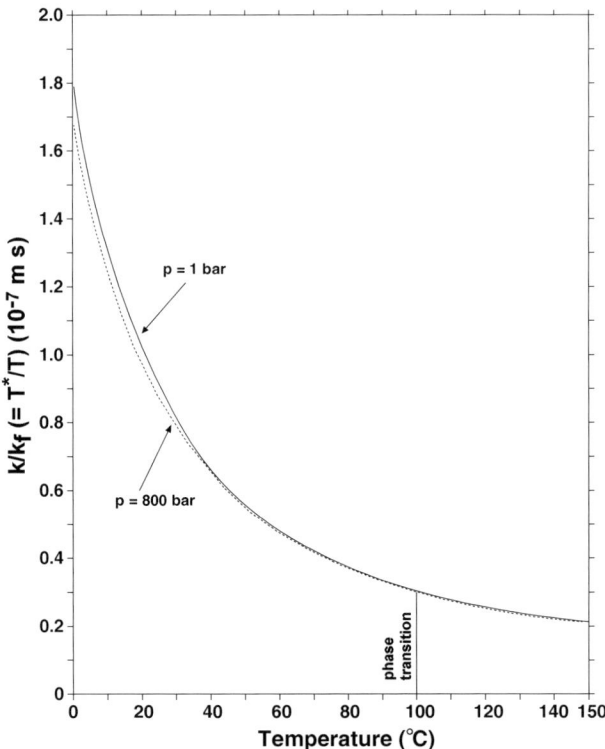

Fig. 8.4 Pressure and temperature dependence of the hydraulic conductivity k_f and transmissivity T^* of pure water. Also shown is the pressure and temperature dependent relationship between the hydraulic conductivity k_f (and transmissivity T, respectively) and the permeability k (and the transmissibility T^*)

The physical and thermal properties of water also depend, in addition to T and P, on its chemical composition. The chemical composition is a complex property and includes the total amount of dissolved solids, the kind of dissolved solids and the amount and type of dissolved gases. The thermal power of a hydrothermal system depends on the chemical composition of the produced fluid and it is imperative to know and monitor the water composition in hydrothermal projects (Sect. 8.6 Eq. 8.6).

Hydraulic conductivity is the central and decisive parameter controlling fluid flow rates in the underground. It is, together with the fluid temperature, the parameter that decides the economic success of the project. Hydraulic conductivity k_f is the material related factor in the phenomenological Darcy flow law (Eq. 8.3a). The total amount of water flowing through the aquifer cross section per unit time Q [$m^3 s^{-1}$] follows from the Darcy equation if the flow relevant cross section [m^2] of the aquifer is known. The amount of hot water available for the hydrothermal system varies with pressure and, mostly, temperature because of the P–T dependence of k_f. Fluid flow Q is clearly higher at high-T than in colder aquifers.

8.2 Thermal and Hydraulic Properties of the Target Aquifer

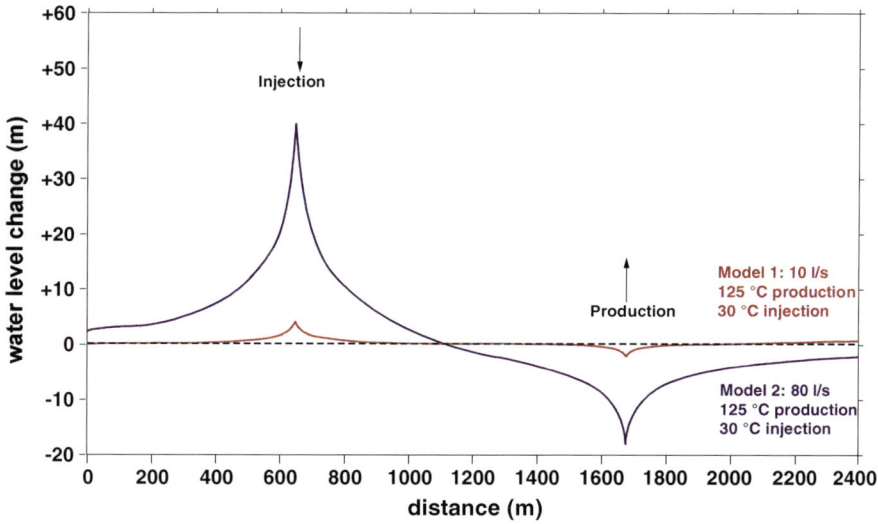

Fig. 8.5 Comparison of the cone of injection and depression at a hydrothermal doublet well. The cone at the injection well is much larger due to the injection of cooled water than the cone of depression at the production well

The Darcy flow law is valid for laminar and linear flow. Other, non linear, flow laws apply to conditions where the geological material has a very low hydraulic conductivity k_f and the hydraulic gradients are very low or of material with extremely high hydraulic conductivity and ultra-high gradients (flow force). Both types of extreme conditions do not normally occur in hydrothermal system setups (Kappelmeyer and Haenel 1974).

Hydraulic well tests provide data on drawdown and relaxation or pressure build-up and decline for given water production and injection rates (pumping rates). The observed gradients i (drawdown, depression cones) and the pumping rates Q permit an insight into the conductivity of the examined (tested) geological unit (Chap. 13). The well test derived conductivity is a property of the tested unit as a whole and describes the conductivity of the aquifer of thickness H [m]. It is called transmissivity T [$m^2\ s^{-1}$] and characterizes the conductivity across the thickness of the tested section. For a homogeneous and isotropic aquifer of thickness H, the hydraulic conductivity k_f can be computed from the measured transmissivity T directly from Eq. 8.4a:

$$T = k_f H \qquad (8.4a)$$

In layered ground, the measured transmissivity T is contributed by several layers i with different thicknesses H_i and hydraulic conductivities k_{f_i}. Thus total transmissivity T is the sum of the transmissivities of all layers i (Eq. 8.4b):

$$T = \Sigma_i\, k_{f_i} H_i \qquad (8.4b)$$

A generalized formulation of the transmissivity-conductivity relation is given by Eq. 8.4c:

$$T = \int_0^H k_f dh \tag{8.4c}$$

The productivity index PI [m^3 s^{-1} MPa^{-1}] proved to be a very useful parameter for characterizing a hydrothermal system. PI is the production rate Q [m^3 s^{-1}] per pressure decrease Δp [Pa]. It can be computed from the obtained production rate Q for a given fixed drawdown (expressed in Pa). The PI is particularly useful if no well test data have been obtained. The PI comprises well specific properties such as skin and wellbore storage in addition to the hydraulic properties of the aquifer (Chap. 13).

The absolute porosity n [dimensionless] is the total volume of all voids and pores of a rock unit per rock volume. A rock contains pores that are interconnected, pores that are isolated, fracture pore space, cavities and any other form of space that is not filled with minerals. The total volume of all these voids is the total pore space. The total volume of a rock contains all voids. The fraction of the total volume of voids and the rock volume is the porosity n (n = V_v/V_r; with V_v total volume of voids; V_r total volume of rock). The porosity n characterizes the storage capacity of an aquifer.

The water conducting features of hard rock aquifers are predominantly fractures and cavities. Conductivity and yield of the aquifer are mostly determined by the fracture network and geometry of the cavity system. Fluid flow is strongly linked to the flow effective porosity n_f [-], which is the portion of the total porosity that is interconnected and where water can be freely transferred from one pore to the next. Water firmly attached to the grain surfaces and water in dead end pores is stagnant water and not part of the flow effective porosity even if this water resides in interconnected porosity. The effective porosity is the relevant porosity parameter for geothermal applications. n_f is always smaller than n. The difference between total and flow effective porosity is particularly significant in shales and marls. Clay and mica-rich rocks may have a high total porosity but a low flow effective porosity and, consequently, a low permeability. Suitable deep hardrock aquifers for hydrothermal system are thus mica-poor rocks with predominantly feldspar and quartz such as granites, sandstones, arkoses and mica-poor gneisses or (karstified) limestone. High effective porosity generally relates to high permeability. However, the relation between n_f and κ is not simple and direct because permeability depends also on the size distribution, shape and connective structures. Both parameters can be derived from tracer tests and pumping tests in the wells (Chap. 13).

Hydraulic well tests may also provide the storage coefficient S [-] characterizing the tested aquifer (Chap. 13). The storage coefficient S is a measure for the volume change ΔV of the stored water (fluid) in response to a change of the heights of the water column Δh per unit surface area A:

$$S = \Delta V / (\Delta h\, A) \tag{8.5a}$$

The specific storage coefficient S_s [m^{-1}] normalizes the volume response to the volume of the aquifer, in contrast to the storage coefficient S where it is normalized to area. The relation between storage coefficient S and specific storage coefficient S_s is analogous to the relation between transmissivity and hydraulic conductivity. The relation is given by Eq. 8.5b for isotropic homogeneous aquifers of the thickness H [m]:

$$S = S_s \cdot H \tag{8.5b}$$

Storage coefficients of layered or inhomogeneous, anisotropic aquifers can be described by equations analogous to the Eqs. 8.4b and 8.4c.

Various methods and tools for hydraulic wells testing and for determination of hydraulic parameters of the deep ground will be presented in Chap. 13.

8.3 Hydraulic and Thermal Range of Hydrothermal Doublets

Hydraulic or thermal short circuits between the production and injection wells must be absolutely avoided during operation of hydrothermal systems (see also Sect. 7.3). Appropriate seals must prevent hydraulic connections to an other groundwater storey. The schematic setup of an injection well is depicted in Fig. 8.6. The distance from the injection well to the production well at connection depth to the aquifer must be sufficiently large for safe operation during a period of about 30 years. During this time the temperature of the produced water should not be influenced by the injection of cooled water into the exploited aquifer. For a specific system, the minimal distance between the two connecting sections of the wells in the aquifer depends on the geology related structure and properties of the ground but also on the technical conditions of the system such as well design and pumping rates. However, the well distance should not be too large because the sustainable yield of the production well depends on the recharge of the aquifer by the injection well.

Dimension and size of the geothermal reservoir is, of course, important for the utilization of the aquifer. The volume of stored hot water and the thermal energy it contains can be computed from the surveyed shape and dimension of the reservoir and the petro-physical reservoir properties including porosity and temperature distribution. For instance, greater aquifer thickness, for all other parameters and dimensions being the same, increases transmissivity and hence also the achievable fluid production rate (Sect. 8.2).

Basis for a three-dimensional geological underground model is data from geophysical, typically seismic exploration and data from existing boreholes in the area (Sect. 8.1). Numerical models optimize the location of and the distance between the two wells. Predrilling models need to assume plausible values for a number of crucial parameters of the target aquifer and the geological structure of the underground and thus are prone to considerable uncertainty. The "known" geologic structure of the underground is based on the interpretation of seismic signals.

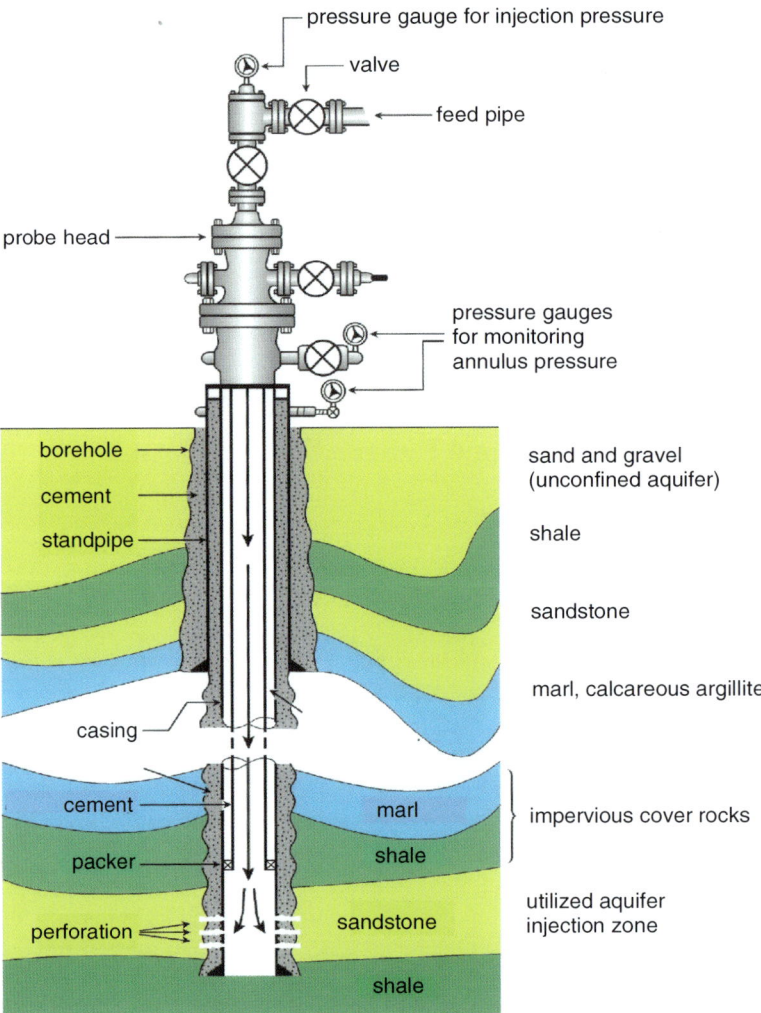

Fig. 8.6 Schematic structure of an injection well (from Owens 1975)

Modeled depths of the target aquifer, details of the stratigraphy, fault and fracture patterns from seismic exploration are combined with data on hydraulic conductivity, temperature and water composition from existing wells and well tests. The reliability of model predictions improves greatly with drilling of the first well of the hydrothermal doublet. Still, the geology of the underground is always good for surprises and the true structure and properties of the target aquifer can only be captured in a very simplified manner by pre-drilling models.

Numerical modeling starts with a conceptual model that bundles all available geological, hydrogeological and thermal data and information about the underground. It is the task of the geologist to construct a three-dimensional geological

8.3 Hydraulic and Thermal Range of Hydrothermal Doublets

model of the structure of the underground from irregularly distributed 1D borehole data and a limited number of 2D seismic sections. The resulting geological 3D model contains the position and orientation of the different stratigraphic layers and their thicknesses. By assigning hydraulic conductivity, porosity and other parameter values to the layers and imposing geothermal gradients to the model area a learned hydraulic model can be developed. This model depicts the basic hydrogeological concept of the underground. The model converts the lithological and stratigraphical units of the geological 3D structure model to hydro-stratigraphic units with associated hydraulic parameters. The thermal structure and properties of the underground follow from the assignment of thermal parameters to the individual strata and lithological units. This still relatively crude and simple model of the underground forms the basis for numerical models of heat and mass transfer. A preliminary stationary groundwater flow model can be further developed from the 3D underground model by assuming plausible geohydraulic boundary conditions and with quantitative estimates of groundwater recharge and vertical components of groundwater flow.

Numerical reservoir models are effective tools for developing, characterizing and optimizing hydrothermal projects (Fig. 8.7a, b). Today, many different numerical 3D models based on finite differences (FD) or finite elements (EF) mathematical techniques are routinely used. Examples are: SPRING by Câmara et al. 1996, FEFLOW by Trefry and Muffels 2007, SHEMAT by Clauser 2003, MODFLOW by Harbaugh 2005.

A numerical thermal water flow model must be calibrated. This is done by varying the geohydraulic aquifer parameters and the chosen boundary conditions until measured and computed groundwater potentials and potential distributions reasonably match and results in plausible groundwater balance. Chosen parameter values must be restricted to a plausible range for the investigated site. The calibration is typically associated with considerable uncertainties because of the limited number of data-backed grid points.

In the following step, the natural temperature field can be modeled using the calibrated stationary groundwater flow model. The typical result of the modeling effort is a rather simplified picture of the temperature distribution in the underground. Normally, the model does not even include heat transfer due to advection because of insufficient database.

In addition to model calibration, a thorough sensitivity analysis is a further important component of project planning. Thus hydraulic well tests producing quantitative test data are essential for a successful model calibration. Using numerical models, it is important to ensure internally consistent mass balance. For example, numerical circulation tests must be run with [kg/s] and not [L/s] because heat extraction changes the density of the hot fluid and hence more mass would be injected than produced when using [L/s]. It is important to carefully follow model instructions and handbooks (although it might appear old fashioned).

A well-calibrated stable numerical model can finally be used for predictions of the later operation phase. Even if the pre-operation distribution of the hydraulic conductivity in the aquifer was relatively homogeneous, the water uptake capacity of the injection well may become problematic during operation (see Fig. 8.5).

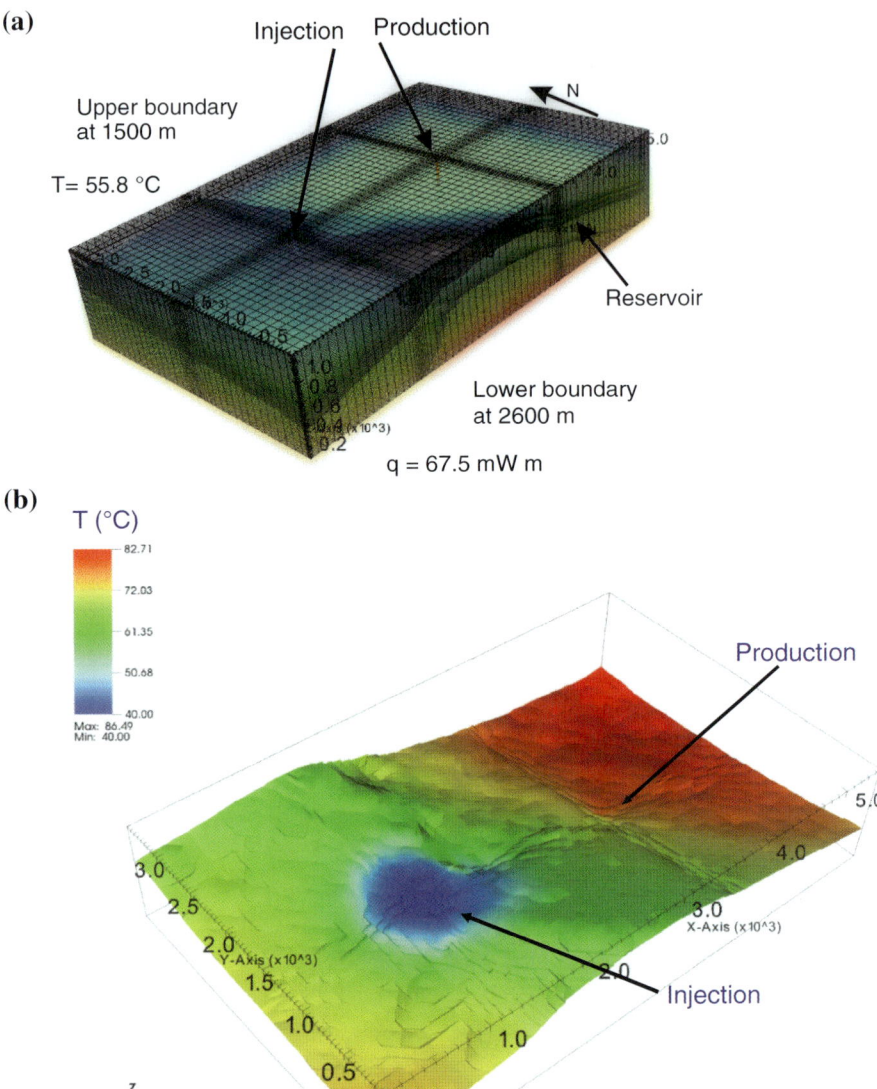

Fig. 8.7 Example of a numerical model for hydrothermal doublet: **a** Geometry of the model. **b** Model predictions for 50 years of operation (figure kindly provided by Geophysica GmbH)

Injection of cool water leads to a strongly lowered hydraulic conductivity in the injection zone, because of the viscosity increase with decreasing temperature (Sect. 8.2). Therefore it is advised to use the borehole with the highest hydraulic conductivity of the rock matrix as injection well.

The derived numerical model is also used to define the license area. A well-structured numerical model shows the likely hydraulic and thermal range influenced by the planned geothermal installation. In many countries, the licensing authorities require a professionally derived model from the petitioner.

With the help of numerical models of geothermal systems, coherent concepts of use can be developed and the planned system can be economically optimized. Recently, numerical geothermal system models have been used to assess seismic risks during construction and operation of geothermal systems (Sect. 10.1.5).

The quality of the model results depends strongly on the number (density) and quality of the measured hydraulic and thermal parameters. Also the scale of the model is crucial for the significance of the model predictions.

A numerical type model of a district heating system supplied by a geothermal doublet near Den Haag (Netherland) is shown in Fig. 8.7. Target horizon of the doublet is the Delft Sandstone unit (upper Jurassic to lower Cretaceous) at 2,200 m depth. Heat flow has been modeled by the three-dimensional finite difference code SHEMAT that solves coupled heat, transport and flow equations. The integrated regional model has the dimension 22.5 × 24.3 km and a depth resolution of 5,000 m. From this regional model local reservoir models have been developed such as the one shown in Fig. 8.7a for a field of 5.5 × 3.5 km and a depth range of 1,500–2,600 m. The number of nodes is about 170,000. Figure 8.7b shows the prediction of the T-distribution after 50 years of operation of the geothermal system at a production rate of 150 m^3/h, a temperature of the produced fluid of 79 °C and a temperature of the reinjected (cooled) fluid of 40 °C. The model predicts a significant decrease in temperature at a distance <1 km around the injection well after 100 years (Mottaghy and Pechnig 2009).

8.4 Hydrochemistry of Hot Waters from Great Depth

The chemical composition of hot water from deep reservoirs has many significant effects on the operation of a hydrothermal system and it may be critical to the economic success of a geothermal project. The produced hot aqueous solutions contain typically a large amount of dissolved solids and gases. In the reservoir the fluids are under a high pressure promoting gas dissolution in the liquid phase. At temperatures below about 110 °C microorganisms are typically present in the fluid-rock systems and add further complications particularly at the fluid reinjection wells. The gas-rich, saline, high-temperature fluids circulating in geothermal systems are extremely corrosive and aggressive and require appropriate materials and suitable anti-corrosion measures. Particularly vulnerable are production pump, heat exchanger, pipe and filter systems (Chap. 14).

Sampling of hot water is challenging. Samples may be taken downhole in the reservoir aquifer or at or near to the wellhead. Downhole samples are taken at P–T conditions of the reservoir. Sample containers can be sealed gas-tight at depth so that no gas-loss or contamination occurs when the samples are brought to the surface. Sampling at the wellhead is often affected by gas-loss and other alterations of the sampled fluid caused by depressurization. Very gas-rich fluids may

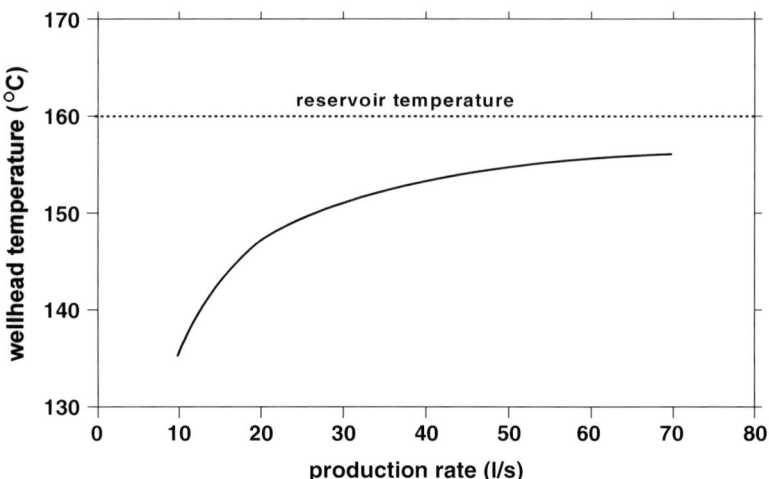

Fig. 8.8 Production temperature as a function of the production rate (models by Ramey 1962)

undergo phase-separation at the surface even in pressurized closed systems. The sample temperature at the wellhead is close to the reservoir temperature if production rate is high. The outflow temperature increases with production rate that is with the flow rate in the well. The production temperature depends on the flow rate (Fig. 8.8). Consequently, a small production rate produces a fluid with a higher density than at high production rates (Fig. 8.2a). These effects influence the productivity index PI defined by $Q/\Delta p$ (Sect. 8.2) also.

Some of the dissolved solids in the produced fluids may precipitate as a result of changing pressure and temperature in the production-injection cycle. The saturation state of the fluid with respect to many minerals may potentially change from undersaturated or saturated in the reservoir aquifer to strongly oversaturated if cooled and depressurized. Particularly critical is the situation if decreasing pressure and temperature is accompanied by gas-loss. As an example, CO_2-loss from the produced fluid very commonly leads to rapid formation of carbonate crusts and deposits (scales) in pipes and in surface installations that may dramatically reduce the efficiency of the entire system (Fig. 8.9). Geothermal systems should be operated under a certain minimum over-pressure and as closed systems in the near surface region to prevent gas-loss and scaling. Because it is the purpose of a geothermal system to extract thermal energy from the produced fluid, the fluid cools and the saturation index of most solids increases (more details in Chap. 14). Particularly problematic is supersaturation of the fluid with respect to several sulphate minerals. Sulphate scales are difficult to prevent, for example with inhibitors or with acidification, and difficult to remove if once formed. Ba- and Sr-sulphate precipitates have been observed in injection wells. Carbonate and iron scales are easier to deal with. Lead salts can be problematic because they accumulate Pb^{210} a ß-ray emitter in addition to Pb^{208} and thus form radioactive scales that require special treatment, handling and disposal. Other challenging scales are briefly described in Sect. 10.3.

8.4 Hydrochemistry of Hot Waters from Great Depth

Fig. 8.9 Carbonate (calcite) scales in a pipe caused by pressure loss and degassing

The chemical properties of the produced geothermal fluid may also make them aggressive to materials. The chemical reactions between the fluids and the mostly metallic system components may cause severe damage known as corrosion. Using corrosion resistant though expensive materials can prevent such damages. Damage analysis of failed submersible pumps showed various, partly massive corrosion related failures partly in combination with scales.

Chemical fluid-rock reactions also will take place in the reservoir aquifer itself. The cooled fluid enters the aquifer from the injection well and its altered saturation status may cause minerals to precipitate in the pore space and so reduce the permeability of the reservoir. Added chemicals such as inhibitors and inorganic acids may, on the other hand, also cause dissolution of minerals in the aquifer thus producing secondary porosity and improve permeability. Dissolution may also occur for some minerals that have a high solubility in the cooled fluid. Some hydrothermal systems are also impaired by microbial corrosion and alteration so that biogeochemical processes must be considered and adequate bio-remediation strategies must be developed (Amann et al. 1997; Dingh et al. 2004).

In any case, each reservoir system is chemically unique and its chemical peculiarities must be carefully evaluated and analyzed.

The hydrochemical properties of the hot fluid also influence the heat content of the fluid. The heat content depends on density and specific heat capacity of the fluid. In highly mineralized water it can be, depending on P, T and composition, higher or also smaller than that of fresh water (Eq. 8.6; Fig. 8.2). Density of water increases, but the heat capacity decreases with increasing salinity (Fig. 8.10).

Hydrochemical analyses and microbiological examinations, also analysis of dissolved gasses should be made in the exploration phase and then in regular intervals during operation of the system. It is important to recognize changes in reservoir conditions as early as possible. The chemical-microbiological monitoring should be accompanied by a hydraulic monitoring that measures production rate, temperature and drawdown. The reaction of the reservoir on continuous system

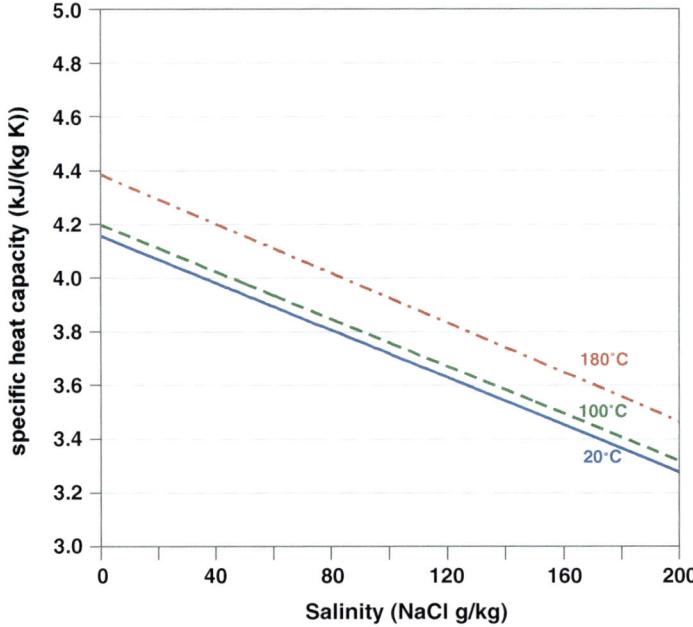

Fig. 8.10 Diagram showing the heat capacity of water versus NaCl-concentration (Sun et al. 2008)

operation must be utterly documented. Purposeful reaction on changes in the reservoir is only possible if these data have been ascertained.

8.5 Reservoir-Improving Measures, Efficiency-Boosting Measures, Stimulation

Ideal sites for hydrothermal systems are deep aquifers with high hydraulic conductivity and high temperature, in contrast to petrothermal systems (EGS systems) where only temperature matters. If the first drilled well of a project does not open, against prediction and expectation, an aquifer with the required hydraulic conductivity for profitable circulation of hot fluid then several possible measures may keep the project within the moneymaking range.

The well can be deepened and additional and deeper aquifers may be accessed that produce fluid at a higher temperature. However, there must be some reliable evidence for success to justify the additional investments. The wells of most hydrothermal doublets are drilled as inclined boreholes resulting in increased contact with the target aquifer and an improved water extraction rate compared with vertical boreholes through a horizontally stratified geology. Inclined boreholes increase the chance of drilling through highly permeable fault and fracture zones. These often steeply dipping structures may function as water conducting channels and much of the water produced from a hydrothermal well is often contributed by high-permeability faults (Sect. 8.1). However, keep in mind that many fault zones have a lower hydraulic conductivity than the rock matrix and are essentially watertight and form hydraulic barriers (Stober et al. 1999).

If the optimal reservoir aquifer has been drilled and its key properties, hydraulic conductivity and temperature, are promising then drilled sidetracks from the wellbore into the target aquifer further increase the yield of the well.

The key parameter for economic success of a hydrothermal project is, in addition to reservoir temperature, the yield of the production well. The yield is the production rate per drawdown that is economically and technically tolerable. Open fractures and connected fracture networks control the hydraulic conductivity and thus the yield in hard-rock aquifers. If the drilled aquifer has a lower than expected hydraulic conductivity, the situation can, to a certain degree, be rescued by specific stimulation measures aimed at improving the hydraulic conductivity of the aquifer.

The classic and prime conductivity improving measure is pumping water to the aquifer under high pressure with the hope to extend existing fractures and so increasing the conductivity of the aquifer. This method is known under various names including aquifer stimulation, conductivity-boosting methods and reservoir-improving measures.

However, reservoir improvement is first attempted by mildly increasing hydraulic pressures in the injection well in order to clean fractures and cavities from conductivity-reducing fine-grained material. Rapidly changing delivery heads may help to cleanse the aquifer rock around the production well. Both techniques are commonly used in near-surface hydrogeology. These are standard methods for obtaining sand-free wells and improving well yield.

Rock matrix treatment with acids is also a common standard method in well engineering for drinking, mineral and thermal water wells. It proved to be particularly successful in carbonate rock aquifers and carbonate deposits on fractures. Acidizing techniques are also routinely used in the oil industry. The reaction of acids with calcite of the carbonate rock matrix or calcite on fractures produces CO_2 gas. Acidizing is also used for wells in silicate rock aquifers such as clay-rich sandstones for example. Commonly used acids include hydrochloric acid (typically 15 %), diluted formic acid, acetic acid, mixtures of HCl and HF and many others (Portier et al. 2007).

The penetration depth of the acid into the rock matrix of the near wellbore depends on the pumping pressure and ranges from a few cm to dm at very low pressure to increasingly greater infiltration radius at higher pressure (usually up to 30 bar). The conductivity-improving effect of pressure acidizing also depends on the amount, concentration and type of acid. Type, intensity and degree of acidizing are adjusted to the type of casing and the well design, which must be efficiently protected from corrosion.

Flushing water conducting structures of the aquifer and increasing their permeability is the sole purpose of increased pumping pressures in combination with acids. The reservoir-improving measures in hydrothermal projects never aim at creating a new fracture network, in contrast to petrothermal systems (EGS). These methods try to establish an improved hydraulic connection from the wellbore to the existing fracture or karst aquifer.

Further aquifer stimulation measures such as massive hydraulic stimulation are not performed in developing hydrothermal systems. These brute force methods need to be applied in developing petrothermal EGS systems (Enhanced Geothermal Systems) for creating the underground heat exchanger (Sect. 9.4).

8.6 Productivity Risk, Exploration Risk, Economic Efficiency

The economic success of a hydrothermal installation depends critically on the geological conditions at great depth (hundreds or thousands of meters underground). Project development rests on data and knowledge from old, often distant boreholes in the area and on data from geophysical exploration tools such as 2D and 3D seismic imaging. The pre-drilling exploration, however, does not provide reliable information on the crucial parameters, temperature, yield and injectivity that determine the future economic success of the hydrothermal project. Therefore, drilling the first borehole for a hydrothermal doublet system is always afflicted with some risk of opening unsuitable ground. The risk of drilling a very expensive borehole into a target horizon that is colder, less permeable and thinner than predicted from the exploration data is inherent in hydrothermal system development. The geological risks associated with hydrothermal projects greatly exceed that of near surface systems. The same geological exploration risks are also known in the oil and gas industry. However, the value of the produced product (oil versus hot water) and the associated rate of return are very different in the two industries and consequently also the economic risk associated with unsuccessful drilling. The geological risk can and must be reduced by an extensive exploration program and a careful, competent and rigorous analysis of the exploration data.

The pre-drilling exploration report is also used as a basis for obtaining exploration risk insurance. Various formats of risk insurances have become increasingly popular in the last years. The insurances may cover different aspects of the total risk associated with the development of a hydrothermal system. The economic overall risk is usually divided into risk groups (geological, exploration, productivity, drilling and so on). The risk groups are analyzed and evaluated separately and contracts are offered for the separate groups as insurance.

Insurance companies generally distinguish five different risk groups: Exploration risk, geological and geotechnical risks, economic risks, environmental and political risks. The individual risk groups cannot always be clearly separated.

The main risk of hydrothermal projects is the exploration risk. It is the risk of drilling one or several boreholes into a hydrothermal aquifer of insufficient thermal productivity and unsuitable fluid composition.

Thermal power P [J/s = W] of a geothermal well is proportional to the production rate (Q) and the temperature (T) and defined by Eq. 8.6:

$$P = \rho_F c_F Q \left(T_i - T_o\right) \tag{8.6}$$

where ρ_F denotes the density [kg/m³] and c_F the specific heat capacity [J/kg K] of the fluid, Q the production rate [m³/s], T_i the production temperature [K] and T_o the injection temperature [K]. The heat capacity of highly saline fluids is about $c_F = 3.9$ kJ/(kg K) and the density ρ_F of the fluid at 120 °C on the boiling curve is near 943 kg/m³. Both parameters depend on P and T and on the composition of the fluid (Sect. 8.2). The total mineralization or salinity of the hydrothermal fluid has a significant effect on the thermal power. An increasing amount of dissolved solids reduces the thermal power considerably. This salinity effect increases with increasing temperature. Although the density of the fluid increases with salinity, the positive effect is overcompensated by the reduced specific heat capacity (Fig. 8.10).

The exploration risk also relates to an unsuitable composition of the hydrothermal fluid. Dissolved solids or gasses may rule out a geothermal utilization or may make it difficult and cost-prohibitive. The fluid may be highly corrosive because of high salinity and high content of hydrogen sulfide. The fluid may precipitate radioactive or highly toxic scales. So far most hydrothermal wells produced fluids that were chemically controllable although at various levels of extra costs.

Accordingly a hydrothermal well is commercially viable if the hot water yield exceeds the project defined lower limit of the production rate Q_{min} at the upper limit of drawdown Δs_{max} and if the produced fluid has a temperature higher than that of the project defined lower T_{min} limit. The project specific values of Q_{min}, Δs_{max} and T_{min} are linked to economic considerations of the operator (Stober et al. 2009). The envisioned products of the hydrothermal system have a controlling effect on the parameter values. If the system should produce electrical power then T_{min} is about 120 °C and production rate Q should be higher than 50 kg/s (limits in the year 2013). General site-independent geological and technical conditions set an upper limit of ~200 K to the exploitable $\Delta T = (T_i - T_o)$ in Eq. 8.6. Likewise, the maximum production Q of hot water from a deep aquifer by a single well is about 150 kg/s. From these limits it follows that the maximum geothermal power for a hydrothermal doublet system is about 50 MW_{th}.

The thermal energy E [J] extracted from a hydrothermal well can be computed from the thermal power P [W] of the system and the time of operation Δt [s] (Eq. 8.7):

$$E = P \Delta t \tag{8.7}$$

During the working life of the hydrothermal plant the crucial parameters production rate Q and temperature of the produced fluid T_i should not decrease noticeably. Precondition for this is a hydrothermal reservoir of a sufficiently large extent. Important is also to exclude impairment by other hydrothermal plants in the surrounding area.

Hydrothermal power plants require very large fluid mass fluxes Q for successful operation. Production rates exceed those common in the oil industry by many times. A production rate of 3 kg/s is considered an excellent well. Hydrothermal reservoirs and the production techniques must satisfy much higher requirements (Chap. 11) than in the hydrocarbon industry. Reservoir properties of a specific geologic unit that is interesting for the oil and gas industry may prove unsuitable for hydrothermal power systems.

8.6.1 Exploration Risks

The hydraulic conductivity of hard rock aquifers and the related yield of a hydrothermal well are controlled by the amount and shape of interconnected open fractures and other water conducting features of the rock matrix and the hydraulic properties of local fracture and fault zones. Hard rock aquifers can be categorized according to their predominant type of pore space into fractured aquifers and karst aquifers (Figs. 8.11 and 8.12).

If the tested hydraulic conductivity of the aquifer in the first drilled wellbore is lower than expected and predicted by pre-drilling exploration stimulation measures need to be carried out to improve aquifer properties (Sect. 8.5). Specific possible actions include acidizing carbonate aquifers (limestone) and hydraulic stimulation, hydraulic fracturing in combination with acidizing. Horizontally

Fig. 8.11 Surface outcrops of typical fracture-dominated sandstone and quartzite aquifers. **a** Lower Triassic Buntsandstein unit in the Voegtlinshofen quarry (Alsace, France). **b** Vendian quartzite with dry fractures, wet fractures and mineral-stained temporarily active water conducting fractures near Lom (Norwegien Caledonides) showing that water conduction depends on detailed fracture properties

8.6 Productivity Risk, Exploration Risk, Economic Efficiency

Fig. 8.11 continued

Fig. 8.12 Example of a water producing karst channel underground in middle Triassic limestone (Muschelkalk) at Talmühle spring near Horb, SW-Germany

drilled sidetracks in the target aquifer may further improve yield (this is common practice in the oil industry).

Conditions for economic success or successful reservoir discovery (strike) are defined at project start by the investor or developer. The anticipated return defines the minimum value for the achievable production rate and the minimum temperature of the produced fluid to make the project an economic success. A hydrothermal well is successful if the parameter limits are achieved or topped.

A partially successful well failed to meet the criteria for strike, however utilization with a revised concept is technically feasible and, financed by the insurance benefit, economically viable.

The exploration risk may not be easily insured in regions of the world where limited or no experience with similar hydrothermal projects exist. Insuring projects that employ new technologies with experimental character or for research purposes such as EGS/HDR projects (Chap. 9) may not be insured at all.

Efficiency and work life of a hydrothermal system depends mainly on the hydraulic, thermal and chemical properties of the aquifer and the stored hot water. These properties need to be explored and investigated best possible and the results, testing methods, and investigation tools should be carefully documented. The operator and investor are making the final decision on the economic efficiency of a hydrothermal plant based on business indicators. The consumer structure of the power consumers is central in the decision making process.

The exploration risk is the controlling factor of all the economic uncertainties. The first successfully drilled and thoroughly tested well greatly reduces the exploration risk of the project. Still, the second well to be drilled, usually the injection well, must be able to take up the produced hot water for a trouble-free hydrothermal fluid circulation (Sect. 8.2). The development costs (drilling, stimulation, hydraulic and other well tests) stand for about 50–70 % of the total costs of a hydrothermal doublet project (exploration, surface installations, and plant being the other costs). Careful project development, a clearly defined agenda of project development phases, milestone scheduling and strict and rigorous termination criteria minimize the economic risk.

Regions with unusually high thermal gradients (thermal anomalies) are potentially attractive and may save investment costs because of short drilling depths. However, high temperature at shallow depth is only beneficial if achievable production and reinjection rates are also sufficiently high for profitable operation of the plant.

Regions with normal thermal gradients produce relatively low temperature fluids even from very deep drill holes (>4,000 m). These low-enthalpy hydrothermal systems supply mostly energy for the heat marked. For profitable operation of a hydrothermal doublet heating system it is necessary for the system to continuously provide heat for local and district heating networks all year around. The successful project provides heat at different temperature levels to a range of different heat clients. The produced thermal energy is distributed following a cascade principle. An example system shall produce hot water of 90 °C. The district heating system of the first user extracts heat and cools the water to 60 °C, which is then used by

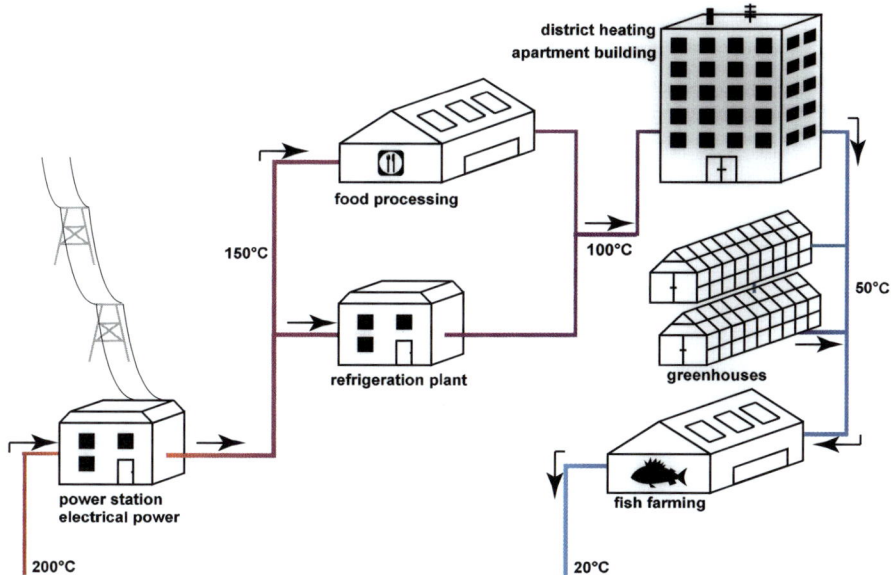

Fig. 8.13 Serial use of thermal energy produced by a hydrothermal doublet system. Cascade system after files of the International Geothermal Association (www.iga.1it.pl)

the second customer to warm a group of greenhouses. The water leaves the second customer with 30 °C and is used for fish farming by a third consumer (Fig. 8.13). The water from the fish farm with a temperature below 30 °C is re-injected into the subsurface reservoir by the hydrothermal doublet system. The uses of thermal energy are not limited to the examples described but further innovative concepts; designs and ideas for heat use wait to be developed by the ingenious entrepreneur. The produced heat can also be used for cooling.

Electricity can be commercially produced from fluids at more than 120 °C with the appropriate technology. The efficiency of electrical power production increases with rising temperature. However, because low-enthalpy hydrothermal doublet systems draw the fluid from aquifers with typical reservoir temperatures rarely exceed 150–170 °C. Very important for the economic success of a system that produces electrical power is the profitable marketing of the thermal energy leaving the electrical power plant (residual heat; ~90 °C). This process heat must be sold and customers must be integrated into the project development. Similar considerations also apply to EGS and HDR systems (Chap. 9).

Geological and geotechnical risks refer to the general method-inherent uncertainty of geological predictions. The anticipated existence of geological structures and underground strata deduced during the pre-drilling exploration phase from geophysical data is prone to intrinsic fuzziness. Geological risks during drilling include the occurrence of unexpected geological strata, unpredicted pressure conditions or large amounts of fluids with potential erosion and collapse of the wellbore. Inadequately handled unexpected geological features usually also cause geotechnical and drilling technical problems.

Drilling risks are related to all technical problems linked to the drilling rig, drilling tools, and the drilling process. Drilling problems include lost tools, damaged casings, defective cementation, pipe sticking, hole deviation, pipe failures, mud contamination, borehole instability, blowout and more (Chap. 11). These drilling problems almost always cause a significant delay in drilling operations. Technical drilling problems may, in the worst case, cause the loss of the wellbore and of all investments up to that day. Drilling risks are borne by the drilling contractor. Drilling risks can be covered by insurances.

Operational risks are related to the power plant technology, the surface installed technical equipment and the plant operation. The operational risk related to the power plant technology can be insured. However, the operational risk related to the hydrothermal reservoir, specifically the constant temperature and production rate during the lifecycle of the system cannot be insured and must be borne by the operator. All variations in temperature and production rate but also in the chemical composition of the fluid during the operation of the hydrothermal doublet are operator risks. This includes all alterations of technical installations that may be caused directly or obliquely by fluctuations in fluid properties. Also part of the operational risk is the changing energy market that causes uncertainties in price trends for electrical and thermal energy over the operation period of 30 years. Planning and developing the project can minimize some of these risks by inserting conservative economic parameters, limiting production rates and distances of the doublet wells in the reservoir.

Hydrothermal systems can be afflicted by a risk to the environment in analogy to the near surface geothermal energy utilization also. Hydrothermal doublets can be hazardous to groundwater resources and to soil. Large projects and massive invasions to the deep underground require appropriate preventive measures and security precautions to minimize hazards. Drilling deep to ultradeep wells requires a mining law approval process in many countries, which also considers the interests of neighbors and local residents. Environmental consequences of and environmental risks associated with geothermal energy installations in general will be discussed in Sects. 10.2 and 10.3.

Reservoir stimulation measures may potentially induce seismic events particularly in areas with natural seismicity. The intensity of the trembling is usually small but in some reported cases strong enough to be felt at the surface. The occurrence of stimulation induced seismicity depends on the geological structure of the ground and the type of rocks present, on existing active tectonic stresses, level of applied injection pressures, magnitude and timeline of injection flow rate during stimulation and the structure and volume of the stimulated fracture system. A monitoring program that measures vibration velocities should therefore accompany deep drilling for geothermal systems. Vibration velocity data characterize the seismic energy of an induced seismic event that reaches the surface. Magnitude data used for characterizing seismic energy released at the site of a natural seismic event are not appropriate for assessing stimulation-induced seismicity in geothermal projects. The occurrence of induced seismicity can, to a certain point, be predicted, evaluated and partially controlled. The keys to seismicity control are continuous measurement and monitoring of the injection pressure and a seismic monitoring program specifically designed for the specific hydrothermal project that measures vibration velocities in the near and far surroundings of the plant site. If the data indicate a growing risk for intolerable seismicity injection pressure and pumping rate must be reduced (Sect. 10.1).

The utilization of low-enthalpy hydrothermal systems for electrical power production and thermal energy uses also depends on governmental subsidies in most countries, which represent a political risk that needs to be considered in project planning. Subsidies comprise for example direct public contributions to the price of electrical energy that is collected from all consumers of electrical energy, direct grants to specific projects, support for research and development, grants for experimental sites and many more. However, at present the majority of the general public and politics favor supporting the expanding use of non-fossil energy in many countries (particularly European countries). Political support for geothermal energy utilization adjusts to arising new technologies, to the evolving global energy market, to the national unemployment situation and other macroeconomic factors and their long-term variations. Low-enthalpy hydrothermal energy utilization is a relatively young player in the global energy market. Substantial progress in improving hydrothermal systems will make the technology more cost competitive in the future. In contrast, production of electricity and heat from fossil resources (coal, oil, gas and others) will inevitability experience long-term cost increases caused by dwindling resources and environmental regulations. The enduring trends on the energy marked make geothermal energy more competitive and finally independent on subsidies in the long run.

8.7 Some Site Examples of Hydrothermal Systems

8.7.1 High-Enthalpy Hydrothermal Systems

The majority of the globally installed 10.9 GW_{el} capacity produce electrical power in active volcanic areas. Excellent site descriptions can be found for (e.g.). The Geysers in California at http://www.geysers.com/, for the power plants on Iceland at http://www.mannvit.com/ and for New Zealand at http://www.nzgeothermal.org.nz/.

8.7.2 Low-Enthalpy Hydrothermal Systems

Only a few geothermal power plants produce electricity with hydrothermal doublets from low-enthalpy areas with reservoir temperatures below 150–170 °C (Bertani 2007). However, the systems have a great potential to supply local communities with reliable base load electricity. Thus we present some specific hydrothermal plants, the experience with development and operation of these plants but also with troubles encountered and the remedies with which the problems have been mastered.

8.7.2.1 Paris Basin (France)

The first hydrothermal doublets for heating residential buildings in the Paris basin (France) have been installed near Melun l'Almont south of Paris in the late 1960s (Ungemach 2001). Subsequently, many more wells were drilled as a result of sharply rising oil prices in the period 1980–1987. Of the total of 63 deep wells only two were complete failures and five wells were only partially successful. The produced thermal energy is used for direct house heating and hot water supply. The produced energy is transferred to secondary loops using heat exchangers and reaches the end-user through separate distribution networks.

Today, 31 hydrothermal doublet systems commercially produce thermal energy in the Paris basin (number from 2012). From the beginning, numerical models optimized the management of the geothermal plants and helped avoiding unwanted interferences with neighboring systems and maximized the life cycle of the installations (Sauty et al. 1980; Antics et al. 2005). All geothermal systems are designed as hydrothermal doublets, which assure sustainability by recycling the produced highly mineralized water into the reservoir.

The Paris Basin is a large concentric geological structure in northern France of several hundred km in diameter. The basin developed after the Triassic by central subsidence of the Paleozoic basement. The basin contains a complete stratigraphic succession from the Permian through the Mesozoic to the Tertiary. Tertiary strata are exposed in the center of the basin structure in the region of Paris. Thermal energy is extracted by hydrothermal doublet systems from several different thermal aquifers. The utilized aquifers are of lower Cretaceous, upper Jurassic, middle Jurassic and Triassic in age. Most of the wells are about 1,700 m deep. The production rates range from 30 to 100 l/s. The temperature of the hot water varies between 65 and 85 °C. The waters are highly mineralized and contain 10–40 g/l total dissolved solids. The hot waters also contain appreciable amounts of dissolved CO_2 and H_2S gas. The waters comprise sulphate from evaporite leaching. Sulphate reducing bacteria produce hydrogen sulfide (H_2S) from sulphate, which caused severe corrosion and scales in the system components including the casing in the first systems installed. Sulfide scales form from the following (net) processes:

1. Sulphate to sulfide reduction

$$2CHO \text{ (organic biomass, bacteria)} + SO_4^{2-} + 4H^+ \rightarrow H_2S + 2H_2O + 2CO_2$$

2. Pipe steel corrosion

$$Fe + 2H^+ + 0.5O_2 \rightarrow Fe^{2+} + H_2O$$

3. Pyrrhotite scale formation from the products of reaction 1 and 2:

$$Fe^{2+} + H_2S \rightarrow FeS \text{ (pyrrhotite)} + 2H^+$$

Today, the entire doublet system, production and injection well and the surface installations are routinely corrosion protected by injecting special inhibitors at bottom hole of the production well (Fig. 8.14).

8.7 Some Site Examples of Hydrothermal Systems

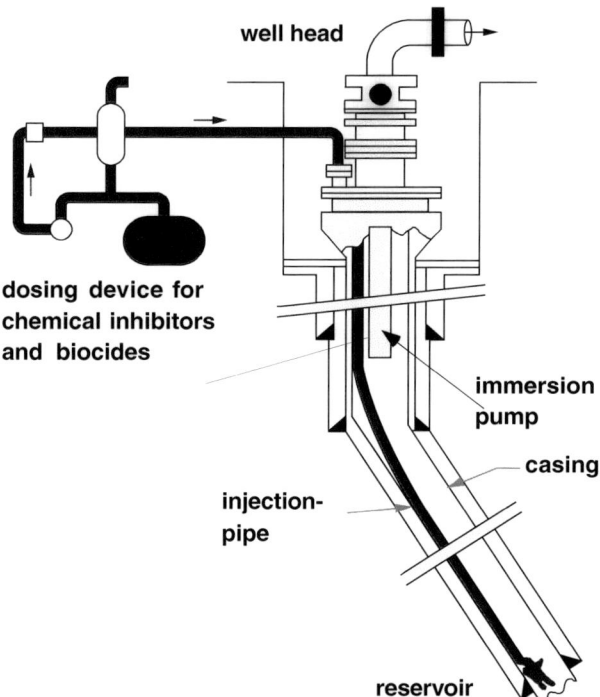

Fig. 8.14 Schematic drawing of a production well in the Paris Basin showing the down hole injection tubing for injecting corrosion preventing inhibitor chemicals at bottom hole (redrawn from BRGM, France)

8.7.2.2 Bavarian Molasse Basin, Unterhaching (Germany)

The geothermal power plant Unterhaching near Munich (Germany) is representative for several similar installations in the Bavarian Molasse Basin. The hydrothermal doublet produces electrical power from hot water pumped from the Upper Jurassic limestone aquifer under the Tertiary Bavarian Molasse Basin. The Permian and Mesozoic cover and the Variscan basement gradually decline towards the Alpine mountain chain in the south, being covered by Tertiary Molasse type sediments of increasing thickness (Fig. 8.15). The strongly asymmetric basin was formed by the Alpine orogeny. Basement and cover reappear in surface outcrops in the Alps and have been incorporated in the complex Alpine structures. The Upper Jurassic limestone (Malm) is the prime aquifer of interest for hydrothermal energy systems. Malm limestones crop out at the surface in the north and its cover increases to several thousand meters in the south (Fig. 8.15). The limestones vary in thickness with a maximum of 500–600 m. At the location of the Unterhaching power plant the Malm limestone aquifer is at about 3,500 m depth and the thermal water has a temperature of 122 °C. The water is stored in fracture and karst

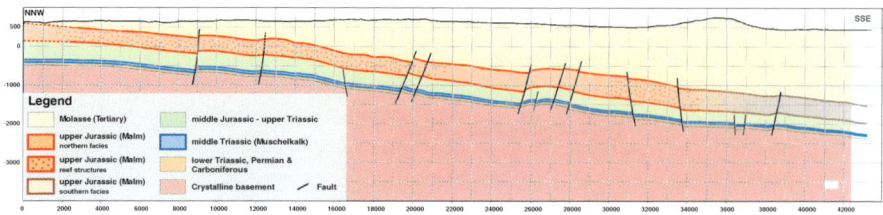

Fig. 8.15 Geological cross section through the Molasse Basin west of Unterhaching showing the southward dip of the basement cover complex with the main aquifer of the upper Jurassic Malm limestone and the minor Muschelkalk aquifer of Triassic age. Note lateral facies change in the Malm unit from highly conductive fracture and karst aquifer in the north and less permeable facies in the south. Particularly permeable are areas with Malm limestone in reef facies. Section from Jodocy and Stober (2009)

porosity. Further towards the south the Malm gradually changes its facies from a fracture/karst limestone aquifer to a dense limestone with low hydraulic conductivity and little potential for geothermal applications.

The project Unterhaching started in 2004 and electrical power production begun in 2009. The capacity of the plant is 3.36 MW_{el} electrical power and in the final stage of development about 70 MW_{th} thermal energy. The yield of the 3,446 m deep production well is 150 l/s. After energy extraction the cooled water is pumped back to the same aquifer horizon using a 3,864 m deep injection well, which is at a higher temperature of 133 °C because of the greater well depth. The two wells are connected by a 3.5 km long thermal water pipe (Fig. 8.16).

The produced thermal water has a surprisingly low content of dissolved solids of only 600–1,000 mg/l. The major dissolved components are Ca^{2+} and HCO_3^- and not NaCl like in most other deep waters from more than 3 km depth. The water also contains appreciable amounts of dissolved nitrogen, hydrogen sulfide and methane. Permanent excess nitrogen pressure imposed to the thermal water cycle prevents chemical precipitation of solids and access of atmospheric oxygen gas. The thermal water pipes are made from glass-fiber reinforced plastic for preventing corrosion problems.

Electrical power is generated by a power plant working on the basis of the Kalina process (Sect. 4.2). The Kalina machine at Unterhaching uses a water–ammonia mixture as heat transfer fluid. The plant supplies electricity for about 10,000 households. In 2006 the construction of the district-heating network was started, which has grown to 35 km length by the middle of 2010 with a total output of 45.7 MW_{th} (BINE 2009). The very successful project Unterhaching has triggered a series of follow-on projects in the Munich area. The positive key properties of the Unterhaching hydrothermal doublet are the extremely high yield of the production well (150 l/s) and a very low salinity of the produced thermal water.

8.7 Some Site Examples of Hydrothermal Systems

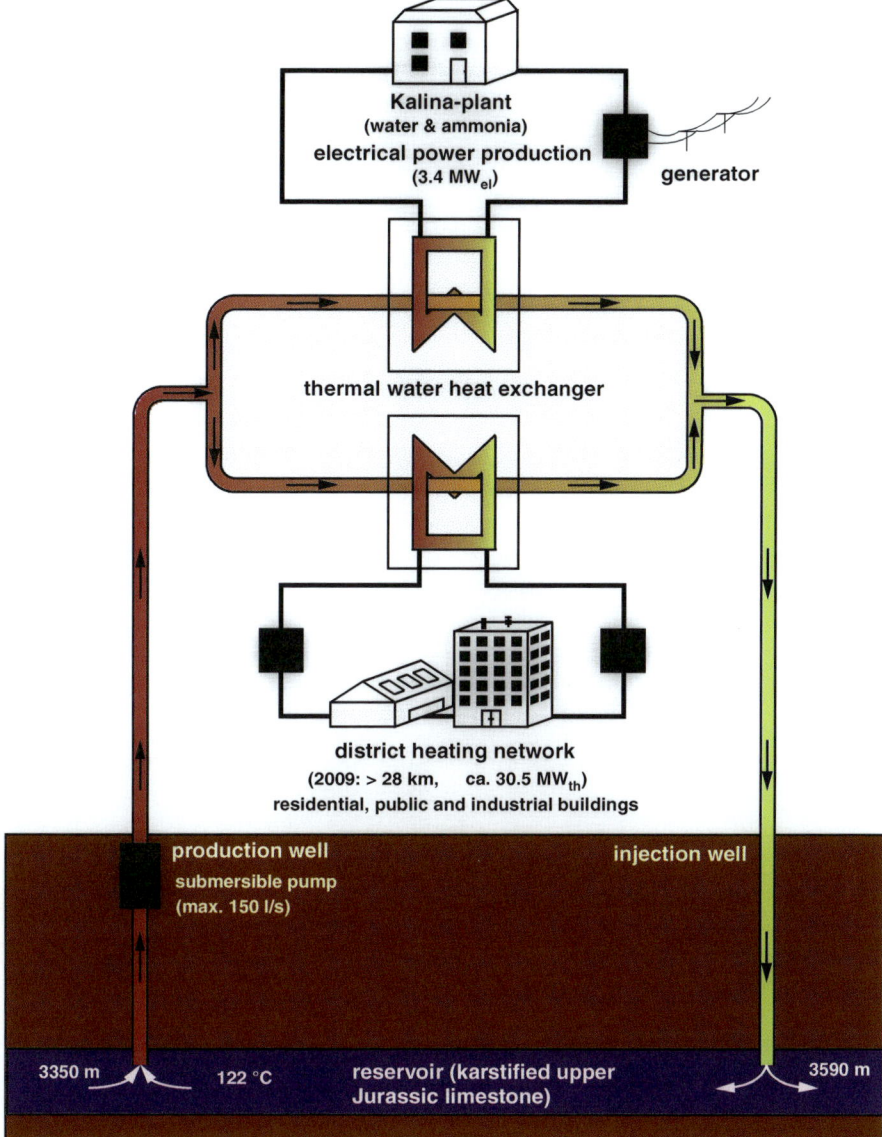

Fig. 8.16 Hydrothermal doublet at Unterhaching, southern Germany (redrawn from documents kindly provided by Geothermie Unterhaching GmbH and CoKG)

8.7.2.3 Bruchsal Research Site in the Upper Rhine Rift Valley (Germany)

In the early 1980s two deep wells have been drilled in the city of Bruchsal (20 km N of Karlsruhe, Germany) with the purpose to produce geothermal energy.

The system was not completed before 2008 because of various economic problems and is running with fairly small power output. Nevertheless, many important technical and other lessons have been learned from the Bruchsal site, which has been a research and test site rather than a commercial power plant.

The site is located in the N–S trending Tertiary (Oligocene) Rhine rift valley near the eastern main boundary fault. The main boundary fault is a major fault zone with a series of listric normal faults with westward dipping fault surfaces. The sedimentary cover sequence of the Variscan basement is downfaulted in the graben interior and several geological units with aquifer properties are present at sufficiently large depth to host thermal water with a great potential for geothermal applications.

The two vertical wells, 1,874 and 2,542 m deep, access the thermal water of the Buntsandstein reservoir, a lower Triassic sandstone (Fig. 8.16). The yield of both wells is nearly identical. The horizontal distance between the wells is 1.4 km and the wells are connected with an insulated thermal water pipeline. Highly saline thermal water is produced from the deeper well. It has a bottom hole temperature of 134 °C. The produced water is reinjected into the shallower wellbore after heat extraction. The transmissivity of the aquifer is $T = 3.6 \times 10^{-5}$ m^2/s (Bertleff et al. 1988). The pumping rate is at modest 24 l/s (2012) limited by the capacity of the present pump (Fig. 8.17). Higher rates seem possible but have not been tested.

Fig. 8.17 Instalation of the submersible pump into the production well GB2 of the Bruchsal hydrothermal doublet (upper Rhine rift, SW Germany)

8.7 Some Site Examples of Hydrothermal Systems

Several earlier pumps with similar capacities broke down due to technical failures unrelated to the specifics of the Bruchsal site.

Water samples have regularly been chemically analyzed from the Bruchsal wells over a period of nearly 30 years. The chemical composition of the highly mineralized and CO_2 rich thermal water is remarkably constant. The main components are Na^+ and Cl^- (Table 8.1). The total of dissolved solids is 127 g/l. The water is effectively a mixture of 1.6 mol/l NaCl, 0.3 mol/l $CaCl_2$ and 0.1 mol/1 KCl. Carbonate alkalinity is very low. However, the pressure decrease during production causes the thermal water to become oversaturated with respect to calcite. In order to prevent

Table 8.1 Chemical composition of water from the Bruchsal well GB2. pH = 5.0 at 134 °C

Component	Concentration (mg/l)
Ca	7,140
Mg	324
Na	37,400
K	3,440
Fe	47
Mn	23
Cl	75,200
HCO_3	350
SO_4	586
SiO_2	83
CO_2 gas	~2,000

Fig. 8.18 CO_2 gas separator for separation of excess dissolved CO_2 from the thermal water before entering the heat exchanger of the Kalina plant at the Bruchsal geothermal plant. The collected CO_2 gas is re-dissolved into the cooled water leaving the plant

Fig. 8.19 Wet cooling tower of the hydrothermal doublet of the Bruchsal power plant

carbonate scales the system is kept under 22 bar pressure in surface installations. The high content of dissolved CO_2 gas requires a special gas separator (Fig. 8.18) that removes a large part of the CO_2 before the thermal water enters the power plant. The gas is re-dissolved in the water leaving the heat exchanger with 60 °C.

The extraction of the thermal energy from the produced thermal water is accomplished by a binary loop Kalina plant using a water–ammonia mixture as a heat transfer fluid. The thermal energy of the hot water is transferred to the working fluid of the secondary loop by plate heat exchangers. The plant is cooled by a wet cooling tower (Fig. 8.19). The electrical power of the plant is 550 kW. Given an annual operating time of around 8,000 h, the plant produces about 4,400 MWh electrical energy.

8.8 Project Planning of Hydrothermal Power Systems

Planning and developing a hydrothermal power plant is a complex undertaking requiring close collaboration of competent experts from many different fields and a clearly structured project management with well-defined responsibilities. The key themes are two fold: (1) the exploration risk, which encumbers all hydrothermal system projects, must be minimized by sound specialist analysis and model-based operation forecasts for the reservoir system. (2) Consumers for the produced thermal energy must be found. A hydrothermal doublet system should never be installed without consumer contracts with buyers for the produced thermal energy. The amount of thermal energy that can be converted to electrical energy is always a small fraction of the total thermal energy produced. Thus most of the energy produced must be sold to heat customers. Consequently, the sites for potential hydrothermal plants are strongly determined by the location of existing or future heat consumers such as district heating grids or industry (in addition to geological conditions). If there are no consumers for the produced heat, the installed hydrothermal plant will essentially be an energy annihilation system (Sect. 8.6).

A hydrothermal doublet project begins with the evaluation of the geothermal potential of a location where nearby heat customers already exist or can be acquired. In low-enthalpy regions and for electrical power plants, there must be plausible potential for the existence of an aquifer at 2–4 km depth with a hot water reservoir temperature of 120 °C or more.

The following checklist in bullet point form summarizes and structures the key work items for the development of a hydrothermal system.

8.8.1 Phase 1: Preliminary study

The preliminary study defines and describes the aim and purpose of the project from which the type of geothermal utilization comes out.

1.1 Objectives of the project
1.2 Geological starting point

- Available data (data compilation; general geology, seismic profiles and wells, hydraulic tests, heat flow data)
- Geological structure of the underground (sections across the study area, interpretation of seismic sections)
- Depth and thickness of aquifers
- First estimate of temperature at potential target aquifers
- Hydraulic conductivity, possible yield
- Chemical composition of thermal water in different target reservoirs
- Local mining regulations, mining concessions, operation licenses

1.3 Energy utilization concept

- Planned and existing district heating (township, communities, local power company, definition of heat to be provided by the geothermal installation)
- Production of electrical energy (optional, if desired)

1.4 Rough technical concept of the geothermal power plant

- Different technical systems (doublet, distance of wellbores, side tracks)

- Well construction design (needed for a first cost estimate)
- Surface installation, power plant concept

1.5 Cost estimate, financial concept, economic responsibility

8.8.2 Phase 2: Feasibility study

2.1–2.4 Points 1.1 – 1.4 of the pilot study in more detail, decision on the alternatives to be planned
2.5 Analysis of costs, investment and financing

- Exploration
- Wellbores, underground installations
- Surface facilities, plant

2.6 Economy

- Operation costs
- Costs and expenditures, revenues
- Profitability analysis, cost-efficiency studies

2.7 Comprehensive risk analysis, quantification of exploration risk
2.8 Ecological analysis, ecological balance study
2.9 Project schedule, project flow

8.8.3 Phase 3: Exploration

3.1 Entrusting a consulting company, assigning the project management
3.2 Applying for the exploration rights at the local mining authority
3.3 Completing geophysical exploration with a competent specialist firm (if necessary)
3.4 Concept for drilling (following the legal requirements of the mining authorities)
3.5 Invitation to bid for the first wellbore, formulation of an operational plan
3.6 Drilling and testing the first well
3.7 If necessary executing stimulation measures
3.8 Decision on strike

8.8.4 Phase 4: Development

4.1 Invitation to bid for the second wellbore, formulation of an operational plan
4.2 Drilling and testing the second wellbore
4.3 If necessary executing stimulation measures

4.4 Building surface installations and a power plant (parallel to 4.1–4.3)
4.5 Securing the licence area at the local mining authority
4.6 Operation of the plant, production of thermal water, thermal and electrical energy

Phases 1 through 3.5 involve all pre-drilling work items for developing a hydrothermal reservoir. Key steps include obtaining the exploration claim from the local mining authorities and deciding on the drilling position. Basis for both is a sound evaluation of the geothermal potential of the site of interest. This is centered around existing geological data and, equally important on developed concepts for the utilization of the geothermal energy to be produced in local and district heating grids and for electrical power production. Also, insurance agreements must be signed and, most important, a financial backer must be found.

Phases 3.6–3.8 decide on the strike of drilling within the means of the project target definition. If the well fails to meet the criteria for being successful according to the project conditions it still may be useful for a substitute geothermal project or other purposes. However, this utilization differs from the original concept because the geological and geothermal situation is different than predicted by pre-drilling exploration. Perhaps, the wellbore struck a gas or oil reservoir! Perhaps, and far more common, the yield or the temperature of the thermal water is lower than hoped for. Still, the produced water can possibly be used for feeding a thermal spa (successfully), however, not for electrical power production as originally intended.

If the first well is successful, the project continues with the invitation to bid for the second wellbore and with further geophysical, hydraulic and hydrochemical research (phase 4). It is important that the production and injection well obey the planned minimum distance in the aquifer so that the produced thermal water maintains its original temperature over the entire lifetime of the system. It is aimed to install a trouble-free working thermal water circulation (primary loop) with sufficient high yield and water temperature. For this, production tests, stimulation measures if necessary, and extensive hydrochemical investigations are needed. Where appropriate, special measures help preventing scaling and corrosion. The challenge of this project phase is also the strain on the drilling equipment and on the drilling technique, the requirements on the well engineering materials and the pump and pump equipment. Decisions on the right technology for the power plant (e.g. ORC versus Kalina) and the engineering of the plant are also made on the basis of hydraulic data and hydrochemical properties of the thermal water.

The multifaceted tasks of a geothermal project require an efficient and unified collaboration of experts from very diverse fields. Engineers, geologists, lawyers, insurance and financing experts must work together hand in hand. Subcontractors must be commissioned and their work must be coordinated. A successful geothermal project requires a well-coordinated and efficient effort of all intermeshed actors.

Chapter 9
Enhanced-Geothermal-Systems, Hot-Dry-Rock Systems, Deep-Heat-Mining

Equipment for hydraulic stimulation

With Enhanced-Geothermal-Systems (EGS) the deep underground is used as a source of heat for the production of electrical and thermal energy irrespective of the hydraulic properties of the deep heat reservoir (Sect. 4.2). In other words the rocks are hot at depth irrespective of whether or not they qualify as aquifers or aquitards. Weakly fractured granites at depth are best described as "hot dry rocks" (HDR systems). However, keep in mind that also the few fractures present are interconnected and filled with hot pore water (Ingebritsen and Manning 1999; Stober and Bucher 2007a, b). The term "hot dry rock" originates from the early-days of geothermal energy utilization where the concept was to drill a deep wellbore into hot but assumingly dry rocks for extraction of thermal energy. Later it was found that the fracture porosity of continental rocks is always saturated with hot water so that the term hot-dry-rock became rather misleading (Stober and Bucher 2007a, b). The upper continental crust is always fractured; its fracture density differs however. A saline, occasionally gas-rich fluid is typically present in the fractures. The geothermal utilization of the hot underground with low hydraulic conductivity is sometimes also referred to as "deep heat mining" (DHM). Because the continental crust is predominantly granitic or gneissic, HDR systems strongly focus on granitoid heat reservoirs. Typical target temperatures for HDR systems are above 200 °C. This means that wellbores of 6–10 km have to be drilled in continental crust of average geothermal gradient.

Recently, it has been proposed to use HDR technology also in deep sedimentary basins (Huenges 2010). Sedimentary series with low hydraulic conductivity at great depth may also provide thermal energy, of course. The systems yet to be developed may also be termed Engineered-Geothermal-Systems (likewise abbreviated as EGS). The HDR technology has originally been developed by the oil- and gas-industry. It is a well-established, decades old method of reservoir engineering and development of low-permeability sedimentary reservoirs, improving the production rates and inflow and hydrocarbon migration conditions (Chap. 13). The concepts have been translated to geothermal energy applications, first expanded to the crystalline basement of the continental crust and recently returned to deep sedimentary successions and thereby closing the circle between the diverse industries.

The desired temperature of the target heat reservoir of EGS (typically > 200 °C) can be reached by adjusting the depth of the wellbore. However, the hydraulic conductivity of typical reservoir rocks is too low for circulating hot fluids for driving a geothermal power plant. For example, crystalline basement rocks at 5 km depth have a mean hydraulic conductivity of only 10^{-9} m/s (Ingebritsen and Manning 1999; Stober and Bucher 2007a, b). The fundamental task of EGS development is the creation of a sufficiently large volume of the reservoir around the wellbores with a significantly higher hydraulic conductivity ($>5 \times 10^{-5}$ m/s). Furthermore, the engineered high-conductivity volumes need to be connected so that sufficient fluid flow rates can be achieved and the fractured rock volume can function as a heat exchanger. Reservoir stimulation techniques are central to EGS development.

EGS development is quite different in distinct types of host rocks because of different natural fracture and porosity patterns and deformation behavior of igneous, metamorphic and sedimentary rocks. The fracture patterns of layered sequences of sedimentary rocks are strongly influenced by post-sedimentary compaction, diagenesis and weak metamorphism. Fracture patterns in coarse-grained granites are mostly controlled by tectonic and thermal stresses. During stimulation measures, the water injected under high pressure migrates predominantly along fractures in quartz-feldspar dominated rocks (e.g. granites), however, in rocks rich in sheet silicates such as shale, slate, mica-schist and mica-rich gneiss the injected water may also migrate along the foliation in addition to fractures. Thus in schistose rocks the injection pulse is damped and consequently the potential for generating seismicity is reduced. Note that at the target temperature of about 200 °C all reservoir rocks classify by definition as metamorphic rocks. Thus granites are strictly sub-green schist facies metagranites, clay and mudstones are low-grade metapelites, limestones are low-grade marbles (Bucher and Grapes 2011).

For improving the hydraulic conductivity of the reservoir rocks EGS are being developed by using both hydraulic and chemical stimulation techniques. These methods have been derived by the oil and gas industry and have been used for decades for improving production from hydrocarbon reservoirs. Stimulation in the context of EGS is a time limited process in the development phase of the system. Once the underground heat exchanger works as planned, no further stimulation measures are needed during regular operation of the plant.

The US Department of Energy (DOE) defines EGS as engineered geothermal reservoirs for the production of economically relevant amounts of thermal energy from low-conductivity or/and low-porosity geothermal resources. The EGS reservoirs need to be stimulated with efficiency-boosting methods irrespective of the type of the reservoir rocks (MIT 2007).

9.1 Techniques, Procedures, Strategies, Aims

EGS produce geothermal energy from weakly fractured hot rocks. The interconnected fractures in crystalline basement rocks such as granites and gneisses are saturated with hot fluids, typically saline brines. The water-conducting fractures in basement rocks of the continental upper crust are complex structures that control the hydraulic conductivity of the basement (Mazurek 2000; Caine and Tomusiak 2003; Stober and Bucher 2007a, b). Stimulation measures typically increase the aperture of existing fractures and thus improve the hydraulic conductivity of the reservoir rocks. High-pressure injection of water irreversibly widens the existing fractures. Rarely new fractures are generated, except perhaps in dense unfoliated metasediments (Huenges 2010).

The created EGS underground heat exchanger is then used in a fashion like in hydrothermal systems. Water is injected into the reservoir through an injection borehole. The water flows through the heat exchanger and extracts thermal energy from the hot rock at depth. After passage through the underground heat exchanger the heated water is pumped to the surface from a production well. The fluid advection is driven by the potential difference between the injection and production well. The subsurface distance between the two wells measures several hundreds to thousand meters.

EGS are primarily built for the conversion of geothermal to electrical energy. Consequently, as mentioned above, the temperature should be 200 °C or higher, which means reservoir depths of 5 km and deeper in areas with moderately warm continental geotherms (38 K/km). However, two 7 km deep wellbores must be drilled in areas with typical average continental geotherms (27 K/km). Since EGS are independent of the presence of highly conductive aquifers at depth, the technology may be installed nearly anyplace. Because of this circumstance, EGS have a tremendous energy potential and the technology can be regarded as the future most important use of geothermal energy (MIT 2007; Lund 2007). EGS technology may have the capability for becoming the prime source of energy in the future.

The method to improve the hydraulic conductivity of the heat reservoir is called hydraulic fracturing. The technique causes an increased conductivity of the rocks by hydraulically fracturing the rock matrix. This always causes seismic noise when the rocks are hydraulically fractured by injecting water into the wellbores under very high pressure. The seismic noise proves that stimulation does what it is supposed to do, opening existing fractures thus increasing the hydraulic conductivity of the rock reservoir. The method runs under different names including hydraulic fracturing and reservoir stimulation. The seismic noise associated with fracturing may be annoying or even scaring to uninformed people in densely populated areas; however, it has never caused damage to surface installations such as homes and other buildings (in sharp contrast to mining activities for example). The key point here is that the population should be educated about the possible "side effects" of possible hydraulic fracturing operations.

When using the hydraulic fracturing method (surface) water is injected into the matrix rock of the heat reservoir with several hundred bar wellhead pressure. The purpose of the effort is to open and widen open fractures and re-fracture old fractures that have been sealed by younger mineral deposits. The fracturing affects volumes of reservoir rock around the uncased wellbore, the open hole or in a packer-isolated section of the wellbore. It is hoped that the efforts permanently increase the hydraulic conductivity of the bedrock. Usually water from a nearby river or lake (or similar) is used as injection fluid. However, the water may be charged with a series of additives such as Na_2CO_3, HCl, NaOH, HF and many others that support the effort by chemically interacting with the bedrocks and the fracture minerals (e.g. Portier et al. 2007).

There are several types of stimulation techniques depending on the sharpness, the duration, the frequency, the chemical additives and the number and dimension of sections to be stimulated. One distinguishes between massive hydraulic stimulation, pulse stimulation, multi-frac, gel stimulation, water-frac, acid-frac and many more. If acid-frac is to be attempted on the basis of the bedrock mineralogy, it has to be decided what kind of acid and at what concentration it will be sufficient to obtain the desired result. The choice of "mild" and "strong" acidizing techniques depends of the mineralogy of the target formation. However, the protection of the casing from chemical attack needs to be considered also.

The minimum volume of heat exchanger rock depends on the temperature and conductivity of the accessed reservoir. However, rule of thumb estimates for an economically profitable heat exchanger volume vary from 10^8 m^3 (MIT 2007) to 2×10^8 m^2 (Rybach 2004). The minimum surface area of a commercially useful heat exchanger is about 2×10^6 m^2 (Rybach 2004). If the open hole is assumed to be about 300 m, it follows that, for a doublet system, the distance between the wellbores at depth should be about 1,000 m.

If the EGS technology ought to become a widely distributed source of renewable energy, mature reservoir engineering is the prime requirement. Many questions are still unanswered and subject to challenging research in the near future. The questions include: How can the hydraulic conductivity of the reservoir be significantly improved using hydraulic and chemical techniques as mild as possible; How can the flow paths of the fluid between the wells be controlled and engineered as desired; How does the hot fluid react with the rocks and what are the chemical consequences of the hot fluid circulation; What is the time evolution of the system and the associated cooling pattern?

9.2 Historical Development of the Hydraulic Fracturing Technology, Early HDR Sites

The oil and gas industry traditionally uses HDR techniques to improve the hydraulic conductivity of sedimentary rocks by fracturing low-conductivity sediments. From early on the industry also used chemical stimulation for reservoir improvement. The HDR has been transferred from the oil and gas industry to the deep geothermal technology. Early efforts to extract thermal energy from the deep underground date back to the early 1970s. The experiments at Fenton Hill (New Mexico, USA) by the Los Alamos National Laboratories have truly been pioneering and innovative research and efforts. The heat reservoir at the Fenton Hill research site was a biotite-granodiorite (Brown 2009). Inspired by the groundbreaking Fenton Hill project, several follow-up projects have been initiated worldwide. The challenging pioneering early projects include: The Urach Deep Wellbore (UDW) in Germany (1970s); Rosemanowes, Cornwall (UK) in the 1980s, Le Mayet (France); Hijiori (Japan); Ohachi (Japan); Soultz-sous-Forêts (France).

The pioneering project Soultz-sous-Forêts in the upper Rhine rift valley started officially in 1988 as an European HDR research cooperation after French-German geological exploration studies. After an extended feasibility study, two boreholes were drilled to 3,500 m (GPK 1 and GPK 2) and a geological heat exchanger has been established by hydraulic stimulation in the years 1993–1997. Hot water could be circulated at a rate of 25 l/s and a temperature of 142 °C. Later GPK 2 has been deepened to 5,000 m and two new boreholes were drilled to 5,000 m (GPK 3 and GPK 4) into the granitic Variscan basement in 2001–2005. The temperature at this depth is 203 °C. Since 2008, the pilot power plant based on ORC technology produces electrical power on the grid. The Soultz-sous-Forêts power plant is primarily a research system that has produced invaluable and important knowledge for the advancement of the HDR technology (Genter et al. 2010, 2012). Immense efforts for further development of the EGS technology are currently undertaken in Hunter Valley in the Cooper Basin, Australia, and in Desert Peak, Nevada and Coso near Los Angeles, California both in the USA.

The common concept of all these projects is the development of a geothermal heat reservoir in the crystalline basement for electrical power production. The fundamental feasibility of the concept has been demonstrated by the groundbreaking Fenton Hill project in New Mexico (USA) in the 1980s. At Fenton Hill the heat exchanger has been created by hydraulic stimulation from two 3,500 m deep wells. The temperature at depth is 234 °C and did not decrease during 11 month of flow testing the system. However, the system was not commercially lucrative (MIT 2007; Duchane and Brown 2002; Brown 2009) mainly because of insufficient flow rates at low pumping pressures. Also, hydraulic experiments showed that fractures did not irreversibly open. Nevertheless, the Fenton Hill project produced unique and valuable data and experience for future EGS projects. The adventure of the Fenton Hill project has been documented and described in great detail by Brown et al. (2012). The EGS interested reader finds a wealth of information on all aspects of HDR system development.

In these early days of the EGS development during the 1970s and the beginning of the 1980s it was still assumed that the crystalline basement of the continents would be mostly free of water (dry) at great depth and that the rocks would be essentially unfractured. Consequently the developing technology has been named "hot-dry-rock" (HDR) and the term "hydraulic fracturing" has been transferred from the oil and gas industry to geothermal system development also. The term hydraulic fracturing is a consequence of the concept that new vertical penny-shaped cracks must be created in the massive rocks (Smith et al. 1975; Duffield et al. 1981; Ernst 1977; Schädel and Dietrich 1979; Kappelmeyer and Rummel 1980; Dash et al. 1981). Today, it is common knowledge that the continental crust is fractured to the brittle ductile transition zone at about 12 km depth and that the fracture system is interconnected and water-saturated (e.g. Stober and Bucher 2005, 2007a, b). Still the hydraulic conductivity of the crystalline basement is normally insufficient for EGS and the geological heat exchanger at depth must be engineered by stimulation methods.

9.3 Stimulation Procedures

Hydraulic stimulation experiments have shown that the crystalline basement normally has a natural fracture network. The existing fracture network has been hydraulically activated during the experiments. No new artificial cracks and fractures have been opened by high-pressure water injection (Batchelor 1977; Armstead and Tester 1987). The prime reaction of granite and gneiss basement on stimulation is that it just modifies the fracture geometry by first increasing the fracture aperture and then displacing the rocks due to shear stresses. The process results in a permanent and irreversible misfit of the two original fracture surfaces

9.3 Stimulation Procedures

Fig. 9.1 Pressure—time curves of stimulation experiments: **a** new crack form, *curve* shows a distinct pressure spike, **b** expanding existing fractures

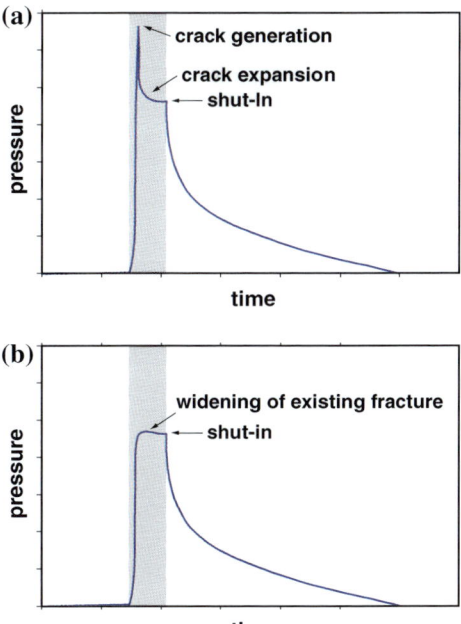

and as a consequence an increased average aperture and hydraulic conductivity (Pearson 1981; Pine and Batchelor 1984; Baria and Green 1989; MIT 2007). This mechanical behavior is different from that of unmetamorphic sediments where stimulation indeed creates new fractures and cracks.

Figure 9.1 shows recorded pressure versus time plots of (a) the process generating a new fracture and (b) the process of expanding an existing fracture. The pressure level and the shape of the pressure—time curve differ distinctly in the two experiments. Generating new cracks requires massively higher pressures and the pressure collapses suddenly when the crack forms. Pressure—time data discriminate stimulations that create new cracks from widening an existing fracture network.

Hydraulic expansion of fractures makes lateral displacement of locked uneven fracture surfaces possible by decreasing the shear strength of the fracture surface. Lateral displacement requires an existing stress component parallel to the fracture surface (Fig. 9.2). Fracture surfaces without a shear stress component open and close elastically without being displaced laterally (Stober 2011). Orientation and magnitudes of the principal stresses at reservoir depth can be derived from in situ stress analysis using data from wellbore breakouts (Zobak el al. 2003) and micro-seismic events. Fractures with a high angle to the normal stress component will not or hardly open and will also not be displaced laterally because of minimal shear stresses. Thus hydraulic stimulation activates only a fraction of the fracture system. Improved hydraulic conductivity only results from fractures that are sheared during hydraulic stimulation. This also means that the fluid flow pattern

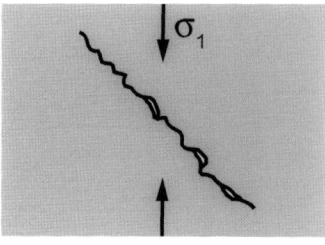

1. fracture with small natural aperture

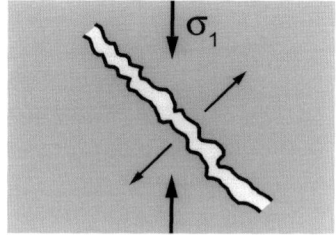

2. fracture expanded by hydraulic pressure during stimulation

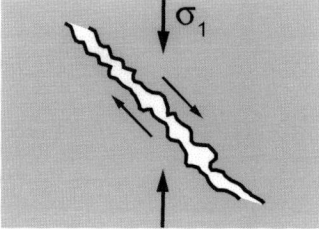

3. displacement of fracture surfaces due to shear stress component

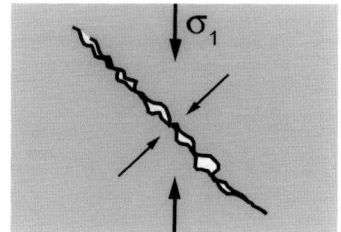

4. Release of hydraulic pressure, results in irreversible increase of aperture and permeability

Fig. 9.2 Permeability increase as a result of hydraulic stimulation

during hydraulic standard well tests, for example pumping or injection tests, may differ totally from the fluid flow pattern caused by hydraulic stimulation. Stimulation generates directed fluid flow.

Shearing of two rough and jagged fracture surfaces causes small mismatches and gaps when the aperture decreases with phasing out the hydraulic pressure. The created misfits cause an increase of the fracture porosity and a permanently enhanced permeability. The effect is known under the term "self-propping".

The function of the injected water is to decrease the shear resistance of the fracture due to pore pressure increase enabling the displacement of the fracture surfaces. After relief of the hydraulic pressure the hydraulic conductivity is permanently and irreversibly improved. However, in the absence of anisotropic stresses in the underground the rocks deform just elastically and hydraulic stimulation does not permanently improve the hydraulic conductivity (Stober 2011) because hydraulic fracture dilatation does not create the required misfits. Furthermore, even if tectonic stresses are present shearing is possible only for those fractures with an appropriate orientation relative to the stress ellipsoid.

The small displacements and shear movements of rock blocks generate mechanical vibrations in the underground. Because of the similarity to seismic tremors the small mechanical vibrations are known as micro-seismicity. The development of EGS power plants inevitably requires reservoir stimulation, which

9.3 Stimulation Procedures

is fundamentally associated with micro-seismicity. Micro-tremors are invariably coupled with successful stimulation measures. The natural stress-state of the underground in particular the magnitude of tectonic shear stresses controls the seismicity of an area. Hydraulic stimulation causes a stepwise relief and decay of naturally present stresses in the underground and micro-seismicity propagates slowly with the advancement of the pressure front of the injected water.

It is the objective of EGS development to keep the micro-seismic tremors at the surface as small as possible but the micro-structural effects in the reservoir maximal and hydraulic stimulation efficient. Chemical methods may be helpful in support of smooth deep geothermal system development. Experimental studies have been recently initiated, using new formulas for the chemical composition of the injection fluid and using various chemical additives to the surface fresh water normally used for reservoir stimulation (Portier et al. 2007). During the 1970s so-called propping (usually quartz sand) have occasionally been added to the injection fluid during stimulation of crystalline basement reservoirs with the idea that the hard solid particles helped to keep fractures open (Smith et al. 1975; Schädel and Dietrich 1979). Today successful stimulation experiments have been carried out with inorganic acids (hydrochloric acid or mixtures of HCl and hydrofluoric acid, HF), complexing substances (nitrilotriacetic acid, NTA) or organic acids (organic clay acid, OCA) (Genter et al. 2010). The purpose of the chemical stimulation is to remove or leach fine-grained mineral dust and carbonates from fracture surfaces. The method has been successfully used by the oil and gas industry for increasing well productivity for a long time. The first acidizing measures have been performed more than 100 years ago in limestone formations.

The hydrocarbon industry distinguishes between slick water stimulation, stimulation with a high-viscosity fluid (additives: polymers or tensides) and acidization (Williams et al. 1979; Kalfayan 2008). Slick water simulation injects a large volume of a low-viscosity fluid (~1,500 m^3) with about 100 t of suspended proppings (quartz sand, bauxite sand) to the reservoir. The wellhead pressures may reach about 700 bar. The proppings help to maintain the permeability improvement after stimulation. Proppings improve the hydraulic properties of fractures that have not undergone shearing during stimulation. The method is mainly used for slate. Stimulation with high-viscosity fluids requires much lower injected fluid volumes (about 400 m^3). Also here about 100 t of proppings are usually suspended in the fluid for injection. The technique is mostly used for sandstones. Acid fracturing also applies up to 700 bar wellhead pressure but the fluid is injected with rapidly and strongly varying rates. The method aims to roughen the fracture surfaces and is used mainly in limestone and other carbonate rocks. The hydrocarbon industry uses stimulation techniques exclusively in sedimentary reservoir rocks in complete contrast to EGS development.

Most recently stimulation of geothermal reservoirs is carried out in sections of the wellbore and in separate isolated zones. The new method helps avoiding hydraulic shortcuts and minimizes the chance to trigger unwanted seismic events.

Hydraulic stimulation measures must be accompanied by a monitoring program that records the micro-seismic signals at or close to the surface. Typically, geophones record ground movement in shallow monitoring wells that are placed around the deep stimulation well. The seismic signals induced by the stimulation can be resolved in 3D and separated into the x-y-z components. The geophones record ground movements quantitatively. Interpretation of the data results in a 3D-image of the activated fracture volume.

This 3D-image of the seismic noise, however, includes only the fractures that experience shear movements. Open fractures without a shear stress component and other cavities that do not suffer shear displacement produce no seismic signal and remain quiet. Thus the geophones of the monitoring wells log merely the seismic active part of the entire fracture volume. The derived 3D-image may correspond to the real fracture volume but not necessarily. This condition is particularly evident in wells that have been repeatedly stimulated. In this case the geophones receive seismic signals mostly from an outer zone around the previously stimulated fracture volume.

The created 3D-image of the activated fracture network is typically not spherical but anisotropic and ellipsoid shaped. The orientation of the created elongated fracture volume determines the target point for the second wellbore. For optimal hydraulic connection the second well is also stimulated (Fig. 9.3). The EGS development makes a great effort creating a zone of increased hydraulic conductivity, the heat exchanger, in the underground that is surrounded by low-conductivity rock. The longitudinal extent of the heat exchanger ranges over several hundred meters, the specific length depending on rock properties, the stress field, and the particulars of the injection procedures.

At the EGS project Soultz-sous-Forêts the natural fracture and fault system of the granitic reservoir rock is oriented parallel to the direction of the principal stress. The orientation of the stimulated fractures as recorded by the geophones excellently matches the direction of the regional stress field. The hydraulic stimulation has been very successful and massively increased the injectivity and the productivity of two wells. The well with the highest initial hydraulic conductivity showed the least response to stimulation. It is thought that the high primary conductivity was related to relatively few highly permeable fractures and faults (Baria et al. 2004; Tischner et al. 2007). The principal stimulation mechanism, fracture broadening followed by fracture displacement, is not efficient in this case.

Widening the natural fracture system for developing the underground heat exchanger requires very high pressures. The so-called opening pressure must be surmounted. Its magnitude depends on the lithostatic pressure and on the orientation of the controlling fracture system. Fluid flow rates do not significantly increase until the opening pressure is exceeded. An opening pressure of 170 bar wellhead pressure was required in gneiss at 4.4 km depth at the Urach EGS site. Slightly lower opening pressures were required in granite at 5 km depth at the Soultz-sous-Forêts EGS site.

At the Soultz-sous-Forêts EGS site maximum wellhead pressures of 180 bar resulted in injection flow rates of about 50 l/s. The maximum seismic response to the stimulation reached a magnitude of 2.9. Stimulation at the EGS site Basel, Switzerland, also located in the upper Rhine rift valley with injection flow rates of

9.3 Stimulation Procedures

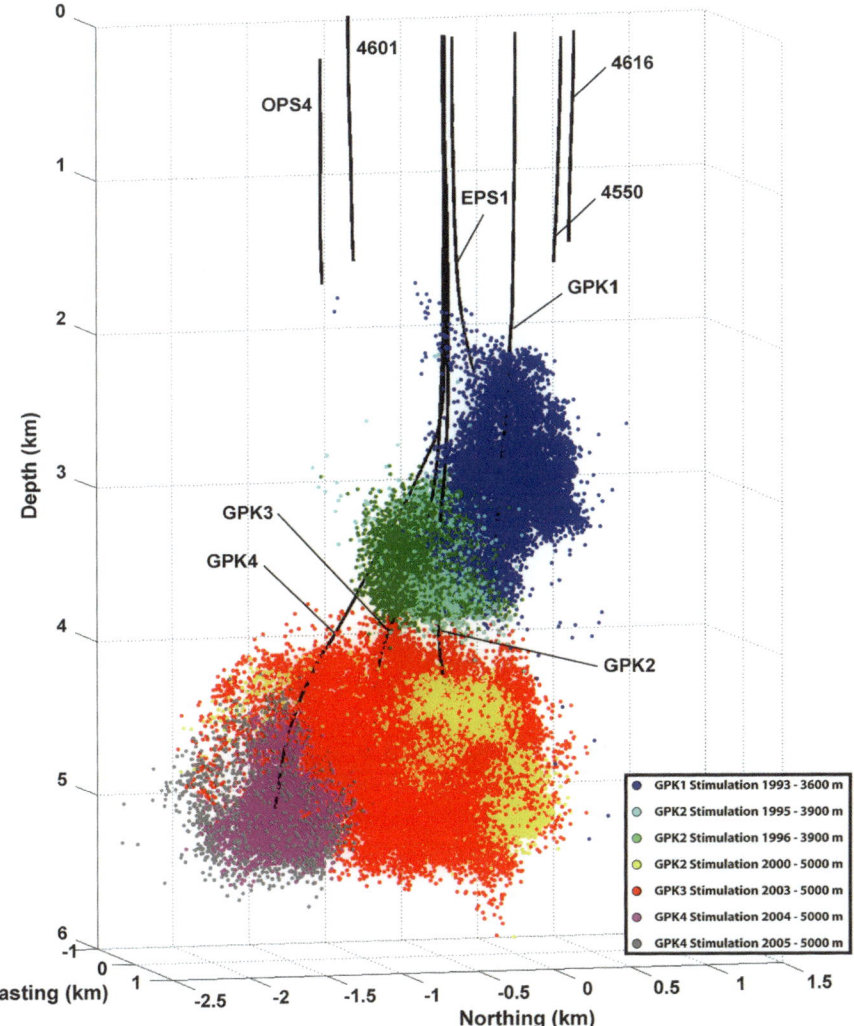

Fig. 9.3 Geophone recordings of seismic noise generated by hydraulic stimulation of the wells at Soulz-sous-Forêts. Note the position of four, about 1,500 m deep monitoring wells (drawing by N. Cuenot, with kind permission by A. Genter)

up to 63 l/s resulted in a wellhead pressure of 300 bar and a seismic response of up to a magnitude of 3.4. Stimulation of the well Habanero 1 at the EGS site in the Cooper Basin, Australia, with 350 bar wellhead pressure lead to injection flow rates of up to 40 l/s. The monitored seismicity reached maximum a magnitude of 3.7.

The summarized experience shows that reservoir stimulation in the deep crystalline basement with injection rates of some 10th of l/s water and wellhead pressures of about 200–300 bar may cause seismic reactions above a magnitude of 3. The magnitude of the seismic reaction to stimulation depends on many factors and parameters including: The injection flow rate, the total volume of injected fluid (water), the properties of the initial natural fracture system, the maximum applied wellhead pressure, the duration of stimulation, the chemical composition of the injected fluid (water), the temperature and, critically, the rates of pressure changes (rates of pressure increase) (Nicholson and Wesson 1990; Shapiro and Dinske 2009; Giardini 2009; Bommer et al. 2006). However, the tectonic condition and the stress state of the underground are the key properties controlling micro-seismicity. How these many relevant properties and parameters interact at any given micro-seismic event is not yet fully understood. However, the concept of seismogenic permeability k_s defines an important derived parameter that relates the hydraulic properties of the reservoir to the chance of inducing seismicity by hydraulic stimulation (Talwani et al. 2007).

9.4 Experience and Dealing with Micro-Seismicity

The stimulation efforts at Soultz-sous-Forêts required about 180 bar wellhead pressure for significantly improving hydraulic conductivity of the granitic reservoir rocks. The same efforts at the Basel EGS project, 150 km south of Soultz also in the Rhine graben, with 300 bar wellhead pressure triggered seismicity that scared the public and caused the abandonment of the project (Sect. 10.1). The EGS project Urach, about 170 km east of Basel, applied up to 660 bar wellhead pressure during stimulation experiments without causing perceptible micro-seismicity (Stober 2011). The three very contrasting examples of seismic response to reservoir stimulation in the same region demonstrate the controlling importance of the existing stress field for the magnitude of triggered micro-seismic events. In the Basel region tectonic stress is high, in the Urach region very low. Therefore macro-seismic events, so-called earthquakes, occur more often and with higher magnitude in the Basel region than in the Urach, or also the Soultz area. This can be seen on earthquake frequency maps, e.g. the European-Mediterranean Seismic Hazard Map.

The early (1993) stimulation works at the Soultz-sous-Forêts site triggered seismicity at 3,600 m as shown on Fig. 9.3 (Cornet et al. 1997). The initial rock permeability was very low 10^{-17} m^2 (Evans et al. 2005). The permeability rapidly increased with increasing injection pressure and the first micro-seismicity was observed when the differential pressure exceeded 5 MPa. Seismicity increased dramatically above 6 l/s injection rate because permeability reached the critical values for seismogenic permeability k_s (Talwani et al. 2007). Further increase of injection rates was followed by decreasing seismicity. Finally the differential pressures stabilized at 9 MPa. The pressure did not further increase with increasing injection rates. The successful stimulation increased the injectivity, defined as flow rate per unit differential pressure, from 0.6 l/s to 9.0 l/s per MPa (Evans et al. 2005). The high injectivity resulted in non-Darcian fluid flow (Kohl et al. 1997).

9.4 Experience and Dealing with Micro-Seismicity

The reservoir at the Basel and Soultz sites is in granitic basement, at the Urach site in metamorphic gneissic basement. The rheology of granite is controlled by quartz and feldspar and it responds to stress predominantly by brittle deformation. The response of the mica-rich gneisses at Urach to stress has a strong ductile component even at temperatures of 200 °C and lower.

Hydraulic reservoir stimulation is a well-established method in the geothermal and hydrocarbon industry that has been applied for decades and worldwide (Bencic 2005). Seismic monitoring of stimulation efforts has not become standard routine for EGS development before the Basel incident in 2006. This is also a consequence of the necessity to build EGS power plants near the potential user and consumer of co-produced thermal energy (in contrast to the hydrocarbon industry that produces oil and gas mostly in uninhabited areas). Thus geothermal EGS project must drill wellbores near cities and other densely populated regions where the public is alarmed and scared if transitory micro-seismicity from stimulation measures can be sensed at the surface.

In regions with natural seismicity necessary stimulation works may interact with natural stress release processes. There induced seismicity is the result of the step-by-step reduction of stored stresses that may not have been released without the high-pressure fluid injection (or at least not at this time). The possibility for induced seismicity and its potential magnitude at a specific site can, to some degree, be assessed, predicted and partly controlled. Efficient seismicity control requires complete and constant readings and supervision of the injection pressure and a seismological monitoring in the close and further surroundings of the site. If observed seismicity increases above a site-specific threshold value injection pressure and flow rates, respectively, must be reduced. However, the details of interacting mechanisms for triggering seismic tremors by hydraulic stimulation are not yet fully understood and the topic requires further basic research (Sect. 10.1).

9.5 Recommendations, Notes

It is a great advantage if, during the reconnaissance survey of the selected site, an exploration well can be drilled to the crystalline basement that later needs to be stimulated. In areas where the reservoir rocks of the basement are covered with a sequence of sedimentary rocks geophysical exploration methods such as gravimetric and magnetotelluric techniques cannot detect the presence, structure and orientation of fracture and fault systems in the basement. Even state-of-the-art seismic investigations may discriminate flow-relevant structures in the underground only vague and uncertain. Very prominent, thick and flat-lying fault zones may give a weak signal and can be imaged under favorable circumstances.

The drilled exploration well may later be used as a monitoring well for seismic signals during stimulation works in the EGS deep wells and further on during operation of the plant to record and observe seismicity. Furthermore, the exploration well can be used for hydraulic tests in the crystalline basement, producing reliable and needed data of the hydraulic conductivity and the storage properties of the basement prior to stimulation. Water (fluid) samples collected in the exploration drillhole give valuable information on the composition of the deep fluid (Bucher and Stober 2010). The chemical data from the uncontaminated basement fluid permits prediction of possible scale formation and corrosion and facilitate development of prevention strategies at the very beginning of the project. Regrettably, most projects drill the first bore as future production well for economic reasons and forgo the exploration well. Unfortunately, after completion of the production well hydraulic reservoir stimulation is often the next work on the agenda. The natural hydraulic and hydrochemical reservoir conditions remain unexplored and unknown. The lack of these crucial data may later threaten the entire project or may cause unnecessary high extra costs. The pretended savings from waiving the exploration well may change suddenly to economic disaster.

The underground in a wide range around the planned stimulation measures needs to be stable and major fault zones should be spaciously avoided. Prominent fault zones may react preferentially on stimulation in regions with increased natural seismicity. They often contain abundant and thick zones of fine-grained crushed rock fragments and clay from cataclasis (brittle fracturing and shearing). Some of these materials may swell during stimulation thus sealing water-conducting structures and resulting in a decreasing hydraulic conductivity.

Drilling engineering and future stimulation measures benefit greatly from a good knowledge of the petrography and the mineralogical composition of the rocks. Deformation properties of rocks depend on rock type and type and amount of rock-forming minerals. The patterns of brittle deformation in granites are controlled by the properties of the dominant minerals quartz and feldspar. Granite is more regularly and more strongly fractured than metamorphic basement, predominantly gneiss. Gneisses are typically rich in mica and hornblende in addition to quartz and feldspar. The deformation pattern in gneiss is strongly influenced by the mechanical properties of mica and the mica-related gneissic fabric. Gneiss tends to form fewer more pronounced fault zones parallel to the foliation (gneiss fabric).

Drilling of the wellbore and the upcoming planned hydraulic stimulation of the reservoir gain much from a profound knowledge of hydrostatic, lithostatic and anisotropic pressures in the subsurface. This requires extensive pressure gauging, investigation of in situ stress indicators (wellbore deformation, wellbore breakouts, hydraulic fracturing) and recording natural pore pressures. These data and information must be ascertained before beginning with the systematic stimulation works. The data are essential for a rigorous assessment of the completed reservoir stimulation and also for the sound appraisal of the observed micro-seismicity.

Oriented drill cores should be taken in the first EGS wellbore from the planned reservoir rocks. The cores give insight into micro-fracture orientation, water-conducting structures and into the macro-fracture network and fault zone properties. The cores can be used for determining mechanical and physical parameters of the rocks (Young's modulus, Poisson number, density, acoustic impedance data, thermal conductivity) and for mineralogical examination. Equally important is a

9.5 Recommendations, Notes

qualified geophysical logging of the wellbore (e.g. gamma-log, temperature-log, electrical conductivity-logging, Caliper-log).

Temperature is of course a central parameter for EGS. Measured temperature data from great depth are scarce. It is thus necessary to extrapolate existing temperature data from shallow ground to EGS reservoir depths. The temperature at depth can be computed from the thermal conductivity of the rocks and the known vertical heat flow. The extrapolation assumes that the hydraulic conductivity of the rock is low enough that advective heat transport by groundwater flow can be ignored. Temperature estimates can be improved by considering the internal heat production of the rocks also.

Water samples for hydrochemical analysis must be collected (with a downhole sampler if necessary) before the start of hydraulic stimulation and the injection of large volumes of fluid. Knowing the chemical composition of the original fluid residing in the fracture pore space of the reservoir is a precondition for planning efficient scaling and corrosion prevention (Chap. 14). If the reservoir is wasted with injection fluid first and the original fluid condition is unknown, system development can react to a difficult fluid composition after damage has already occurred. Water in the crystalline basement is normally highly saline at typical reservoir depths of some km. Total dissolved solids (TDS) range from some tens to some hundreds of g/l. The major solutes are sodium, calcium and chloride (Bucher and Stober 2000, 2010). Some fluids contain appreciable amounts of dissolved gasses (e.g. CO_2, N_2, CH_4, H_2S). Planning and construction of the surface installations require knowledge of the chemical composition and properties of the produced thermal fluid for efficiently coping with scaling and corrosion and handling the aggressive, saline and possibly gas-rich hot fluid.

The initial hydraulic tests should be of short duration and should be run with low flow rates. They should not cause substantial pressure changes. Therefore the slug test is the ideal choice of the first test method (Chap. 13). In the next step constant rate pumping or injection tests, depending on hydraulic conductivity, supply the necessary information on natural flow regime and the quantification of the natural hydraulic conductivity (Chap. 13). Pressure variations in these tests should be kept relatively small. A short step-drawdown test (step test) may follow. It gives first data on the well efficiency. Then pre-simulation tests can be designed using these derived test data. Pre-simulation tests provide extensive information and experiences on the reaction of the underground on fluid injection. With these hydraulic experiments the system developer is slowly and carefully approaching the concept of the actual site-optimized stimulation methods and putting them into practice.

The success of stimulation measures depends on the existence of a suitable natural fracture network and an appropriate stress field. If nothing can be stimulated in a stress free reservoir an EGS project will crash. If stimulation was possible, the final hydraulic properties of the reservoir decide on the economic success of the project.

Hydraulic short-cuts and an extreme stimulation of singular fractures or fault zones can be prevented or the danger minimized by section-by-section stimulation

Fig. 9.4 Examples of a single packer and a double packer. A packer consists of a central pipe and rubber mantle. The rubber mantle can be pressed to the borehole wall after it has been positioned in the wellbore by inflating or by compressing it

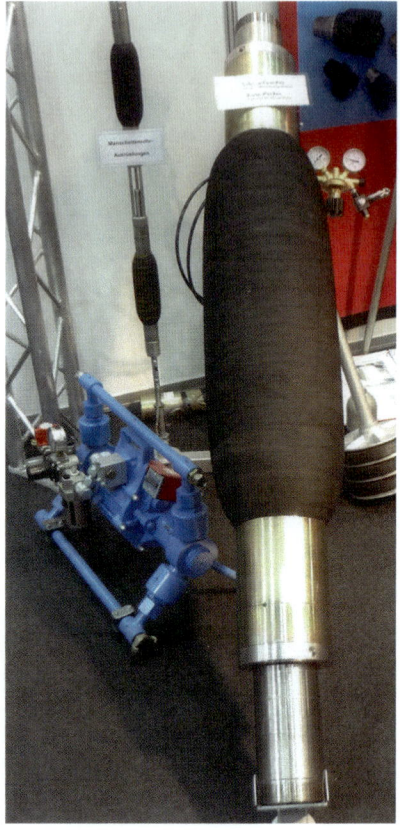

of isolated portions of the reservoir. Packer and cement bridges (concrete plugs) are used for the hydraulic isolation of defined rock volumes (Figs. 9.4, 13.2). A wide range of packer systems are available in all dimensions for different wellbore diameters and special applications and different needs.

Target points of the wells are planned in accordance with the natural stress field. It can be expected that the stimulated rock volume, the future hot rock heat exchanger, develops in direction of the stress ellipsoid. The definition of the optimal distance of the bottom holes of the two wells, the quantification of the thermal range and the prediction of the life cycle and ageing of the system require knowledge of thermo-physical rock parameters (thermal conductivity, density, heat capacity, heat production rate).

If the EGS project uses exclusively vertical wellbores, the surface installation will be spread over distances of several hundreds of meters and these space requirements must be taken into account. An unrelated note: Vertical bores have the advantage that the handling of thermal casing expansion during production of thermal water is considerably simpler.

9.5 Recommendations, Notes

It is strongly recommended to set up a seismic monitoring network at a very early stage of EGS project development. The network should continuously record all seismic signals above a magnitude of 1.0 (Richter scale) in a 10 km range around the site of the planned power plant. The measurements should be continued during drilling of the wells, stimulation of the reservoir and operation of the plant. If induced seismicity occurs, it most likely occurs at the initial stages of hydraulic simulation. During continuous operation of the plant induced seismicity is unlikely. However, plant shutdown for maintenance or the subsequent resumption of operation may cause induced seismicity.

Chapter 10
Environmental Issues Related to Deep Geothermal Systems

Drilling rig for deep boreholes

The conversion of geothermal energy into electrical power or useful heat produces no CO_2 and no flue gas emissions such as soot particles, sulfur dioxide and nitrogen oxides. The operation of a geothermal power plant is deeply friendly to the environment. The risk for harmful environmental effects is extremely low during normal operation and even during accidents. The low-risk systems result from the use of high-quality structural materials and from the mature technology with numerous safety precaution installations.

Construction of geothermal systems and power plants causes CO_2 emissions related to manufacturing construction materials, transport of materials and equipment and service traffic, no different as with other types of power plants. Careful planning of logistics helps to minimize these emissions.

Developing the underground heat exchanger of an enhanced geothermal system (EGS) (Chap. 9) involves hydraulic stimulation measures that cause microseismicity. Rarely the induced microseismicity can cause irritation if physically sensed at the surface. Problems related to microseismicity are dealt with in Sect. 10.1.

The geothermal fluid circulates in a closed system and cannot cause any damage to the environment. If a leak occurs in the surface installations, the fluid circulation can be stopped and the leaking section replaced. The working fluid in the secondary loop of the power production circulates also in a closed system. If leaks occur here, precautions on the construction and technical side help to minimize environmental pollution.

However, it may be necessary to test the thermal fluid circulation during the system development phase when the primary loop is not yet completely closed at

Fig. 10.1 Steam clouds at the Landau power plant after successful development of the primary loop

the surface. It is only during these tests that the success of the entire effort can be visually seen from the rising white steam plumes (Fig. 10.1).

In binary geothermal power plants the working fluid of the secondary loop must be cooled below condensation conditions after leaving the turbines. The excess heat is released to the environment (Sect. 10.3), like in any other thermal power plant. The thermal emissions of geothermal plants are, however, orders of magnitude lower than those of thermal large-scale coal- and gas-fired plants and nuclear power plants. The emissions are considerably lower even if normalized to the power output of the units.

Several countries recommend the concept of combined heat and power or the technology may even be mandatory. Combined heat and power reduces energy loss related to the fluid cooling process and optimizes the use of the extracted geothermal energy. Integrated and cascade use of geothermal low-enthalpy resources also uses produced thermal energy rather than wasting it in the fluid cooling process (Sect. 8.6).

This book does not cover all potential environmental hazards that can be associated with the construction and operation of a low-enthalpy geothermal power plant. In the following sections, we present several environmental problems selected partly because of the topicality of the problems, partly because the hazards are sometimes unjustifiably ignored. Also geothermal projects can be plagued by very diverse failures and troubles during system development and later during operation. Careful planning, management, monitoring, educated and experienced personnel and adequate quality of used materials and equipment help reducing potential troubles and thus also minimize environmental effects. Finally, an open and straight communication philosophy that integrates all active parties in the project and a trusting working together is the best insurance against troubles and hazards.

10.1 Seismicity Related to EGS Projects

Rock deformation caused by massive hydraulic stimulation of the planned EGS granite reservoir at 5 km depth has been physically sensed and acoustically recognized in the city of Basel (Switzerland). Ground shaking and thundering noises scared the local population (Kraft et al. 2009). The incident had disastrous effects for the public acceptance of deep geothermal projects, not only EGS projects but also for hydrogeothermal doublets. The "Basel incident" had extremely negative consequences for the further development of enhanced geothermal system. The question must be asked by anyone interested in geothermal energy development: Why did it happen? How can such incidents be avoided?

We explained in the previous Chap. 9 that the development of the underground heat exchanger of HDR systems requires massive hydraulic (and chemical) stimulation of the reservoir. The success of the stimulation process is expressed by microseismicity in the underground, which can be monitored and recorded by seismic monitoring programs. The data provide a 3D view of the stimulated rock volume that can be expected to have a significantly enhanced hydraulic conductivity compared with the original situation. The data derived from the observed microseismicity are fundamentally relevant for the location and detailed path of the second borehole.

The stimulation related microseismicity is indispensable for the development of HDR/EGS project. However, it is absolutely essential to avoid any surface effects that can be sensed or heard (Chap. 9). We are firmly convinced that this is possible with prudent planning and conservative reservoir development strategies. It also should be kept in mind that in several documented geothermal projects massive hydraulic stimulation did not or barely result in recognizable and recordable microseismicity, but also making reservoir development difficult.

Microseismic incidents are not exclusive to petrothermal systems (EGS, HDR). Deep geothermal systems that produce hot water from fault systems or inject fluids into faults have been exposed to problematic microseismicity. There is also a known incident from one hydrothermal system. A selection of known incidents is listed in Table 10.1.

There are several possibilities to express the intensity of microseismic events. A widely used conventional parameter is the logarithmic Richter magnitude scale, which expresses the energy released by the (micro) seismic event. By definition any seismicity below magnitude 2 is microseismicity. Minor seismicity covers the magnitude range of 2–4. The infamous "Basel incident" reached a magnitude of 3.6 (Table 10.1). By definition of the Richter scale magnitude 3.6 characterizes an event that can be felt at the surface by most people but rarely causes structural damage. The listed magnitudes of seismic events related to geothermal projects (Table 10.1) refer to events caused by massive hydraulic stimulation of the geothermal reservoir. In some of the listed projects there were no predefined acceptable maximum magnitude, in other projects including Basel an upper limit has been part of the concession contract. The listed projects utilizing fault zones have

Table 10.1 Seismicity related to geothermal projects

Geothermal project	Type	Magnitude[a]
Unterhaching, D	H	2.2
Landau, D	S	2.7
Insheim, D	S	2.3
Riehen, CH	H	–
Paris Basin, F	H	–
Gross Schönebeck, D	P	–
Horstberg, D	P	Minimal
Soultz (3.5 km depth), F	P	2.2
Soultz (5 km depth), F	P	2.9
Basel, CH	P	3.4
Urach, D	P	1.8
Fenton Hill, N.Mex. USA	P	Minimal
The Geysers, Calif. USA	D	4.0
Cooper Basin, AUS	P	3.7

[a] Maximum magnitude: below detection limit, minimal < 1. System type: Hydrothermal (*H*), petrothermal (*P*), fault system (*S*), dry steam (*D*): Country abbreviations: *D* Germany, *F* France, *CH* Switzerland, *USA* United States of America, *AUS* Australia

massively stimulated the fault systems or injected the cooled fluid of the primary loop under high pressure into the fault systems.

The EGS related seismic incidents caused discussions about the cause for the unwanted seismicity above the magnitude two level and the potential risks for even stronger triggered seismicity that may cause major structural damage at the surface. Also social issues have been intensely discussed including improved communication with the local population at the project site. The general acceptance of an EGS project for the citizens vary strongly depending on complete transparency of the project and all its development phases. The perception of projects by the public depends a lot on the use of terminology. Instrumentally monitored microseismicity should definitely not be communicated to the public as "earthquakes". For the layman the term earthquake is loaded by pictures of damage and death and prompts fears that are not backed by the physical nature of stimulation related microseismicity.

In Germany, the seismic incidents cause major insecurity at the involved authorities resulting in widely varying approval practices and processes. A major difficulty in this context is that a sound pre-drilling risk assessment of induced seismicity is impossible because of lacking data.

International research projects analyze induced seismicity and model geomechanical processes behind the seismicity. The results of the projects such as Geothermal Engineering Integrating Mitigation of Induced Seismicity in Reservoirs (GEISER), Physics and Application of Seismic Emission (PHASE) and others will hopefully lead to improved strategies for minimizing induced seismicity.

10.1.1 Induced Earthquakes

Induced earthquakes are earthquakes that occurred as a direct or indirect consequence of human undertakings. The stress buildup in the underground can be entirely related to constructional measures in the subsurface directly causing an earthquake. On the other extreme high sub-failure stresses may exist in the underground that any form of construction related actions releases as an earthquake. The difference is the different portions of pre-existing natural stresses ("autochthonous") and added anthropogenic stresses ("induced"). This distinction is often difficult to quantify and a continuous range of direct and indirect causes exist.

Earthquakes are caused by spontaneous release of stored elastic deformation energy by friction-based sliding along pre-existing fault planes. Earthquakes are events where stored stresses are relieved by the displacement of rock masses. However, the terminology is strictly scale dependent. The displacement of the rock packets of the Earth crust occurs in general parallel to existing fault systems and may be active across the fault system on a cm to m scale.

The deformation processes in the earthquake hypocenter generate ground movements at the Earth surface. The intensity with which an earthquake is felt at the surface depends on the total strain energy released at the hypocenter (magnitude), the focal depth of the hypocenter below the epicenter and properties of the local hard ground and soil. The surface effects of a seismic event can be predicted and described from parameters including seismic intensity, vibration velocity and vibration acceleration (Sect. 10.1.2). These parameters characterize the earthquake at the surface; magnitude is a parameter that characterizes the earthquake at the hypocenter at depth.

Earthquakes suddenly release stored elastic deformation energy in rock masses at depth. Failure occurs when the stress state exceeds a certain threshold value (Nicholson and Wesson 1990). The high-pressure injection of fluids during reservoir stimulation does not significantly increase stored deformation energy. However, fluid injection lowers the threshold value for failure by reducing the effective frictional resistance on the critically stressed and loaded fault planes.

The hydrological and geological properties and the locally existing stresses at and around the injection zone of the reservoir control the reaction on fluid injection, in some situations an earthquake may be triggered in others high-pressure fluid injection may not result in an earthquake. The hydrological conditions for triggering earthquakes can be related to the seismogenic permeability (Talwani et al. 2007). It should be distinguished between factors that make earthquakes possible and factures that release them. It is not possible to quantitatively predict the probability for a triggered earthquake that may result from increased fluid pore pressure by fluid injection into a deep well (Nicholson and Wesson 1990).

The magnitude of an earthquake is proportional to the logarithm of the length of the activated fault (Wyss 1979; Wells and Coppersmith 1994). A magnitude 8 earthquake activates a fault system of several hundred km length and displaces the moved fault blocks by several meters. A magnitude 3 earthquake activates about some tens of meters of a fault and displaces fault blocks by some cm.

Anthropogenic seismicity is not exclusive to geothermal project development. It is also known from the oil and gas industry, from dammed lakes used as reservoirs for hydroelectric power systems, from underground storage of gas and compressed air, from the high-pressure injection of liquid waste in deep wells and also from underground and open pit mining (Nicholson and Wesson 1990; Shapiro et al. 2007; McGarr 1991; Rutledge et al. 2004; Segall 1989; Cook 1976). Increasing seismic activity can also be triggered by torrential rainfall events (Husen et al. 2007).

Production from oil and gas reservoirs may cause seismicity by decreasing pore fluid pressure from fluid extraction and increased loading. Isostatic compensation after massive reservoir exploitation can cause earthquakes even at large distances (Grasso 1992). As an example: High oil production from the Wilmington oilfield in the Los Angeles Basin (California, USA) subsequently caused subsidence of locally up to 8.8 m with subsidence rates of up to 0.71 m per year. Resulting from these ground movements several damaging earthquakes occurred from the 1940s to the 1960s with a maximum magnitude M_L of about 5.1 (Kovach 1974). At the end of the 1950s massive water injection was started with the aim to bring subsidence to a halt and also to improve oil recovery from the field. The operations were successful. However, water injection triggered a series of small quakes ($M_L < 3.2$) (Nicholson and Wesson 1990). The example shows that water injection into an oil reservoir can reduce the effective load and may trigger quakes. Other examples of man-made, subsidence-related seismicity in oil and gas fields include the Goose Creek oil field quakes from 1925 in Texas (Davis and Pennington 1989), gas-extraction subsidence in the gas field of the Pau Basin near Lacq (SW France) (Segall et al. 1994) caused >1,000 seismic events, 44 reaching magnitude >3.0, two of them magnitude >4. Further examples can be found, for example, in Grasso (1992).

The largest reported earthquake related to reservoir infill is probably the magnitude 6.5 earthquake of Koyna in India (Gupta and Rastogi 1976). Many earthquakes reported from dammed lakes have been summarized and analyzed by Talwani et al. (2007).

Liquids or gasses can be injected to the underground from deep wells for very diverse reasons including leaching salt from evaporites, disposing liquid toxic wastes, improving production from ageing oil and gas fields or fracturing reservoir rocks for improved hydraulic conductivity. Injected fluid volumes have been enormous at some sites so that fluid injection induced earthquakes. An infamous case of induced seismicity occurred at the Rocky Mountain Arsenal injection well near Denver (Colorado, USA). The 3,671 m deep well was used around 1962 for disposal of hundreds of millions liters of toxic waste. The liquid was injected with wellhead pressures of about 72 bar into a reservoir that is cut by several sub-parallel faults. The resulting several hundred earthquakes reached a magnitude of 5.5 (Hsieh and Bredehoeft 1981; Nicholson and Wesson 1990).

A worldwide compilation of seismicity related to geothermal energy projects did not report any loss-producing damage quakes during drilling or during reservoir stimulation (Majer et al. 2008). During operation of a geothermal system fluid injection is counterbalanced by fluid extraction, which is a fundamental difference to the other examples of injection-induced seismicity described above.

10.1.2 Quantifying Seismic Events

Seismic events can be characterized by the two quite different parameters magnitude and intensity. Magnitude is related to the strain energy released at the source of the event, the hypocenter at depth. The intensity of a seismic incident describes the effects and consequences of the event at a specific location at the surface. The empiric logarithmic magnitude scale (Richter Scale) derives the released energy from the vibration amplitude recorded by a seismometer. The amplitude relates to the seismic energy released by the ground rupturing process. Seismic events below magnitude 2 are referred to as microseismicity. It is not sensed at the surface. Magnitude 2–3 events are extremely light earthquakes and can be sensed at the surface only at very special conditions. A magnitude 3–4 event is a very light earthquake. Many people can feel such quakes, however, structural damages are very unusual. The Richter scale is open-ended and each succeeding scale unit represents ten times more seismic energy released. Since 1990, the planet Earth experienced five earthquakes with magnitude >9, magnitude 10 earthquakes have not been recorded yet.

The earthquake magnitude is not a measure for consequences and damages at the surface, the epicenter, above the hypocenter. The ramifications at the surface are influenced, as mentioned above, by the focal depth below the epicenter, by the type of rocks and soil and the structure and type of buildings.

The Gutenberg-Richter law relates the magnitude and the number of earthquakes in a given defined region during a certain time period with at least that magnitude (Eq. 10.1).

$$\log N = a - bM \quad (10.1)$$

where N is the number of events with at least magnitude $\geq M$ and a and b are regional specific constants. The parameter b is close to 1 for tectonic earthquakes. Fluid injection may increase b, resulting in more but weaker seismicity. The Gutenberg-Richter law can be used to explore the possibility of high-magnitude events that have not occurred so far.

The macroseismic intensity is an empirical classification of local effects of an earthquake on residents and natural and man-made structures at the surface. The local near-surface structure of rocks and soil strongly influence the ground motion and thus the macroseismic intensity. The ground motion at the surface of loose soil can be massively higher than on solid basement bedrock. Consequently an earthquake with the same magnitude and focal depth may have much higher macroseismic intensity on loose soil than on solid rock. The Mercalli scale expresses the macroseismic intensity by roman numerals from I (lowest) to XII (highest). Its highest value defines the location of the epicenter above the earthquake focal region. It declines with increasing distance from the epicenter.

In the context of induced seismicity related to geothermal projects robust quantitative predictions of local surface ground motions that may result from reservoir stimulation measures would be very helpful. The ground vibrations, ground movements and accelerations can be quantitatively characterized using methods of petrophysics and soil physics. The vibration velocity relates directly to the structural effects. Vibrations exceeding 5 mm/s may cause smaller damage and cracks of the plaster. Severe structural damage is not caused until vibration velocity exceeds much higher values.

The lower limit of sensing a seismic event at 5 km focal depth is about magnitude 2–2.5. Under special circumstances a magnitude 3.5–4.5 event may cause light damages to buildings at the epicenter. The US Geological Survey supposes that clear damages occur at magnitude above 4.5 and ground vibration velocity clearly above 34 mm/s. Magnitude 5 earthquakes may cause scattered severe structural damages to buildings. It follows that a maximum magnitude of 2–2.5 should not be exceeded during massive hydraulic stimulation or operation of a geothermal system.

10.1.3 The Basel Incident

The city of Basel is located at the southern termination of the upper Rhine rift valley. The local high geothermal gradient and the geological structure of the rift valley and the potential for local buyers of produced thermal energy make Basel an optimal site for an EGS project. The valley has an infill of a thick succession of Mesozoic and Tertiary sediments covered with Quaternary deposits. The top of the crystalline

10.1 Seismicity Related to EGS Projects

basement beneath the sediments is between 2,640 and 2,750 m below surface. The knowledge about the geological structure was limited to relatively shallow depth before project start. The Basel 1 wellbore reached 5,009 m and drilled through 2,300 m of basement rocks mostly granite. The basement and the older part of the cover are cut by numerous steeply dipping faults and form a complex pattern of fault blocks. The drilling site is located about 4.5 km from the surface outcrop of the eastern main border fault of the Rhine rift valley. The exact position of the border fault system at 5 km depth is not known. However, the fault system undoubtedly dips to the west and could be at close distance to the Basel 1 wellbore at 5 km depth. The natural seismicity in the Rhine rift valley decreases from Basel towards the north.

December 2nd, 2006 was the beginning of massive hydraulic stimulation of the granite at 5 km depth by injecting river water at high pressure into the fractured granite at depth. The drilling site of Basel 1 is located within the city area of Basel (Fig. 10.2). Six days later (Dec 8th) a seismic event with a magnitude of 3.4 and the epicenter at the drilling site of the geothermal project occurred. The measured ground vibration velocity was 9.3 mm/s. Three further seismic events with magnitude >3 occurred at January 6th, January 16th and February 2nd in 2007. The ground vibrations were accompanied by acoustic noise (bangs) and were felt and heard by many frightened people in the city.

The tremors have been caused by the hydraulic stimulation works by the geothermal project and are termed induced earthquakes following seismological terminology. Most of the monitored seismic events were in the range of microseismicity (Fig. 10.2). Microseismicity is the desired response of the fractured rocks

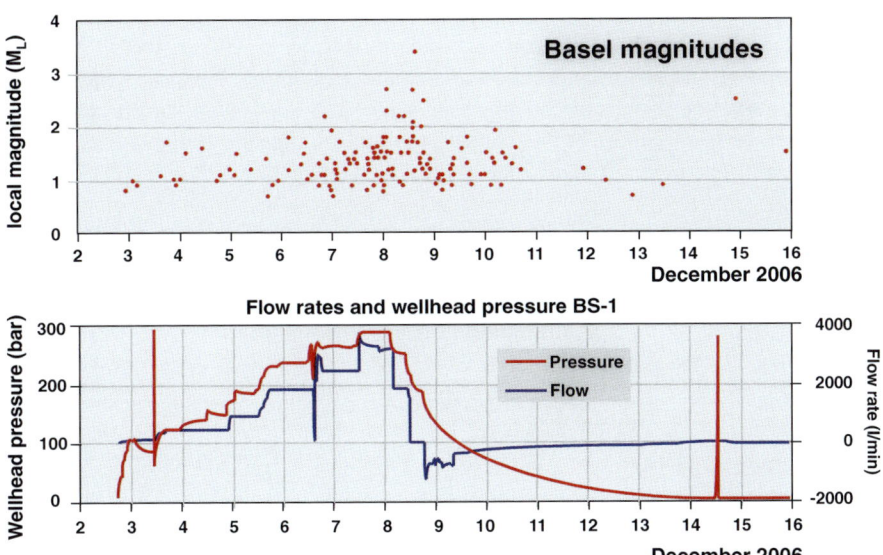

Fig. 10.2 Recorded seismic events induced by hydraulic simulation of granite at 5 km depth of the EGS project Basel (Kraft et al. 2009)

to widening the aperture of joints and fractures for generating the underground heat exchanger. Some of the seismic events (Fig. 10.2) were above the tolerable microseismicity and represent accidentally induced stronger partly noticeable quakes. The transition between the two types of seismicity is continuous.

In the run-up to the stimulation works a pre-stimulation test had been carried out at November 25th, 2006. During the test water was injected to the wellbore at increasing rates starting with 3 l/min, then 6 l/min and finally 10 l/min. In the process the wellhead pressure increased from the initial 15 to 33 bar, 52 bar and finally 74 bar. The test reflects the natural hydraulic response of the granitic bedrock at conditions below the opening pressure and suggests that the hydraulic conductivity is about 10^{-10} m/s. This is a relatively low hydraulic conductivity compared to other locations in crystalline basement at similar depths (Ladner et al. 2008; Stober and Bucher 2007a, b).

Based on the results from the pre-stimulation test hydraulic stimulation of the granite started December 2nd, 2006. In the run-up, the procedural details have been hotly debated. Stimulation works ended December 8th, 2006. A total of 11,566 m^3 water was injected. The injection rate had been increased in 5 steps to a maximum of 3,750 l/min. Towards the end of the injection procedure the wellhead pressure reached a maximum of 296 bar. Until about 2 o'clock local time of December 6th the injection rate of 1,800 l/min and the generated wellhead pressure of 250 bar was not exceeded. Until then monitored seismic events were below magnitude 2 (Fig. 10.2). After increasing the injection rates to 3,000 l/min and a corresponding wellhead pressure of 275 bar the first events with magnitude larger than 2 occurred. A further increase of the injection rate to 3,750 l/min and pressures of close to 300 bar resulted in an increasing number of events with magnitudes >2.5 (Fig. 10.2).

Injection has been gradually reduced and stopped at the 8th of December. The opened wellbore spewed a large amount of water and after a few hours the magnitudes of the quakes decreased considerably below 2.0. One event with a magnitude above 2.0 has been recorded at December 14th shortly after the well has been apparently closed ("shut-in") and the wellhead pressure has again increased to 285 bar.

The data (Fig. 10.2) suggest a correlation between injection rate or injection pressure and the magnitude of seismic events. This observation in turn implies that properly adjusted hydraulic parameters may control induced seismicity. Critical for answering the question of controllability of quakes by hydraulics is the understanding of quakes of magnitude 3 and more that followed in January and February 2007 at the Basel site.

The stronger earthquakes observed at the EGS project Basel (magnitude > 3) were presumably caused by releasing tectonic shear stresses stored in existing shear zones by hydraulic stimulation. It is very unlikely to detect such a loaded shear zone during pre-drilling exploration or with hydraulic tests. However, during the stimulation works is perhaps the best chance to detect potentially dangerous seismogenic shear zones in good time and sensibly react on the threat.

The induced Basel earthquakes may have had a stabilizing effect in the sense that they may have prevented a much stronger natural earthquake in the future.

However, the injected large volume of water that still resides in the ground may also have an opposite destabilizing effect. Fluid migration along pressure gradients may induce further quakes in the future, though probably with decreasing magnitudes. The hydraulic and seismic details of the Basel events suggest actually a destabilizing effect of the residual injection fluids (Langenbruch and Shapiro 2010). The observed late post-stimulation seismicity mirrors earthquake aftershocks following the law of Omori (1894).

The induced seismic events at the Basel EGS site did not cause special or unusual effects at the surface. The recorded seismic vibrations and the macroseismic perceptions were characteristic and typical of earthquakes with these magnitudes and focal depths. Also the sensed acoustic signal (bang) is a typical near-field earthquake phenomenon if high-frequency seismic signals are involved. Unique for the Basel incident was, however, that the epicenter was located in an extensive urban agglomeration, that the tremors were man-made and that the population has been caught off-guard and completely unprepared by the earthquakes.

10.1.4 Observed Seismicity at Other EGS Projects

High hydraulic pressures are necessary to expand the natural fracture network for creating the underground heat exchanger. The so-called opening pressure depends on the lithostatic pressure and the orientation of the controlling fracture and fault system and must be surpassed. Significant increases of flow rates begin above the opening pressure. Much higher pressures are needed if a natural fracture network is not present and must be created by hydraulic methods. Prior to 2000, EGS projects created and stimulated the subsurface heat exchangers without seismological monitoring.

Stimulation work has activated fractures extending for several hundred meters. The strongest induced earthquake caused by geothermal system development reached magnitude 3.7 (Richter scale). Personal or ponderable material damage is not known from any of the incidents.

At the EGS project Urach (SW Germany) the first stimulation experiments have been carried out in the late 1970s in a 3,300 m deep wellbore. 640 bar wellhead pressure and an injection rate of 1,200 l/min have been reached. Some years later maximum wellhead pressure even reached 660 bar during the continued stimulation experiments. Seismicity has not been monitored at that time. However, no reports on sensed tremors, acoustic signals or earthquakes have been recorded.

The thermal spa Bad Urach, in direct proximity of the EGS wellbore, produces hot water from the Triassic Upper Muschelkalk strata at 650–700 m depth. No evidence for any impairments, damages or irregularities have been observed by the spa. During a further period with stimulation experiments (2002) wellhead pressures of about 350 bar and injection rates of 600 l/min have been applied. A seismological monitoring network has been established and observed microseismicity as a response to stimulation. Microseismicity was of low magnitude and reached a maximum magnitude of 1.8 in one single event. The opening pressure in the gneiss basement of the EGS project Urach is 176 bar.

Injection experiments with very high wellhead pressures of 420 bar in the drillhole Horstberg in the North German plains produced no impairments and caused no tremors with a measurable magnitude.

Stimulation experiments have also been carried out in the 4,421 m deep wellbore Habanero 1 in the Cooper Basin of Australia in 2003. More than 20,000 m^3 water have been injected at a flow rate of 40 l/s and a resulting overpressure of 350 bar. The resulting vigorous microseismicity has been interpreted to originate from a sub horizontal 2.0×1.5 km sized structure of 150–200 m thickness. The seismic monitoring recorded a total of 12 macroseismic events with magnitudes between 2.5 and 3.7. The spatial distribution of the seismic events suggests a shear slip mechanism on an existing fault zone that has been released by fluid-reduced normal stress (Baisch et al. 2006). The interpretation has been confirmed by a new deep drill hole with a target 500 m distant point from Habanero 1. Habanero 2 drilled through a highly conductive fracture zone at 4,325 m depth.

Habanero 1 has been stimulated again in 2005. A total of 22,500 m^3 water was injected at rates of up to 31 l/s and a resulting overpressure of 270 bar (Baisch et al. 2006). This test was run with lower injection rates and consequently with lower overpressures at the wellhead. Only three stronger seismic events occurred and reached magnitudes of 2.5, 2.9 and 3.0. The early events of the 2005 stimulation occurred at the rims of the previously stimulated volume whereas the volume close to the wellbore has not been seismically reactivated and remained inactive.

Stimulation measures at Soultz-sous-Forêts in the Rhine rift valley reached a maximum wellhead pressure of 180 bar at injection rates near 50 l/s (Baria et al. 2006). A maximum magnitude of 2.9 resulted from the fluid injections. In comparison Basel: flow rate 63 l/s, pressure 300 bar, and magnitude 3.4.

In the Soultz EGS area there are 2 exploration drill holes (GPK1, EPS1), 3 bores for seismic monitoring (4550, 4601, OPS4), 3 deep (5,000 m) geothermal wellbores (GPK2, GPK3, GPK4) and a large number of wellbores of the oil and gas industry. About 25 km of the total length of all these bores have been drilled in granite. In none of all these drillholes signs of seismicity induced by the drilling process have been observed.

The underground heat reservoir at the Soultz EGS site has been hydraulically stimulated during several measures. The "upper reservoir" at 3,500 m has been stimulated in the wellbore GPK1 in 1993 and in the wellbore GPK2 in 1994 and 1995. The "deeper reservoir at 5,000 m has been stimulated in the borehole GPK2 in 2000, in GPK3 in 2003 and in GPK4 in the years 2004 and 2005 (Gérard et al. 2006). Hydraulic stimulation produced several thousand seismic events with magnitudes <2 up to a 2.9 maximum. All events with magnitude ≥ 2 occurred during the shut-in phase (Genter et al. 2010). Some data on the volumes of injected water and resulting magnitudes of seismic events are summarized for the stimulation works in the deeper reservoir in Table 10.2. The monitored seismicity can be related to shear slip on existing fault and fracture surfaces. No indication for extensional fractures was ever found.

10.1 Seismicity Related to EGS Projects

Stimulation experiments in wellbore GPK4 in 2005 with an incrementally increasing injection rate caused only about 200 seismic events.

The induced seismicity during chemical stimulation is also clearly weaker than during purely hydraulic stimulation (Table 10.3). Three different types of chemicals are usually applied in chemical stimulation (Portier et al. 2007; Genter et al. 2010):

- Regular Mud Acid (RMA) is a chemical dissolving silicate minerals such as clay, feldspars and micas
- Nitrilo Triacetic Acid (NTA) dissolves calcite and some other carbonates
- Organic Clay Acid (OCA) is temperature resistant and is applied in clay-rich formations

No seismic events occurred during the hydraulic circulation test of 4-month duration between GPK1 and GPK2 in the upper reservoir. Hydraulic circulation of several months duration in the deeper reservoir induced measurable seismicity, which was significantly weaker than during the stimulation phase. During circulation tests both flow rates and pressures were much lower, of course, than during stimulation. A few slightly stronger events (Table 10.4) with higher magnitudes have been recorded. They occurred always during the shut-in phase. All observed seismic events occurred within a distinct zone of the reservoir.

Table 10.2 EGS Soultz-sous-Forêts, hydraulic stimulation of the deeper reservoir (5 km)

Wellbore (Year)	Injected volume (m^3)	Flow rate max. (l/s)	Wellhead pressure max. (bar)	Number of induced seismic events	Magnitudes
GPK2 (2000)	~23,400	50	130	~14,000	75 × ≥1.8 2 × 2.4 1 × 2.6
GPK3 (2003)	~3,400	50; 60 & 90	180	~22,000	43 × ≥ 1.8 2 × 2.7 1 × 2.9
GPK4 (2004)	~9,300	45	170	~5,800	3 × ≥ 1.8 1 × 2.0
GPK4 (2005)	~12,300	45	190	~3,000	17 × ≥ 1.8 1 × 2.3 1 × 2.6

Table 10.3 EGS Soultz-sous-Forêts, combined chemical and hydraulic stimulation of the deeper reservoir (5 km)

Chemicals used	Date	Flow rate max. (l/s)	Number of induced events	Magnitudes
RMA[a]	May 2006	28	~20	M ≤ 1.9
NTA	October 2006	40	–	–
OCA	February 2007	55	~80	M ≤ 1.5

[a]*RMA* regular mud acid, *NTA* nitrilotriacetic acid (chelatant), *OCA* organic clay acid

Table 10.4 EGS Soultz-sous-Forêts, observed seismicity during circulation tests in the deeper reservoir (5 km)

	July–December 2005	Jul–August 2008	Nov–December 2008
GPK2 production rate	~12 l/s	~25 l/s	~17 l/s
GPK3 injection rate	~15 l/s later ~20 l/s	~23 l/s	~12 l/s later ~27 l/s
GPK4 production rate	~3 l/s	–	~12 l/s
GPK3 max. wellhead pressure (bar)	40 later 70	73	28 later 86
# seismic events	~600	~190	53
Max. magnitude	2.3	1.4	1.7

10.1.5 Conclusions and Recommendations Regarding Seismicity Control in Hydrothermal and Petrothermal (EGS) Projects

Seismic events may also occur at very shallow focal depths of 1–2 km in sedimentary sequences. This has been well documented in many studies. In sensible areas seismic events can be induced by very minor excursions of the stress field or the hydrogeological conditions. However, these kind of seismic events are relatively rare. Anyhow, the possibility for induced seismicity can never completely be excluded even if very smooth procedures are applied in developing a geothermal system.

Seismic risk assessment must strictly distinguish between the several phases of geothermal system development: (a) the drilling phase, (b) wellbore cleaning and hydrogeological efficiency-boosting measures, (c) massive hydraulic stimulation of the reservoir, (d) operation phase of the system.

The method of massive hydraulic reservoir stimulation is used routinely by the oil and gas industry (Sect. 9.2). During the many decades of wellbore drilling by the hydrocarbon industry induced seismicity has never been observed during the actual drilling process. To our knowledge, drilling related seismicity has not been reported or documented in the international literature.

Induced seismicity is unlikely to occur in typical hydrothermal systems at shallow depth (<1 km) and low temperatures. Microseismic events may inherently occur in hydrothermal reservoirs at greater depth and elevated temperature due to cooling by injection fluid or caused by pressure changes resulting from varying operating conditions. The seismicity originates on local fracture and fault zones. Fluid reinjection into the reservoir may alter the existing stress pattern in the underground and induce microseismic events. These events release very little energy, are of very short duration, have a high vibration frequency and a very low magnitude (Majer et al. 2008). Consequently the tremors remain unnoticed at the surface.

Production of thermal water from an aquifer may potentially induce seismicity after a long operating time. Fluid extraction reduces pore pressure in the aquifer and increases the effective load. Production of hydrocarbons in oil and gas fields is often associated with massive local subsidence at the surface. The problem is insignificant in geothermal doublets because the heat transfer fluid is re-injected to the aquifer in a closed loop. Even if the transfer fluid circulation is not completely

10.1 Seismicity Related to EGS Projects

closed the potential volume losses are small and insufficient to induce seismicity or subsidence.

Induced seismicity resulting from thermally induced stresses generated by injection of cold fluid into hot rock is potentially possible but has never been observed. It is also unknown in the oil and gas industry.

Well improvement measures in hydrothermal system development include well deepening, inclined drilling, deflected drilling (side tracks), directional boreholes, buildup of hydraulic overpressure, shocking and (pressure) acidizing (in limestone). The mature methods are routinely used in drinking water, mineral water and thermal water well development. In contrast to EGS development, the application of overpressure has the purpose to improve the hydraulic connection of the borehole to the aquifer and the fracture network respectively and not to generate a fracture network for the underground heat exchanger (Sect. 8.5). Consequently the applied overpressures are much lower for improving hydrothermal wells.

In recent years massive hydraulic stimulation has been used in some hydrothermal and fault system projects. Usually the stimulation attempts have been made as a last-ditch effort when the expected highly conductive aquifer proved to be an aquitard or when the drilled fault zone had a lower hydraulic conductivity than it was hoped for.

Assessing hydraulic stimulation in an aquifer is very different from considering stimulation in a fault zone. Massive hydraulic stimulation of a major fault and fracture zone bears the potential danger of spontaneously releasing stored stress that may be possibly cause a significant seismic event. Failure of seismogenic faults causing a natural earthquake requires that shear stress exceeds a certain threshold value, which is given by the normal stress, the friction coefficient of the fault system and the shear strength of the rocks. Fluid injection into a stress loaded fault zone may trigger a natural earthquake because the fluids alter the controlling parameters, cohesion, friction and normal stress and the system may prematurely fail. It is therefore highly advised to avoid extra sensitive zones or treat them with great care and prudence. Continuous seismic monitoring and real-time modeling of the underground is strictly mandatory.

Slow migration of injected water in the shear zone and slow pressure build-up by heating of the injected cold water may favor retardation of abrupt stress release in the shear zone.

Massive hydraulic stimulation is the key method for developing a plant based on EGS that has been used since the 1970s. The method has also been the subject of considerable research efforts. Still many physical processes and controlling parameters that influence the abundance and distribution of induced seismic events and their magnitude are not yet fully understood (Kraft et al. 2009). During massive hydraulic stimulation injection rates of some thousand l/min water and wellhead pressures of several hundred bars are commonly applied. Critical for inducing seismicity is, however, the local tectonic shear stress at reservoir depth. The released seismic energy originates predominantly from stored natural deformation energy. The specific interactions between the controlling parameters are not precisely known, neither qualitatively nor quantitatively. Therefore, it is inherently impossible to reliably and quantitatively predict the risk for inducing seismicity already in the planning stage of an EGS project.

The widening of fractures and joints in the geothermal reservoir by stimulation measures needs to be irreversible. The improvement of the hydraulic conductivity must be permanent and stable. This fundamental demand requires shear slip during stimulation; pure elastic deformation is reversible (Stober 2011). Chemical stimulation and the use of proppings (Sect. 9.3) may keep fractures open in the absence of shear slip deformation.

Geothermal reservoirs with high hydraulic conductivity and good storage capacity are capable of taking up injected fluids at relatively low wellhead pressures. Consequently they are generally less susceptible to strong induced seismic events. Prominent geologically young fault zones in the vicinity of injections bear a significant potential for a strong induced seismic event, particularly if they have a low conductivity and a low storage capacity. The injected fluid then migrates preferentially along the fault zone and unlocks stored shear stresses. The likelihood of triggering an earthquake decreases if major fault zones are absent in the neighborhood of the wellbores. Generally regions with high natural seismicity are often also strongly faulted and fault zones are abundant. Thus such regions have clearly a higher risk for inducing seismic events during EGS development (Nicholson and Wesson 1990).

It is recommended to develop a geothermal energy project following methodically a structured succession of phases. The objectives of the phases include creating: A risk assessment study, a risk mitigation and response plan, a seismic monitoring plan, an emission measurement network and a monitoring system for the hydraulic procedures. It must be strictly distinguished between planning for a geothermal project that requires hydraulic stimulation (petrothermal, EGS) and projects that don't such as hydrothermal systems.

It is very important to provide the public with comprehensive, honest, serious and professional information about the project, particularly if it is near villages or towns. Information must be continuous from the initial planning, through development and operation.

10.2 Interaction Between Geothermal System Operation and the Subsurface

Settling and subsidence effects in the land surface are occasionally associated with extraction of large amounts of fluids from the underground. The effects are well known in the hydrocarbon industry and also from drinking water abstraction areas. Land surface subsidence is a slow process that usually affects a large area and is often partly reversible.

Several meters of surface subsidence resulted from high drinking water abstractions in the San Joaquin Valley near the city of Mendota in Fresno County (California, USA). In the period 1925–1975 subsidence added up to nearly 9 m. Subsidence rates of up to 5 cm per year have been observed in several areas in the USA with high groundwater abstraction. Massive subsidence may go together with faulting (Johnson 1991). Decimeter to meter scale subsidence is commonly observed in the major oil and gas producing regions of the USA. An example is the Diatomite oil field in Kern County, California (Bondor and Rouffignac 1995). A further example is the Slochteren gas field in the Nederlands, which is in production since 1960. Settling of 30 cm in an area of 250 km^2 resulted from the long-term gas extraction.

One of the most sizable subsidence occurred in the Wilmington oil field, Long Beach (California), which came to nearly 9 m. Highest subsidence rates have been measured with >40 cm per year in San Joaquin Valley, California (Fielding et al. 1998). Subsidence is often accompanied by self-sealing of the wellbores and a resultant reduction of the production rate. Both effects are a consequence of the lowered pore pressure caused by fluid extraction. From this the subsequent increase of load reduces the pore volume (porosity), which results in subsidence. Subsidence is an unwanted incident. It has been combated in oil and gas fields by injecting water or gas (usually CO_2) into the reservoir. The method also stopped self-sealing of the wellbores and partly reversed the process. Subsidence halted and was partly compensated by subsequent uplift.

Subsidence is a potential threat also to geothermal systems where large amounts of thermal water are extracted from a deep reservoir for geothermal utilization. However, most geothermal systems are being planned as doublet systems with a quantitative recycling of the produced thermal water by reinjecting the fluid back to the reservoir. The closed primary fluid cycle regenerates the reservoir and prevents leaking of highly saline deep fluids to the surface environment. The closed primary fluid cycle requires a hydraulic connection between production and injection well. Therefore production-related pore pressure drop and injection-related pore pressure increase is typically restricted in doublet operation to the immediate vicinity of the respective wellbores. All together the hydraulic consequences are clearly more moderate than for pure production or injection only operation.

Radioactive elements are globally present in many minerals and rocks. Particularly granites and granite-derived gneisses contain minerals with abundant naturally radioactive elements such as uranium, thorium and potassium. In granites the zirconium silicate, zircon, incorporates U and Th in its structure (a favorable circumstance that allows isotopic dating of the granite). Decay of uranium in the structure of accessory zircon in granite produces a series of decay products including ^{226}Ra, ^{210}Po and ultimately a series of stable lead isotopes. Typical subsurface heat exchangers in deep HDR systems are developed in granitic reservoir rocks. Interaction of the fractured granite with the heat transfer fluid also transfers some of these "naturally occurring radioactive materials" (NORM substances) to the fluid in addition to the thermal energy. Consequently the thermal water pumped to the surface through the production well is naturally radioactive to variable degrees. In some cases the concentration of NORM in the fluid is critically high. Inappropriate disposal of NORM waste implies a significant risk for health.

Natural radionuclides are present in all deep waters of the upper continental crust. The total activity of the waters depends on the rock types building up the reservoir and varies over a wide range. As mentioned above granites and granite-derived rocks are typically relatively rich in radionuclides. The principal radioactive isotopes are: ^{226}Ra, ^{210}Pb, ^{228}Ra, ^{224}Ra, ^{40}K (Faure 1986). Radium 226 has a half live of 1,600 years. From an environmental point of view the isotope is relatively long lived. It decays to radon, a radioactive noble gas, which also may cause environmental troubles.

The risk of radioactive materials leaking to the surface environment is small in geothermal systems operated with a closed system primary cycle. Extracted radioactive elements either dissolved in the heat transfer fluid or as suspended solid particles in the fluid are reinjected into the reservoir in doublet systems. Precipitation and deposition of hazardous materials in the surface installations can be prevented by optimized operation of the system and by precipitation inhibitors added to the produced fluid. However, there remains the potential that radioactive solid deposits and scales may form in surface installations, particularly in pressure shadows of pipe junctions and pipe bends, also in the heat exchanger, filters and pumps.

Barite—celestite scales are relatively widespread (Chap. 14) and the ((Ba, Sr)SO$_4$) solid solution crystals may exchange Ra on the Ba sites of the crystal structure because of chemical similarities between the two elements. Thus deposition of low solubility barite may co-precipitate radioactive radium isotopes (^{226}Ra, ^{228}Ra, ^{224}Ra). Scales of galenite (PbS) and native lead (Pb) are also common and may incorporate the chemically identical radionuclide ^{210}Pb. To sum up, radioactive solutes of produced thermal waters can be considerably enriched in scales and become a serious disposal problem.

Scale deposits should be given serious attention. Scales in replaced pipe sections, pumps, filters and heat exchangers need to be analyzed and appropriately treated if necessary and responsibly disposed of.

10.3 Environmental Issues Related to Surface Installations and Operation

Geothermal system development considers potential environmental hazards well in advance and takes the necessary and suitable precautions. For instance, drilling mud containing commonly problematic organic components must be adequately and safely disposed. For collecting the commonly highly saline or even toxic deep fluids that have been brought to the surface during pumping tests appropriate collecting tanks must be made ready if they cannot be reinjected to the deep reservoir immediately. Cooling fluids contain softener, biocides and anti-corrosion chemicals and must be properly disposed.

During drilling but also later during regular operation the geothermal plant generates noise emissions. These must be carefully considered and mitigation measures should be planned particularly if the plant is close to inhabited places. Effective noise abatement measures reduce noise emissions efficiently. However, the improvements require investments into appropriate technical installations at extra costs.

Drilling with electrical motors is more quiet than with fuel driven motors. Noise protection barriers and walls around the drilling site reduce noise efficiently.

Air fin coolers are noisier than water-cooled systems (Fig. 10.3). If air-cooling is necessary, low-speed air fin coolers are more favorable with respect to noise emissions. Also running turbines make noise. Noise-induced annoyance can be reduced by thoughtful structural engineering and by creating green areas. Of some environmental concern is also the heat released by the heat transfer fluid during re-cooling of the fluid before reinjection. Anticipated noise emissions are one of the principal concerns of the public facing plans for a geothermal project.

Fig. 10.3 Example of an air cooling system: **a** Overview, **b** Rotor, view from below

Impact on landscape by pipelines can be avoided by putting them underground. This is more expensive than surface pipelines but it considerably increases the acceptance of the geothermal project in the public.

In the past dry-steam high-enthalpy geothermal power plants released steam to the atmosphere after the turbine unit with considerable odor nuisance. Today, the condensed and cooled steam is re-injected into the heat reservoir. This reduces or completely prevents obnoxious odors from reaching the air and improves the productivity of the reservoir by maintaining a sufficiently high fluid pressure in the reservoir. Reinjection of the produced thermal fluids is standard today in all geothermal power systems.

At temperatures below 200 °C special working fluids are used in the secondary cycles for the production of electrical power. Systems on the basis of the Organic Rankin Cycle (ORC) utilize organic working fluids, pentane for example. Kalina systems use an ammonia-water mixture as working fluid. It is necessary to be prepared for a damage event and the appropriate security concepts, installations and equipment must be operational to prevent environmental hazard and hazards on the facility's grounds. Precautions against chemical hazards are routine in the chemical industry.

Chapter 11
Drilling Techniques for Deep Wellbores

Top drive of a deep drilling rig

Drilling costs stand for about 70 % of the total costs of a deep geothermal project. The drilling technique used in deep geothermal projects has been adopted for the most part from the oil and gas industry. The drilling technique used in geothermal projects, however, must satisfy higher requirements because of the combination of high temperatures, high volume fluxes and typically high concentrations of aggressive and corrosive solutes in the produced fluid. Borehole diameters are larger because of the high volume fluxes. In contrast to oil and gas wells, wellbores in the geothermal industry must provide evidence for an operation life of 30 years. Geothermal wells pump hot salty fluids directly along the casing to the surface. In contrast, oil wells produce hydrocarbons along a liner protecting the casing. The costs for a deep drillhole in the geothermal industry is higher by a factor of 2–5 compared to boreholes in the oil and gas industry (Teodoriu and Falcone 2009).

Drilling and lining deep geothermal wellbores is a very complex and demanding affair. It requires interaction and cooperation of many different high-quality professionals and specialized service companies. Also the requirements to the pumping technique and equipment are extremely demanding because of the high temperature saline fluids that normally need to be pumped in geothermal systems (Sect. 14.3). In this book we try to give a brief overview over the subject only. Specialized textbooks dealing with drilling technique include Bourgoyne et al. (1986) and Aadony (1999).

Deep boreholes are normally drilled in shift operation 24 h/d without interruptions. An optimized logistics at the wellsite makes provisions for sufficient storage areas for drill pipes, casing, spare parts, cuttings, drilling mud and consumables. If the wellsite is located near inhabited buildings noise protection must be organized.

The type of lining of geothermal wells depends on the type of geothermal system (deep geothermal probe, hydrothermal well, EGS borehole) and on the actual lithological and hydraulic conditions. The drilling process of deep geothermal wells is subdivided into a series of drilling phases marked by the installation and cementing of the casing. The diameter of the wellbore and the casing decreases step-by-step at each new drilling phase (Fig. 11.1).

The structure of all deep wells is thus tapered. The planned final diameter of the wellbore at depth depends on the required or aspired flow rate. The needed final diameter at depth, the geological stratigraphy and the planned final depth define the initial diameter which needs to be started with at the surface. For a bore with four drilling phases typical diameters are for example: An $18^{5/8}$ inch surface casing is drilled with a 23 inch drill bit, a $13^{3/8}$ inch casing is drilled with a 16 inch drill bit, a $9^{5/8}$ inch liner with a $12^{1/4}$ inch drill bit and a $8^{1/2}$ inch drill bit is used for the open hole section. If the reservoir consists of stable rocks a casing is not needed in the reservoir section (open hole). Unstable rocks or rock fragments in the fluid require the installation of perforated liner or filter pipes (cased hole) (Devereux 2012; Bourgoyne et al. 1986). It is very important that the hydraulic conductivity is not permanently and irreversibly reduced (skin) by the drilling operation or the drilling mud in the reservoir section (Chap. 13).

At the surface, drilling begins with a large diameter. The diameter successively tapers from phase to phase downward. The needed fluid flow rate, as mentioned,

11 Drilling Techniques for Deep Wellbores

Fig. 11.1 Well design of a deep bore (example)

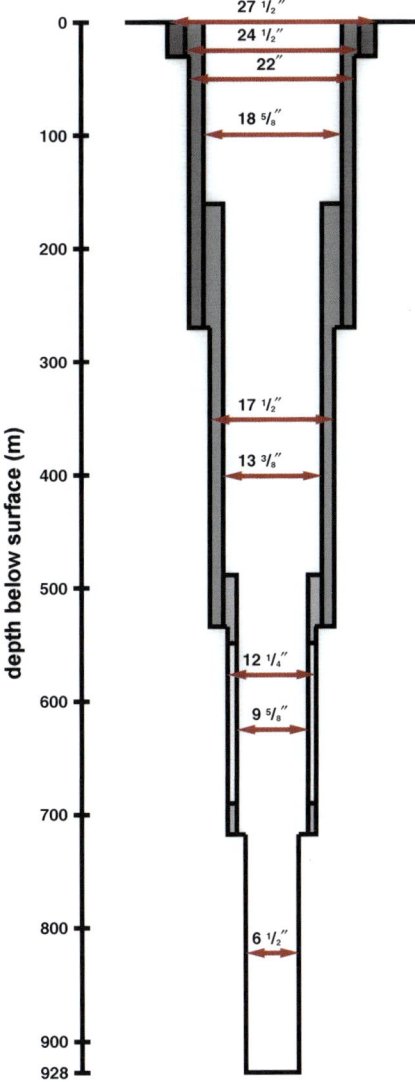

defines the minimum diameter of the open hole section and thus the overall structure of the wellbore. However, within these conditions some other aspects must be considered. The diameter must be large enough for trouble free installation of the casing and that sufficient space remains for a good quality cementation. Also the diameter should be large enough to minimize friction losses. On the other side, large diameter wellbores require larger drilling rigs and more energy and material resources. In short, they are more expensive. Drilling costs are roughly proportional to the volume of drilled rock.

Surface casing, conductor casing, casing and liner are needed to prevent instability of the wellbore. Liner are not installed to the surface but are mounted at a landed casing. The casing protects the wall of the bore and seals the well, together with the cementation, from fluid-bearing layers other than the target reservoir. This also makes it possible to separate possible layers with different hydraulic potential. The pipe material and the connectors must be pressure-resistant and must have a high tensile strength at the same time. Both properties drop off with increasing temperature. Therefore, careful pre-drilling consideration and engineering calculations of expected pressures on the casing (external and internal pressure), the anticipated loads such as the weight of the casing string, torsional loads from directional drilling and sidetracks, grinding loads, compression loads and others are mandatory. The high temperature at the base of deep geothermal wells causes the casing of the production well to expand. The casing of the injection well tends to shrink during plant operation. The well engineering must consider the effect of the thermal parameters on the casing to were the casing is not damaged during expansion or contraction.

After fitting the casing cement suspension is pumped into the annulus from bottom to top replacing progressively the drilling fluid. Cementation of a deep drill hole requires special cement properties and special preparation efforts. Cement is a powdered, hydraulic mineral binder that if mixed with water hardens to concrete (in air and in water). Deep drilling uses cements with fine powdered inerts of similar grain size as fresh-water or salt-water suspensions. Bores of more than 3,000 m depth require special mixtures of cement, inerts and chemical additives and advanced recipes (Smolczyk 1968). The dry raw materials are delivered ready mixed to the drilling site. At the drilling pad the tempering water is prepared by adding NaCl, delayer and chemical additives in blending tanks. Then the powder mix is injected into the mixing water under continuous stirring.

The described mixing procedure results in a homogeneous cement slurry with minimal air bubbles. The appropriate recipe for the cement grout and the proper technique of cement injection under the conditions of the bore are crucially important for the quality of the cementation. Quality criteria for deep well cementation are: high early strength, chemically resistant, impermeability to chemically aggressive fluids and excellent binding to both the casing and the rock. This also means that the cementation needs to remain at constant volume. Casing and cementation are the key elements for a safe and long-lasting plant operation. Very useful guidelines have been released by the American Petroleum Institute (API) (www.api.org).

The cemented casing also helps to safely proceed with the next phase of drilling and allows controlling geology-related over- and under-pressure relative to hydrostatic pressure. The position of the casing during cementation must be strictly centric in the bore thereby ensuring that it is completely surrounded by cement. Else the cementation may remain incomplete and cavities with drilling fluid persist. The centric positioning of the casing is brought about by centralizer elements if the annulus is sufficiently large. If the annulus is narrow centralizer fins made of e.g. carbon fibers are being mounted or painted directly on the pipes before placing in the bore. Special centralizing efforts are necessary for directional drilling and sidetracks.

The casing requirements are high; it must resist all stresses imposed by the regular operation of the plant for the entire life cycle of the system. It also must resist the enormous strains caused by chemical and hydraulic reservoir stimulation. If necessary, the casing of the production well must be protected from aggressive and corrosive fluids by using (expensive) corrosion-resistant materials. At $T < 120\ °C$ and $p < 250$ bar, corrosion-prove fiberglass-enforced plastic pipes can be an option.

The hook load describes the size of operation of a drilling rig. It is the total load that can be held or lifted with the rig. The hook load limits the maximum drilling depth and the drilling and finished diameter of the bore. Hook loads of 150–500 t are required for 2,000–6,000 m deep wellbores. The essential components of a deep drilling rig are shown on Fig. 11.2. Figure 11.3 shows the some of the components of Fig. 11.2 from a specific drilling site. Figure 11.4 shows an example of drilling operations from a very remote place in western China.

Modern drilling rigs (Bjelm 2006; Hole 2006; Binder 2007) use the rotary drilling technique. Rotary drilling can only be used for drilling deep wellbores. Diesel-electric power supply drives the rotary table, the drill pipe and the drill bit at depth. The driving motor is located high up in the rig and drives the drill pipes from top.

Another drilling technique is turbo drilling. In this technique a turbine drives a drilling bit at depth in the wellbore. This technique is mainly used in directional drilling.

The drilling string consists of drilling pipes of about 9 m length each (fitted by special connectors). The drilling string is permanently adjusted to the appropriate pressure by a qualified operator (Fig. 11.4c). Immediately above the drill bit heavy weight drill pipes are typically mounted to increase the weight onto the drill bits (Fig. 11.5).

Roller cones and diamond bits are used as cutting tools (Fig. 11.6a) in rotary drilling. The tools must be optimized for removing and transporting the cuttings in the respective geological formation. For coring the formation different bits must be installed (Fig. 11.6b). Diamond bits have a longer lifetime and no movable parts compared to roller bits. Fine grained sediments must be drilled with different bits than granitic basement (Fig. 11.6c).

Directional drilling is required if the wellbore needs to reach a well-defined target area such as a major fault zone that is expected to be highly water conducting. Typically drilling begins vertically. The wellbore deflection begins at a certain depth, at the so-called kick-off point (KOP). The slope then gradually decreases and the deviation from the vertical increases progressively. The length of the drilled section and the vertical depth differ steadily. The total length of the drilled section after completion of drilling is designated as "final depth". This may cause confusion in regards to interpretation of temperature data, hydraulic tests and other parameters and it needs to be correctly recognized that this is not a vertical depth (if drilling is not exclusively vertical). One must be aware that modern deep wellbores may have a drilled section that exceeds the vertical depth massively.

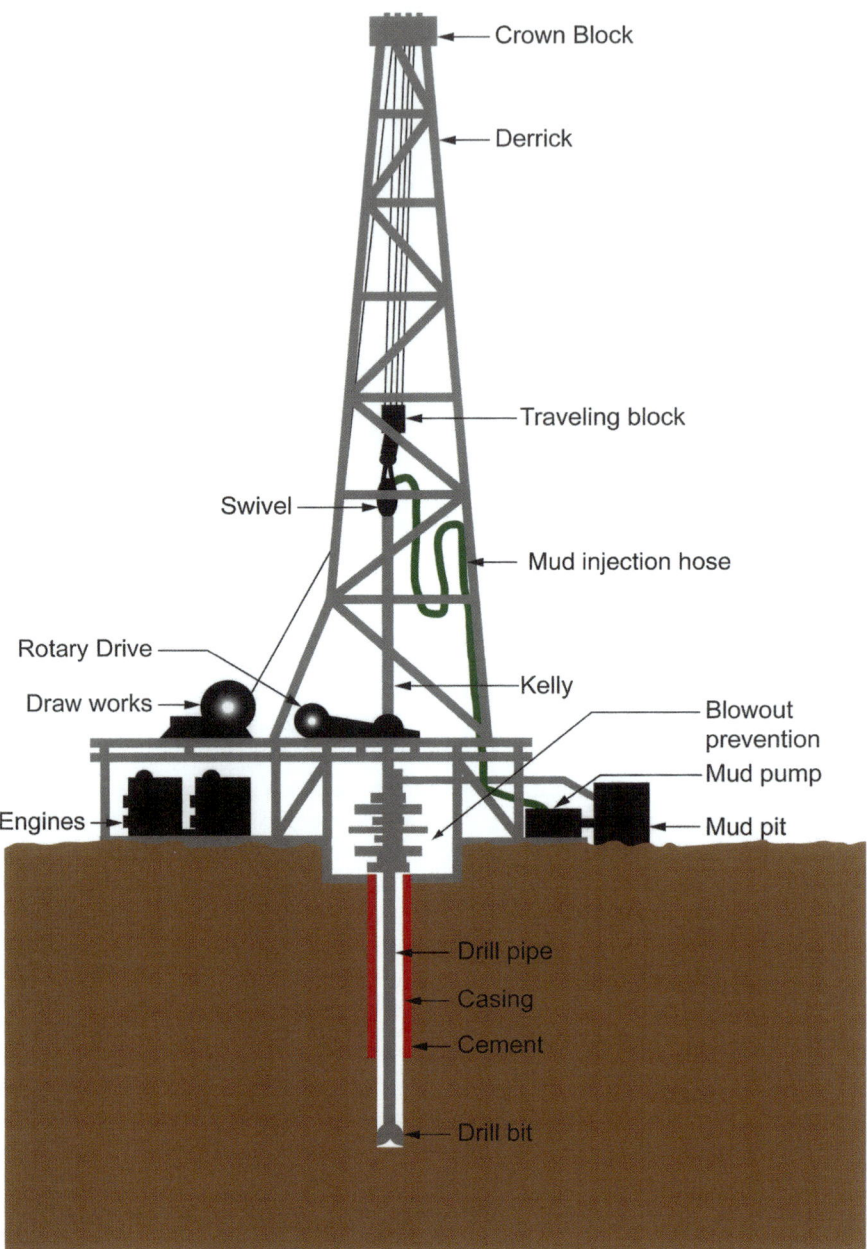

Fig. 11.2 Schematic view of a deep drilling rig

The drilling drive is placed below surface, as a down hole motor directly above the drilling bit. Directed drilling motors have an adjustable bend at the case. Deflected drilling drills operates without turning the entire drill string. The drill bit slowly and continuously deviates from the vertical drilling axis and it describes, because of the

Fig. 11.3 Rig with hook (*yellow*)

bend in the housing, a curved trajectory. Drives for directed drilling have a strong wear at the knee joint. The position of the drill bit is continuously recorded during drilling and "measurement-while-drilling" devices (MWD) transmit the data to the surface (Reich 2011). If the data analysis reveals a discrepancy between the actual and the planned drilling path, appropriate corrections are made. It is also possible to combine two motors: the downhole motor for the steering head to optimize the borehole placement and an additional second motor at the surface to optimize drilling efficiency. This relatively new technique enables to drill faster and deeper.

Rotary drilling turns the entire drill string including the drive. The drillmaster may influence the drilling path somewhat by varying the pressure on the drill bit. The drilling path can also be influenced by activating hydraulic fins. The control electronics sets hydraulically driven fins in motion and presses them against the wellbore thereby deflecting the drilling path into the opposite direction.

Planning the well drilling considers many bits and pieces including: planning details for directional drilling and defining a detailed drilling path, assigning appropriate casing materials, tailoring pipe wall thicknesses, choosing pipe connectors, specifying special thermally resistant cements, choosing appropriate down hole tools, reservoir-conserving drilling fluids, disposal of drilling mud and cuttings and choosing the right size (hook load) of the rig. The drilling location and

Fig. 11.4 **a** Drilling rig in the Qilian Mountains (West China). **b** Details of the platform of the rig. **c** Drilling camp in a remote area without noise-sensitive neighbors. **d** Safety-first applies to drilling operations worldwide

Fig. 11.4 continued

the target point in the reservoir set the frame for the drilling path and the geologically controlled casing scheme.

In the upper 500 m of the bore, provision must be made for sufficient space for mounting the production pump. Continuous operation establishes a thermal steady state with a hot environment in the production well and a cooler setting in the injection well. Frequent operation interruptions create significant thermal stress on the material (casing, cementation, etc.). This must be considered in the planning stage and appropriate thermal resistivity of the installations

Fig. 11.5 Drill collars support the load on the drill bit

Fig. 11.6 Drilling tools: **a** Rollercones for deed drilling. **b** Diamond bit for coring. **c** Drilling for drilling fine grained sediments (clays). **d** Spent drill bits

is well-invested money. Generally, the casing in geothermal wells must resist extremely high mechanical, thermal and, as will be discussed later, chemical stresses. Compressive strength can be critical for large-caliber pipes, whereas tensile strength is limiting for small-caliber casing. Typically, however, the connectors are the vulnerable part for the tensile load (Australian Drilling Industry 1997).

Friction losses (Δp) in wells depend on flow rate and diameter of the casing (Fig. 11.7). The friction related pressure drop (Δp) is proportional to the flow rate (Q) squared. Thus doubling the flow rate (Q) leads to a four-fold pressure drop (Δp). Similarly, a 15 % reduction of the flow cross-section amplifies Δp by a factor of two. Friction losses in wells with a $9^{5/8}$ inch casing and typical geothermal flow rates of 150 l/s are normally smaller than p = 20 bar. 7 inch casings, however, can be associated with friction losses of $\Delta p > 90$ bar, which may significantly threaten the profitability of the system. For flow-rates Q of about 100 l/s, expected Δp is correspondingly lower (10 and 45 bar for the considered casing diameters, respectively).

Large diameter drilling, directed drilling and securing problematic and difficult drilling sections require the detailed study of active tectonic stresses, gas influxes and reservoir-conserving resource connection. The key devices preventing uncontrolled discharges of fluid and gas are blow-out preventer, choke-manifold and drilling-spool (Fig. 11.8).

Fig. 11.7 Friction-losses Δp (bar) in a geothermal well as a function of flow-rate Q (l/s) and internal casing diameter ID (m) (BGR)

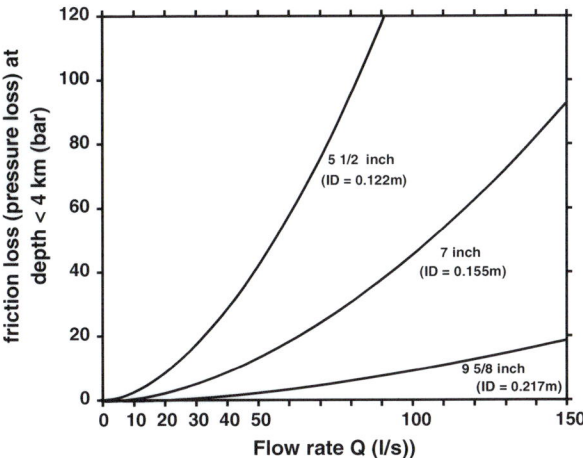

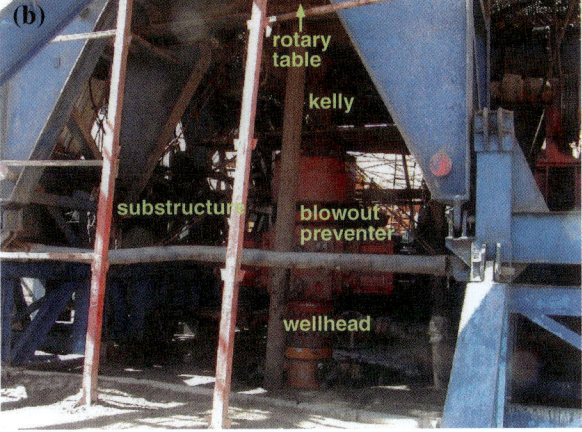

Fig. 11.8 **a** Blow-Out-Preventer waiting for installation. **b** Blow-out preventer of the Qilian Rig shown in Fig. 11.4

The very first action for the planned deep drilling is the selection of a drilling site (well site). It typically requires an area of 3,000–5,000 m^2 and access to water and electrical power. It is important that groundwater threatening fluids may not contaminate groundwater. Fluid- and waste-handling must be designed. Both standpipes for the planned doublet and the rig cellar should be defined decision of the drilling site foundation. Collection pools for drilling mud and cuttings need to be designed and brought into function. Collection tanks for the storage of highly mineralized and potentially toxic deep thermal fluids must be provided for hydraulic tests of several days of duration (Fig. 11.9). In most countries, the design and setting up of the drilling site must be coordinated and tuned with the responsible authorities.

Drilling mud performs many tasks. It cools the drilling tools, lifts the cuttings to the surface, stabilizes the bore and serves many other purposes. Deep geothermal drilling typically uses water-based drilling fluids. Special conditions may require oil-based drilling fluids or special foams. The complex site-specific requirements on drilling fluids require expert knowhow. Dedicated firms provide materials and chemicals that are mixed and added to the drilling fluid. An expert engineer supervises the operation and is responsible for optimizing the formula and monitoring changes in the produced mud. The appropriate formula for the drilling fluid also depends on the type of rock to be drilled. Clay-suspension in water is often used in the near-surface top hole because of groundwater protection concerns and because of optimal drilling progress with low-density fluids. Polymers are commonly added to increase the viscosity of the drilling fluid while

Fig. 11.9 a Storage pond for highly mineralized and potentially toxic deep thermal reservoir fluids during hydraulic tests. **b** Storage pond with inflowing thermal fluid

Fig. 11.9 continued

drilling highly permeable formations. The filter cake from this drilling fluid helps to seal the permeable formation. Special inhibitor additives significantly reduce swelling of clay in drilled shales and mudstones. Still the viscosity of the fluid remains low and the clay-bearing cuttings are rapidly removed from the bore. Barite ($BaSO_4$) is a standard stabilizer for drilling formations under pressure. Drilling fluid reduces friction between drilling string and the bore in long and particularly in directional drilling sections (Huenges 2010; Putra 2008).

The drilling mud is pumped to mud tanks (pools) above ground, cleaned and reconditioned for reuse in the borehole. The drilling fluid is also used for transmitting pressure signals and data for the "measurement-while-drilling" (MWD) technology. The mud logging firm records important parameters, such as drilling progress, current depth and position of the drill bit, load on the drill bit, torque, speed, mud load, flow rate of drilling fluid among others. During drilling, geologically trained personnel examine the cuttings and identify the current drilled formation and compare the current data with the predicted profiles from the pre-drilling exploration phase. Specialized service firms carry out the precise maneuvers of the directed drilling. Other

Fig. 11.10 Well head

expert firms handle the fitting of the pipes and the subsequent cementation. Pipes, cement and additives must be ordered, delivered and stored on site in just sufficient amounts at exactly the right time. After drilling and completion the wellbore is securely sealed with a well head (Fig. 11.10).

During drilling and after completion of the wellbore, systematic hydraulic tests provide data on the hydraulic conductivity of the target horizon and the thermal reservoir (Chap. 13). Repeatedly collected fluid samples supply the necessary information on the hydrochemical properties of the formation fluid (Chap. 14) and methodical geophysical well logging makes physical and structural rock and formation properties available (Sect. 12.2). The type and duration of hydraulic tests depend on the typical prevailing hydraulic conductivity of the formation. If the initial natural conductivity is lower than required for a successful project it may be increased with engineered hydraulic and (or) chemical methods (Sect. 8.5). If the two wellbores of a doublet and the reservoir design have been successfully completed then the two wells can be tightly connected above ground. The following circulation test of several weeks duration reveals the properties and conditions of the entire system. After switch-off of the production pump the well commonly keeps producing thermal water (Sect. 8.2). Injecting high-density salt water into

11 Drilling Techniques for Deep Wellbores

Fig. 11.11 Blending device for high-density brine. The brine is injected into the production well to stop artesian over-flow after pump-switch-off

the well may stop the unwanted artesian water. Some geothermal plants have been equipped with special devices for this purpose (Fig. 11.11).

Production pumps (submersible pumps, line shaft pumps) belong to the mechanically most strained assemblies of a geothermal power plant. Pump failures cause a long-standing unplanned outage of the entire system. Pump and motor are combined in one unit in submersible pumps. The pump is submersed in the geothermal fluid at relatively large depth in the production well (several 100 m below surface). The pump operates in a chemically aggressive fluid at high pressure and temperature. The pump itself produces additional waste heat. The electrical power supply requires a mechanically and chemically resistant cable leading through the wellbore from the surface to the pump at depth. Line shaft pumps have a pump unit that is placed in the production well at depth, the motor unit, however, is located above ground. The motor is thus not exposed to the high temperature of the produced fluid. Consequently, line shaft pumps can produce fluids with very high temperature (Fig. 11.12). Because pump and motor units are mechanically connected with a rod assembly the maximum depth for operating of line shaft pumps is about 300 m. Most geothermal plants produce geothermal fluid with submersible pumps. However, motor-failures of line shaft pumps can be fixed and repaired immediately, which is a significant advantage. Damaged submersible pumps must be recovered from depth, dismantled, repaired and finally re-installed.

Fig. 11.12 Line shaft pump

The repair is likely more time consuming and thus causes a longer full stop of the plant if no replacement pump is hold at the site.

Submersible centrifugal pumps (Sect. 4.2) have been in use for short operation periods down to 3,000 m depth and volume flows of up to 280 l/s by the oil industry. Special submersible centrifugal pumps have been operated at temperatures of up to 232 °C and pumped highly viscous fluids, fluids with dissolved carbon dioxide, hydrogen sulfide and suspended solid particles. Such extreme conditions, however, reduce the working life of the pumps dramatically (Sect. 14.3).

The efficiency of the submersible pumps depends not only on the critical components in the borehole including motor, seals and fittings, pump, sensors and cables but also on the above ground control systems. A diligent choice of fitting components optimized for the specific site conditions are crucial for the efficient, reliable and during operation of the pump system. Especially resistant materials must be used for pumping strongly corrosive fluids.

At given hydraulic conditions the well yield decreases with increasing temperature. Most pump manufacturers define 180 °C as maximum temperature at which their pumps can be operated reliably and constantly. The actual operating conditions may later differ significantly from the predicted conditions forming the basis for the chosen pump design. If so, the efficiency of the system and pump

11 Drilling Techniques for Deep Wellbores

Fig. 11.13 Injection pump

durability may be appreciably reduced. Today most of the practical technical experience with submersible pumps draws on their use in oil fields. However, in the last years experience from specially designed pump systems for use in geothermal power plants is steadily growing (e.g. Ichikawa et al. 2000).

Injection pumps (Fig. 11.13) for re-injecting cooled heat transfer fluid back to the underground reservoir can be placed above ground, in contrast to production pumps.

Chapter 12
Geophysical Methods, Exploration and Analysis

Seismic exploration

Geophysical sounding and investigations provide an indirect view into the underground. Geophysical investigations collect data using instruments at the surface or placed in boreholes. Borehole geophysics and geophysical well logging can probe and research in cased and in uncased bores. In this chapter, we briefly present a selection of geophysical investigation methods. Detailed accounts on geophysical methods have been presented by e.g. Sheriff and Geldart 2006; Telford et al. 2010.

12.1 Geophysical Pre-drilling Exploration, Seismic Investigations

Applied geophysics uses physical measuring systems for sensing the geological structure and the properties of the subsurface. The methods determine parameters and properties from diverse domains including the gravitational field, magnetic field, electrical conductivity, propagation of acoustic and electromagnetic waves. The methods are indirect, meaning that geological structure must be deduced from the collected data. The data must be geologically interpreted. The interpreted geophysical data portray the structure of the underground, the lithostratigraphy, the thickness and depth of the target formations and the position, orientation and thickness of fault zones. Under favorable circumstances the data even give evidence on the specific type of sedimentary facies of the target formation.

The very first step in geophysical exploration for a new geothermal power system is the careful and thorough search for all available geophysical data and direct and indirect information on the geology of the underground from previous exploration campaigns in the wider region of the new site. The compiled "old" data must be reviewed, evaluated and re-interpreted if necessary. The effort may make a new expensive seismic campaign obsolete or it can be significantly downsized. Digital reprocessing of "old" seismic data may significantly increase the conclusiveness of such data.

Reflexion seismology is a particularly important method for the geophysical investigation of deep geological structures. Other methods including gravimetry, geomagnetics, geoelectrics, magnetotelurics or combinations of these methods can be useful for cost-efficient reconnaissance investigations. They are sometimes needed for solving special local problems. For example for describing and evaluating recognized local seismic anomalies or seismic barriers.

Gravimetry measures the strength of the Earth gravitational field. A gravimeter is the standard instrument used by this geophysical method. It is an accelerometer specially designed for measuring the downward acceleration due to gravity. The local variations of the acceleration due to gravity reflect the density structure of the local geological underground but also the latitude and altitude of the instrument position. The geological structure and nature of the underground can thus be deduced from the detected density variations (Telford et al. 2010). For example, the seismic exploration may have discovered an intrusive body crosscutting a well-stratified sequence of presumably sedimentary rocks. Gravimetric analysis can with certainty distinguish between intrusions of high-density igneous rocks such as gabbro and low-density salt domes. Gravimetry may also discover and localize larger caves in karstified formations in the subsurface. The method may also localize the depth of the basement-cover contact surface which separates dense basement rocks from lighter sedimentary cover rocks. There are many more potential useful applications of the powerful method.

Geomagnetic surveys measure deviations from the undisturbed geomagnetic field. The data relate to the magnetic susceptibility of the geological material forming the underground. Measurements can be made from the Earth surface or airborne. The magnetic susceptibility of near surface rocks (some km) and soil deform the natural magnetic field that has its origin in the deep Earth. The local deviations from the undisturbed magnetic field, so called magnetic anomalies, result from induced magnetization of Earth materials by the magnetic field. The size of the anomalies depends on the strength and direction of the Earth magnetic field at the location in question and on the magnetic properties of the geological material, the size of especially magnetizable geological units, strata or bodies. Typically, pronounced magnetic anomalies are caused by discrete perturbing rock bodies in the underground. The classic example: Lenses of iron ore cause strong magnetic anomalies. Iron oxides, iron sulfides and similar material display an inherited permanent remanent magnetization, independent of the present day magnetic field. Geomagnetic measurements try to relate observed anomalies to the nature, shape, size and depth of magnetic perturbations (Telford et al. 2010). Geomagnetic measurements have been successfully used to locate and characterize fault zones in the crystalline basement.

Magnetotelluric surveys for deep exploration use alternating electromagnetic fields for investigating the underground. Natural or technical sources can be used as stimulating primary magnetic fields. The geophysical method can be used to explore a wide range of depths from the near surface area to the deep underground because the range of periods of the alternating fields is very large and the penetration depth depends on the frequency. The applied magnetic fields induce electrical currents in conductive geological units and these currents in turn generate electromagnetic fields. From the recorded temporal variation of magnetic and electric fields follows, after appropriate data processing, the distribution of electrical conductivity in the geological units in the crust and down to the uppermost mantle. The method has been significantly refined during the past years. However, research and development is still needed particularly in noise-reduction during data recording and refining the processing methods (Huenges 2010).

The prime exploration method for depths greater than 1,000 m is reflection seismology. Compared with other geophysical methods, reflection seismology provides the clearest and most precise picture of the structure of the subsurface. The method sends acoustic waves from the surface to the underground. The acoustic waves are generated at the surface with truck-mounted vibrators (metal mass vibrating at high-frequency), falling weights or with detonating explosives in boreholes (Fig. 12.1). The waves propagate through the rocks and are being reflected and refracted at the boundaries of rock units with (highly) contrasting density (Fig. 12.2). A relatively small fraction of the wave energy reaches the surface where it is recorded by complex arrays of instruments, so-called geophones (Fig. 12.3). Geophones are highly sensitive microphones which record the wave echoes from the various geological discontinuities underground. The reflected

Fig. 12.1 Reflection seismology. The illustrated vibrator is fitted to a truck (see front page of this chapter)

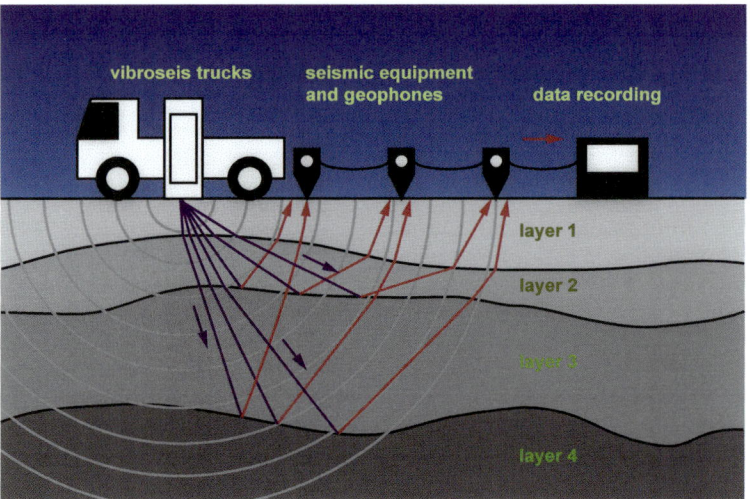

Fig. 12.2 Propagation of acoustic waves in the geological material underground

wave has traveled from the source at the surface to the reflecting geological surface underground and back to the geophone at the surface. The position of the reflector at depth is thus recorded as two-way travel time (in seconds). The two-way travel time depends on the depth of the reflector and the acoustic properties of all rock units that the wave has passed through.

Launching a seismic campaign starts with obtaining the necessary official permissions. Target areas and depth of the potential reservoir must be defined next.

12.1 Geophysical Pre-drilling Exploration, Seismic Investigations

Fig. 12.3 Data recording during a reflection seismic survey

The layout of the seismic lines is fixed on the basis of topographic maps and on-site inspections. Experimental parameters such as point intervals; exsiccation energy, equipment, transducer channels and other must be determined. The details of the grid are finalized during the experimental fieldwork and the geophones are laid out and the vibro-lines are started to be run. The distance between the geophones controls the minimum resolvable wave length. The total length of the line array defines the selectivity. The seismic resolution gradually decreases with depth. Higher frequency signals are being lost with increasing depth. The resolution at greater depth can be improved by signals from higher source energy. The measuring instruments, recording truck, the vibroseis trucks, the personnel and all necessary infrastructure must be moved to the exploration area. Preliminary data processing can be continuously performed during fieldwork. The major part of the data analysis and processing with sophisticated software is done post-fieldwork. The final travel-time to depth conversion requires a model of the geological structure of the underground or a known geological profile from an earlier nearby wellbore (Fig. 12.4).

Geological material boundaries (lithological boundaries) refract and reflect acoustic waves and partly convert them to other types of waves. The details of these processes at lithologigal boundaries bear valuable information on the geological nature of the boundary (Fig. 12.3). The processed data of the wave field result in a graphic display of the material (density) discontinuities along the seismic section.

The acoustic impedance Z controls the traveling velocity of seismic waves in a geological formation. The seismic wave velocity V is the ratio of the acoustic impedance Z and the density of the rock ρ ($V = Z/\rho$). The contact surface between two geological units can only be "seen" if there is an impedance contrast between the two units. If not then the reflection coefficient R at the interface is zero.

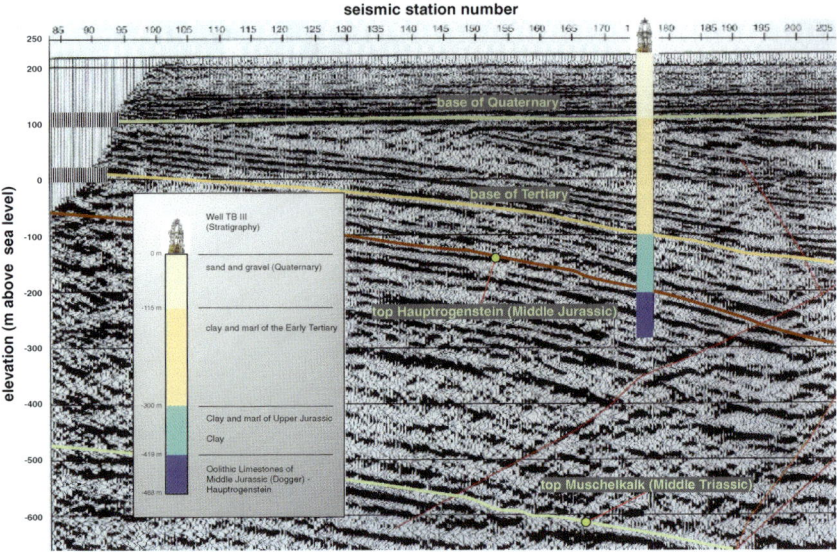

Fig. 12.4 Example of an interpreted seismic section. Calibration of the recorded discontinuities using an existing drillhole

Seismic data processing copes with enormous quantities of data. The data are loaded with noise and must be filtered to make primary reflections visible. The information from single lines is then stacked. Skillful stacking of selected single lines enhances the visibility of important reflections from the underground. Processing distinguishes relevant geological signals from interfering signals and selectively eliminates the later. Interfering signals include for example multiple reflections of primary signals.

Common-Midpoint staking (CMP) is the most widely used technique for seismic data processing. The installed geophone array receives energy from multiple source positions and the recorded tracks are then reordered according to common midpoints between source and geophone. After travel time correction the seismic signals appear directly above the geological reflector. Further numerical data engineering improves the quality of the final seismic section. The final result is a graphic display of a seismic model section showing the geological reflectors at two-travel-time depth. The CMP technique distorts inclined or curved reflectors. In order to convert travel time in seconds to depth in meters specific seismic velocities need to be assigned to the different geological units and rock types. The final seismic section is a simplified lithostratigraphic model of the underground, showing lithological boundaries between geological units with an impedance contrast (Sheriff and Geldart 2006; Telford et al. 2010; Shaw et al. 2005). The model can be verified if accessible data from older drillholes exist from the same area. Processing and geological interpretation of the data from a seismic section (2D seismics) ultimately produce a geological cross-section (Fig. 12.4). If data from a number of intersecting seismic sections are available a 3D seismic model can be constructed. This can be further developed to a 3D block model of the geological structure of the underground, which may also be used as a basis for a geothermal simulation model. The 3D model can also be used for virtual planning of the bore course (Chap. 11).

The hydrocarbon industry uses sophisticated and mature computer based techniques for reservoir characterization and optimization based on seismic data. Standard software applications such as **PETREL** and **ECLIPSE** (both from the company Schlumberger) help to develop an optimized picture of the geological underground and the potential hydrocarbon reservoir. PETREL is used to develop a geological and structural model from 2D/3D seismic data and for preparing a simulation model. ECLIPSE helps to analyze and predict the dynamic behavior of the reservoir over time and to model the evolution of pressure, temperature and flow rate during long-time operation. The economically critical decision on the exact position of the drillhole and the exact drill course for hitting the reservoir at the optimal location is ultimately based on the concept and picture of the geology of the underground derived from seismic data.

Refraction seismic surveys analyze refracted seismic waves, in contrast to reflection seismology. The method uses travel times of waves from seismic sources that have been refracted along interfaces of geological layers with density contrasts. The wave that travels along the geological contact generates seismic signals that can be recorded by the geophones (Telford et al. 2010). The seismic source is, like in reflection seismology, a shot (blasting explosives), vibrators (vibroseis trucks) or other sources. The sensors (geophones) are arranged in a regular spacing along a profile line and record the propagation of the wave field. Processing of the data is done using travel time diagrams. The resulting model shows the geological structure of the underground with layers with different seismic wave velocities (different densities of the material). The recorded travel time data relate to the depth of interfaces of geological units. Refraction seismology can be used to investigate the underground to a depth of about one third of the length of the geophone array. The velocity of acoustic wave in a layer is an important parameter because it relates to the density of the material. It can be derived directly from refraction seismology. Thus the rock material and geological identity of the layer can be directly deduced from refraction seismic travel time data (Stark 2008; Avseth et al. 2010; Reynolds 2011).

12.2 Geophysical Well Logging and Data Interpretation

Geophysical well logging examines the wellbore, its immediate vicinity and the surrounding area of a wellbore. Well logging is the essential data collecting method in deep well drilling. A wide variety of physical principles are used for measuring methods including geoelectrical, magnetic, and acoustic techniques but also radar and methods using radioactivity. The measurements in wellbores produce data on geological, lithological, petrophysical, reservoir-relevant properties of rocks and materials. Also structural, textural and drill technical data can be routinely measured. State-of-the-art geophysical well logging has replaced or greatly reduced the necessity for time consuming and cost-intensive coring. Well logging produces in situ rock and material data in the natural environment, which is a great advantage over laboratory derived rock data (Ellis and Singer 2007).

The measurement of hydraulic and rock mechanical properties of the drilled rocks requires very robust instruments (borehole logging devices) that must resist high temperature, high pressure and chemically aggressive fluids. The probes and sensors are lowered or hauled in the borehole and record the data along the axis of the wellbore. The resulting depth versus parameter data and graphs are called logs. Cables connect the probes (instrument units) with a registration station above surface. The probes can be stopped and fixed at any depth and can so be used to measure parameter variation with time.

Well logging measures different groups of parameters: (a) physical parameters that characterize the immediate vicinity of the borehole, (b) measurements regarding the geometry and morphological details of the borehole, (c) properties of the fluid in the borehole (drilling fluid, formation water). The desired physical and chemical parameters can be collected with passive or active methods. A passive measurement reacts on an external forcing, e.g. an electrical self potential, magnetic field, natural radioactivity. Active measurements use engineered signals penetrating the rocks, e.g. electrical currents, radioactivity or acoustic waves. The active methods measure the interaction of the engineered external forcings with the rocks. The measurement of parameters of the wellbore geometry include: Borehole diameter, borehole cross section, borehole inclination and azimuth. The most important measurable parameters of the fluid phase present in the borehole are: Temperature, electrical conductivity (with a complex relationship to the salinity or the total mineralization of the fluid) and pH of the fluid (that can be measured to about 150 °C and 150 bar pressure, Midgley 1990).

The borehole cable serves as a mechanical mounting support for the probe. However, it also supplies the probe with electrical power, transmits the data to the surface recording unit and registers the vertical (depth) position of the probe and thus makes the log (depth vs. parameter relation). Measurements in deep drillholes must consider and correct for the cable extension due to weight and temperature. The registration unit controls the measurement procedure, supplies the probe with energy via the cable, records measured parameters, stores the data and produces a real time graphic presentation of the log. The system also records formation parameters, the driving speed of the probe, the tensile stress on the cable and other crucial parameters. Most of the deployed probes are multi-channel probes that are able to transmit more than one parameter per probe run.

The temperature log records the temperature of the fluid in the wellbore. The drilling disturbs the thermal conditions around the borehole for some time. The undisturbed steady state temperature distribution can be accessed by repeated recording of the temperature log or measuring after longer periods of shut-down time. Significant changes in the steady state temperature versus depth profile may indicate major water inflow or outflow structures (Fig. 12.5). Consequently, temperature logs may also detect a leaking casing or back fill. Temperature data

12.2 Geophysical Well Logging and Data Interpretation

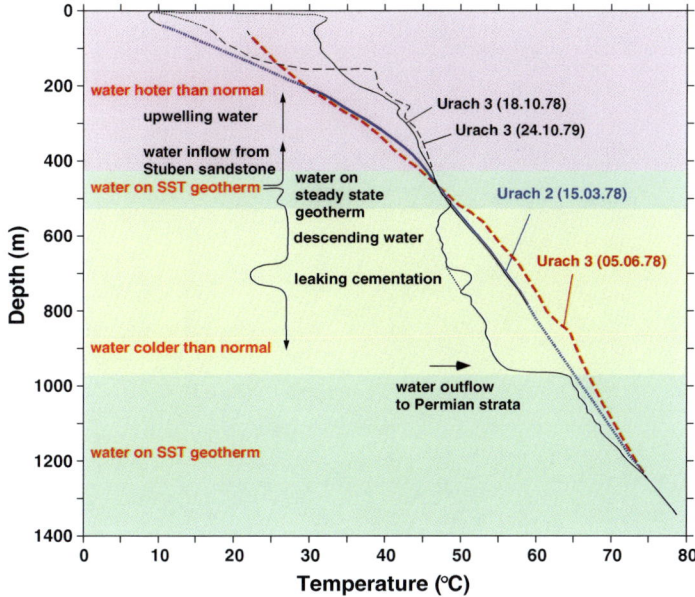

Fig. 12.5 Temperature log indicating water inflow and outflow structures. Urach 3 deep well, SW Germany (Stober 1986)

collected during a hydraulic pumping or injection test can be used for deriving thermal parameters of the subsurface materials (rocks). In addition, the temperature data can even be used to derive hydraulic conductivity of individual geological strata (Sect. 13.2 in Chap. 13).

The electrical conductivity log is important and interesting because it correlates easily measurable electrical conductivity (EC) with the total amount of ions present in the borehole fluid. The electrical conductivity of a fluid is related to the total amount of charged species (ions) dissolved in the fluid. The dissolved ions transport the electrical charges. Thus the EC can be used as a proxy for salinity (dissolved salt). The variations measured by an EC log correlates with the total mineralization of the fluid in the wellbore, it also helps to identify influx and outflow structures, and leakages in the casing and the cementation. Similar to the temperature logs, EC logs can recognize inflow and outflow structures only if they result in EC contrasts, which does not have necessarily to be the case. In other words, if for example the inflow water has the same temperature and salinity as the fluid present in the wellbore, the inflow cannot be recognized by the two methods. The electrical conductivity log is usually collected together with the temperature log. Because the EC depends on temperature, the EC data must be adjusted using the T data from the temperature log.

The caliper log is a geophysical well logging tool with extendable sensors that measures the geometrical details of the cross section of a wellbore along a depth profile. It is also used to measure the inside diameter of the casing. It

Fig. 12.6 Caliper log being removed from a deep well

shows wellbore breakouts and cavities and gives hints about the mechanical properties of the borehole wall. It is also an excellent tool for detecting mineral scales, corrosion of the casing or any other damage and deformation of the casing (Fig. 12.6).

Gamma ray logging measures the naturally occurring gamma radiation of the drilled geological formations in a borehole. The gamma rays originate from the radioactive decay of unstable isotopes such as ^{40}K in the structure of K-bearing minerals such as clays, micas and feldspar. Gamma rays are also emitted from uranium and thorium isotopes incorporated in various minerals including zircon, monazite and others. Consequently, the gamma ray log detects lithological boundaries in the drilled profile if the fraction of minerals containing radioactive isotopes abruptly changes. Gamma ray logs are very useful in combination with cutting analysis for creating geological drill logs. They are helpful for locating the existence and position of planned clay barriers.

The density log measures a continuous record of the bulk density of the drilled formations (also gamma–gamma log). It uses an active gamma ray source and measures the density-dependent absorption and scattering of the gamma rays. The measured bulk density of the formation is related to the density of the rock matrix, the volume of pores and fractures and the density of the pore fluid. Thus knowing bulk density from density log and rock density from lab measurements the porosity of the formation can be deduced from: $\phi = (\rho_{matrix} - \rho_{bulk})/(\rho_{matrix} - \rho_{fluid})$. For simple monomineralic rocks such as limestone, dolomite and sandstone ρ_{matrix} can be directly taken from tables.

12.2 Geophysical Well Logging and Data Interpretation

Sonic logging measures the travel time difference Δt of a compressional wave in a borehole generated by a transmitter and picked up by two receivers at different distances to the transmitter (p-wave velocity). This time difference depends on the lithology, rock structure and the porosity of the formation. In the same rock matrix, Δt increases with porosity. Sonic logs can be used to generate a continuous porosity profile of the drilled section.

A borehole-imaging log provides a detailed image wellbore from micro-resistivity or acoustic measurements. It shows rock structural dip, fractures, faults, breakouts and other conditions of the borehole. The borehole-imaging log can provide detailed information on the quality of the cementation and the casing. The images can be used to create a 3D picture of the geological structure in the wellbore.

A variety of further logging tools are routinely used for solving specific problems. Water inflow points and structures can be detected with flowmeter logging. The method uses a device with a spinner whose rotation relates to water flow rate in the borehole. The quality of cementation of the wellbore and its bonding to the formation and to the casing can be assessed by cement bond logging.

An important method for measuring hydraulic properties of the formation is fluid logging. It combines a hydraulic well test with geophysical well logging in an uncased wellbore. The resulting fluid log represents a hydraulic conductivity profile of the drilled formations. The method is particularly useful for aquitards (Sect. 13.2 in Chap. 13).

The so called fishing tool is a useful piece of equipment for recovering objects that have been lost in or fallen into the bore. There are many differently designed devices for the purpose of fishing lost instruments and other tools from the drillhole (Fig. 12.7).

In addition to the highly developed well logging techniques a new development started in the 1970s and 1980s of the last century, logging while drilling (LWD). The methods measure relevant parameters during drilling by sensors integrated in the drill string. Also drilling paths can be measured directly during drilling

Fig. 12.7 An example of a fishing tool that has been used for recovering a probe that crashed because of a cable brakeage

(MWD) (Chap. 11). The data are being transmitted to the surface by pressure signals through the drilling fluid column. However, networked and wired drill pipes that transmit high-definition downhole data to the surface are increasingly used also. The real-time parameter data allow for instant decisions about the drilling direction and other well management attributes. The LWD tools tolerate formation temperatures of about 150 °C, some also up to 170 °C. For pure directional drilling procedures MWD tools for up to 200 °C are available. All measuring systems have a very limited formation penetration depth. A special seismological technique has been developed which is used to explore the geology ahead of the bit whilst the well is being drilled. The technique is known as "seismic while drilling SWD" (Poletto and Miranda 2004).

Chapter 13
Testing the Hydraulic Properties of the Drilled Formations

Vortex on the river Rhine

Hydraulic tests provide the key data on the hydraulic conductivity of the reservoir formation and permeability structure of the reservoir. These hydraulic properties are fundamental for the success of a geothermal project. The first hydraulic tests are already made in the hanging wall of the intended reservoir formation during drilling of the deep well. After completion of the wellbore, the hydraulic properties of the reservoir formation must be extensively tested. This includes long-term tests, circulation experiments, or tracer tests. Chapter 13 gives a brief overview over some standard hydraulic testing methods, the practical conductance of the tests and the processing and interpretation of measured data.

13.1 Principles of Hydraulic Testing

Hydraulic tests may solve very diverse problems. Therefore, the appropriate testing procedures depend on the specific data needed to answer the current question. However, all test methods monitor water pressure changes that result from an incurred excursion from the undisturbed pressure distribution in the reservoir. The excursion is being imposed by the testing method. A large variety of hydraulic testing schemes are currently used in groundwater exploration and by the oil and gas industry (Kruseman and de Ridder 1994; Witt 2009; Nielsen 2007). One type of test, the pumping tests and the production tests, produces water, whilst another type of test, the injection tests, introduces water into the formation to be examined. The details of the pressure to flow rate relationship are controlled by the sought-after hydraulic properties of the reservoir. Some tests use pressure pulses to get response signals from the reservoir. Some test require only a few minutes, other tests run for days. The duration of a hydraulic test depends on the type of test, the type of needed data and the hydraulic properties of the tested formation (Fig. 13.1). Some methods test the entire open-hole or the entire screened section of the wellbore, other experiments test specific sections of the formation that are of particular interest by separating the sections with packer or other systems (Fig. 13.2). Some tests continuously record the water pressure in the tested section of the bore, other tests measure water pressure near the surface, and some just monitor the groundwater table as a pressure response to extracting or injecting water. The tests also differ with respect to the water extraction rate. Some tests pump water at a constant rate from the reservoir and continuously monitor the pressure response. In other tests, the pumping rate increases stepwise. Tests may also be run at constant pressure so that the water flow rate gradually decreases during the test. Some tests are carried out concurrently with geophysical well logging (Sect. 12.2) others not. Some tests monitor the water temperature at the wellhead or near the pressure gauge in the tested section of the wellbore alongside with the pressure recordings.

The hydraulic properties of the reservoir measured and analyzed with the testing methods provide the conclusions on the yield of the well. This is the prime goal of hydraulic tests. Yield and temperature determine the commercial success

13.1 Principles of Hydraulic Testing

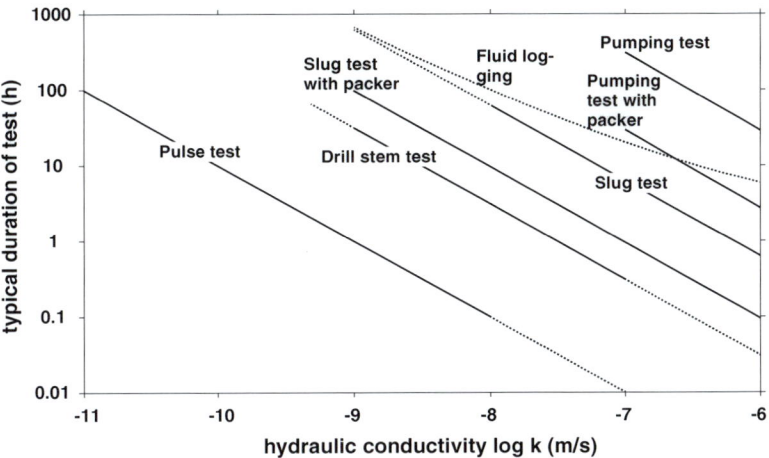

Fig. 13.1 Duration of various types of hydraulic tests in formations with different hydraulic conductivity (Stober et al. 2009; Hekel 2011)

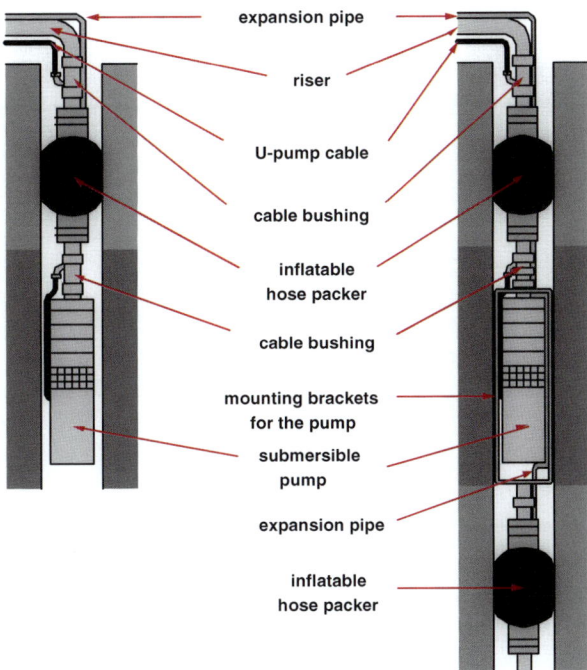

Fig. 13.2 Schematic illustration of single and double packer systems in hydraulic tests

of the geothermal project. However, hydraulic tests also provide water samples for the necessary hydrochemical analysis and isotope studies (Chap. 14). The well yield depends not exclusively on the hydraulic properties of the reservoir formation (hydraulic conductivity, storativity) but to some degree also on the hydraulic properties of the wellbore itself (skin, wellbore storage). Some testing methods are suitable to separate properties of the formation and properties of the wellbore. Appropriately designed well tests provide data on the hydraulic potential, the pre-testing pressure of the tested formation and give clues on the structure of the thermal aquifer (or aquitard) and the flow properties of the formation. The hydraulic response to the tests may indicate hydraulic interaction and communication with the formations above and below the reservoir (leakage). The tests may provide insights into the interaction between fractures and porous rock matrix or into the hydraulic significance of major fractures and faults. The tested volume of the reservoir formation increases with the duration of the test. Long-term tests may therefore provide information on the extent of the reservoir and the nature of hydraulically active borders (Stober and Bucher 1999a, b, 2005; Kruseman and de Ridder 1994).

The known hydraulic signal or stimulus imposed on the reservoir formation during the test by extracting or injecting water or by sending a pressure pulse triggers a response or reaction of the unknown hydraulic system. The reaction, a pressure drop or pressure increase (or water table changes) are continuously recorded. Thus input and response signals are known and must be analyzed and interpreted in the context of the known geological situation, the structure of the underground known from seismic studies and plausible hydrogeological properties of the tested formation. Solving the mathematical inverse problem and finding the sought-after hydraulic parameters of the formation requires a sharp model concept of the tested formation or the wellbore that is as close as possible to the real structure and properties of the system. This well-defined and well-founded model of the underground is required to react with the same response signals on the input signals like the tested system (Fig. 13.3). Figure 13.3 shows six different geological model concepts of the hydraulic situation in the tested formation. The six systems respond characteristically on the imposed external signal, in this case pumping water at a constant rate from a well. The hydraulic reaction of the system is monitored as drawdown and plotted against time elapsed since pumping started. The data are conventionally represented on double-log or semi-log plots. After pumping stops the drawdown slowly recovers and the recovery versus time after pumpstop also reflect the hydraulic properties of the tested underground. The recovery data can be displayed on so-called Horner plots (Fig. 13.3). The detailed shape of the drawdown versus time curves in pumping tests depends on many possible geological structures and features that may influence the hydraulic behavior of the tested formations. The graphical evaluation of the measured drawdown versus time data requires the choice of a model concept that best represents the hydraulic situation of the underground as stated above. The choice must be made among many model concepts, which may prove difficult because of vague knowledge of the real situation. The number of feasible concepts can be drastically reduced by careful planning and implementation of the hydraulic tests. Critical is particularly the duration of the test. Very short tests may trigger response signals from the immediate vicinity of the wellbore (skin, wellbore storage). The hydraulic conductivity of the formation close to the wellbore is typically severely altered by

13.1 Principles of Hydraulic Testing 237

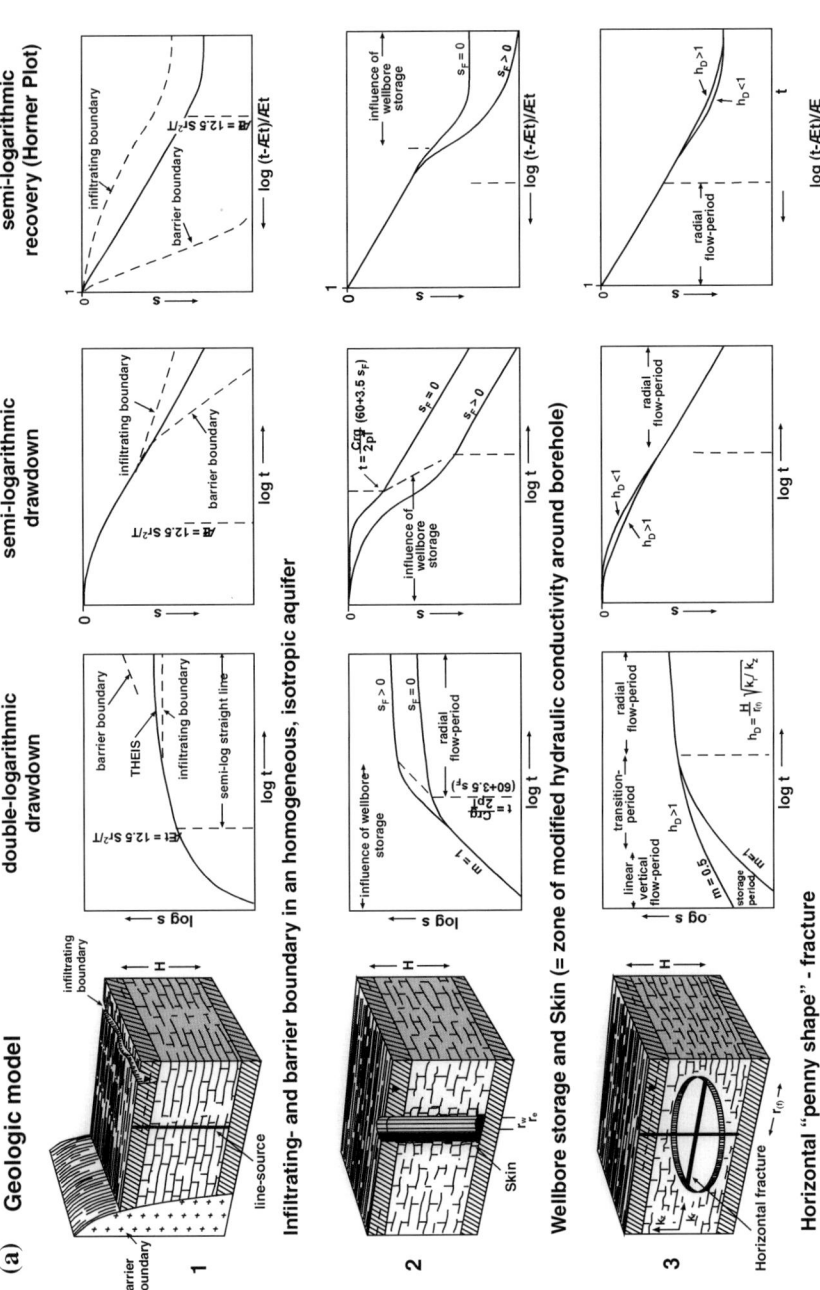

Fig. 13.3 a and b Six geological model concepts and the associated drawdown response to pumping at a constant rate, or the recovery of the drawdown after pumping stops (Horner plot at the right hand side of the figure). The figure illustrates the reaction of the system to a constant input signal depending on formation and wellbore related details of fracture shape and orientation, skin, impervious hydraulic boundary, recharge by infiltration and other features (Stober 2011)

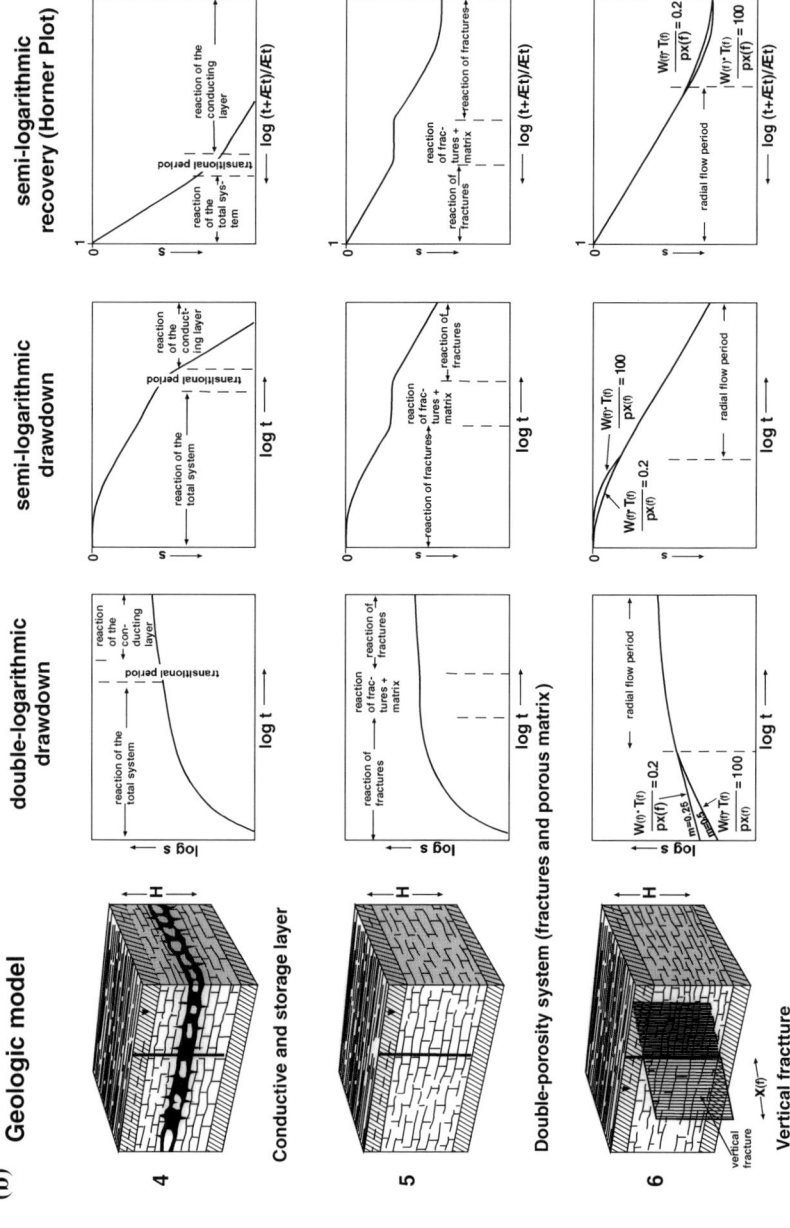

Fig. 13.3 continued

13.1 Principles of Hydraulic Testing

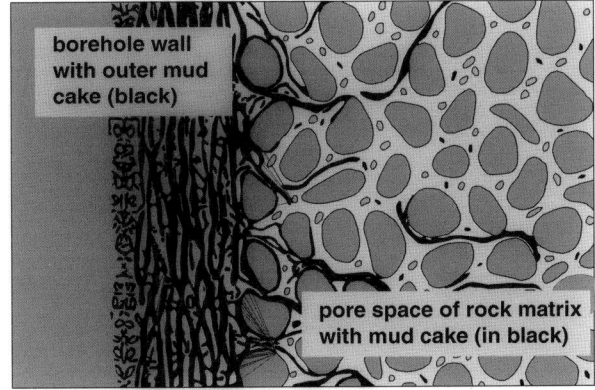

Fig. 13.4 Rock structure across a wellbore: next the wall of the bore a damage zone with severely altered hydraulic conductivity (so-called skin) typically exists, the consequences of the drilling operations reach beyond the skin, however (mud cake in the porosity of the rock matrix)

the drilling operation (drilling mud, fracturing, acidization a.o.) and by technical efforts in the borehole (Fig. 13.4). Many experiments in crystalline basement rocks prove the presence of a zone of increased hydraulic conductivity near the wellbore (Stober 2011).

The skin, the zone of altered conductivity near the wellbore, influences the pressure response in the course of a hydraulic test. If the conductivity of the skin is lower than that of the tested formation pumping is accompanied by an additional pressure drop (increased drawdown in Fig. 13.1). The pressure response on pumping is smaller (decreased drawdown) if the skin has a higher conductivity than the formation. The skin related additional pressure change can be expressed as contribution to the drawdown Δs_{skin} (in meters):

$$\Delta s_{skin} = s_F Q / (2\pi T) \quad (13.1)$$

where s_F denotes the skin factor (dimensionless), Q is the production rate (pumping rate) in m^3/s and T [m^2/s] stands for the transmissivity. The dimensionless skin factor can be positive or negative depending, as mentioned, on the hydraulic conductivity of the altered zone around the wellbore compared to that of the undisturbed formation (van Everdingen 1953; Hawkins 1956; Agarwal et al. 1970). For fully impervious wellbores $s_F = +\infty$ and for highly stimulated, acidized or fractured zones near the wellbores s_F may be as low as $-\infty$. A simple procedure to derive a dependable value for the skin factor from pressure–time test data can be found in Matthews and Russel (1967).

At the beginning of a hydraulic pumping test the fluid in the wellbore is produced. Later fluid from the formation flowing to the wellbore due to the imposed pressure gradient by pumping is being produced gradually also. The reaction of the tested formation is delayed. This effect is called wellbore storage (C). It corresponds to the volume change (ΔV) in the wellbore per pressure difference (Δp), thus it has the dimension m^3 Pa^{-1}. The wellbore storage can be computed from Eq. 13.2:

$$C = \Delta V / \Delta p \quad (13.2)$$

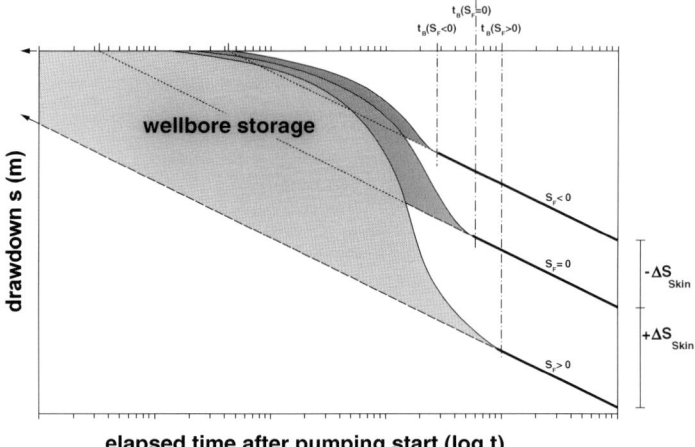

Fig. 13.5 Drawdown versus time plot showing that the early phase of the test is absolutely dominated by the wellbore storage, it gradually decreases (middle of diagram) and fades out at different times depending on the skin factor s_F. Note the constant slope of the straight-line relationship of the drawdown—time data (heavy solid lines) reflect the "true" hydraulic conductivity of the tested formation. The slope of these straight-line sections is a direct measure of the conductivity of the aquifer (this is what one wants to know!). Therefore, it is evident from the figure that tests running for a too short period of time will result in erroneous (high) hydraulic conductivities of the formation

Equation 13.2 illustrates that the wellbore storage depends on the diameter of the borehole (2 r_w) controlling ΔV. The duration of the wellbore storage t_B [s] is furthermore controlled by the transmissivity (T) of the tested formation and by the skin factor.

$$t_B = \left[r_w^2/(2T) \right] \left[60 + 3.5 s_F \right] \quad (13.3)$$

It follows from Eq. 13.3 that the duration of the wellbore storage t_B increases with the skin factor and the square of the radius of the well and that it increases with decreasing transmissivity of the formation (Fig. 13.5). Consequently, tests in large-caliber wells drilled in aquitards are affected by the wellbore storage for the longest period of time.

If hydraulic tests in confined aquifers are carried out with dedicated testing tools, then the size of the tool and the compressibility of the fluid are the important parameters controlling the wellbore storage.

A network of monitoring wells is needed to record the spatial pressure distribution during pumping or injection in the tested well. However, in deep drilling a network of monitoring wells is not available. Evaluation of hydraulic tests in the first well is totally restricted to pressure data (drawdown) collected in that single deep borehole. In the early phase of the test the pressure signals are dominated by the geometry of the wellbore, the casing and cementation (if present) and the size of the testing tools. The hydraulic properties of the formation become visible in the data after a minimum duration of the test that must be longer than the duration of the wellbore storage (Fig. 13.5).

13.1 Principles of Hydraulic Testing

When testing non-thermal groundwater wells, the measured water table corresponds directly to the hydraulic pressure in the tested formation. Testing thermal water reservoirs, the pressure data must be corrected for the temperature related density differences. Because the density of water depends on temperature and pressure, water columns of equal weight but at different temperature have different length. The relatively small density difference results in length differences of several meters if the water columns are several hundred or even thousands of meters in a deep well (Sect. 8.2).

Under quiescent conditions, the water column thermally equilibrates with the rock. It is cool near the surface and hot at depth. If water is pumped from the well, warm or hot water from depth flows upward and the entire water column increases in temperature controlled by the pumping rate, duration of pumping, the thermal conductivity of the rocks and other parameters. Because of this thermal effect, it can be frequently observed in pumping tests that in the initial phases of the test the water table increases instead of reacting with the expected drawdown. After shut-in the water column reacts with a thermally induced drawdown instead of the expected water table recovery. The inverse effects can be observed in injection tests pumping cool surface water into a deep thermal well. For the evaluation of pumping or injection tests in thermal reservoirs the measured drawdown (or recorded near surface pressures) must be normalized to a reference temperature. Each data point for the length of the water column (drawdown) in the wellbore must be density corrected to a reference temperature and pressure.

Straightforward and unproblematic is the direct pressure measurement in the wellbore at the depth of the tested formation. This avoids troubles with intricate and error-loaded temperature and density corrections that ignore further complications caused by temperature anomalies, increased salinity or high gas concentrations in the fluid.

The water conducting structures of hard rock aquifers are typically single fractures or fracture zones, in contrast to rocks with a porous matrix. Thus the distribution of water conducting structures is heterogeneous. The orientation and geometry of these structures varies widely in fractured hard rock formations. In contrast to porous aquifers, they represent hydraulic discontinuum by nature. Many diverse model concepts have been developed especially by the oil and gas industry for the quantitative analysis and interpretation of hydraulic well test data (type curves, approximate solutions, specialized software). The models can be grouped into the following categories (Kruseman and de Ridder 1994):

Type 1 models: The water conducting fractures are randomly oriented and regularly distributed in the formation. On a sufficiently large scale the formation behaves like homogenous continuous aquifer. The hydraulic properties can be modeled and interpreted with the concepts of Theis (1935). In this case the test data also can be interpreted by the approximation of Cooper and Jacob (1946). Typical example is regularly fractured crystalline basement (Fig. 13.1).

Type 2 models: The tested formation contains local domains (zones, horizons) of preferred high conductivity and high fracture porosity (Fig. 13.4). There are two endmember types of such domains: Conductive zones dominate the flow properties of the entire formation and have minimal storage capacity. Storage zones, in contrast, behave hydraulically the opposite way (e.g. Berkaloff 1967). Most discontinuities are mixtures of both endmember type domains. Classic example is a karstified zone in a limestone formation.

Type 3 models: The tested formation can be comprehended as double-porosity system, matching a fractured porous rock (Fig. 13.5). This model concept assumes the existence of two continuous homogeneous regimes of flow property, one characterizes the pore space of the rock matrix the other the regular random fracture pore space like in model type 1 (e.g. Barenblatt et al. 1960). Classic example: Fractured sandstone with matrix porosity.

Type 4 models: In the tested formation a prominent vertical fracture of limited extension strongly influences the hydraulic behavior of the system (Figs. 13.3, 13.6). In strongly stimulated wells, it was found that hydraulic test data required the presence of a fracture (Dyes et al. 1958). The effects of fractures of different orientation on the pressure–time data in well tests have been further explored by e.g. Russel and Truitt (1964), Gringarten and Ramey (1974), Cinco et al. (1975).

To what extent such models adequately represent the geological and hydraulic structure of the tested formation needs to be decided for each tested formation and each well anew. It is principally impossible to assign a certain model a priory to the formation of interest because the geometric details of the voids (general pore space) and their hydraulic interaction cannot be predicted beforehand. The right way for finding an appropriate model compares the measured with theoretical model pressure–time data (Fig. 13.3). Figure 13.3 presents six common example situations. It proved to be helpful for the model-finding process to also consider graphs of the derivatives of the pressure (drawdown, water table) time data (Fig. 13.6) and other special functions (e.g. Bourdet et al. 1989). It is immediately obvious from Fig. 13.6, that derivative plots are graphically much more distinctive than pressure—time plots alone.

For finding an appropriate hydraulic model that best describes the properties of the formation, the data (pressure, drawdown) are plotted against time on a log-log graph or on a linear-log graph as shown on Figs. 13.3 and 13.6. The pressure recovery behavior after shutdown of the pump can be represented on a Horner-plot (s vs. log (t + t')/t') (Fig. 13.3). If many data have been recorded per unit time, the derivative of the pressure (drawdown) per unit time can be plotted (e.g. Fig. 13.6). The graphical representations of these data (log [($\delta s/\delta t$) t] vs. log t) are a very powerful tool for the diagnosis of the hydraulic behavior of the tested geothermal formation as illustrated by Fig. 13.6.

Radial flow towards a well (linear sink) results in a straight line relationship of the data on a s versus log t plot (radial flow period). Volume flow towards an imperfect (heterogeneous) well can be recognized on a linear relationship between s versus $1/t^{0.5}$ data. If fluid flow from the fracture pore space is followed by fluid flow from the porous matrix, the bilinear flow behaviour can be identified by a linear relationship on a s versus $t^{0.25}$ plot.

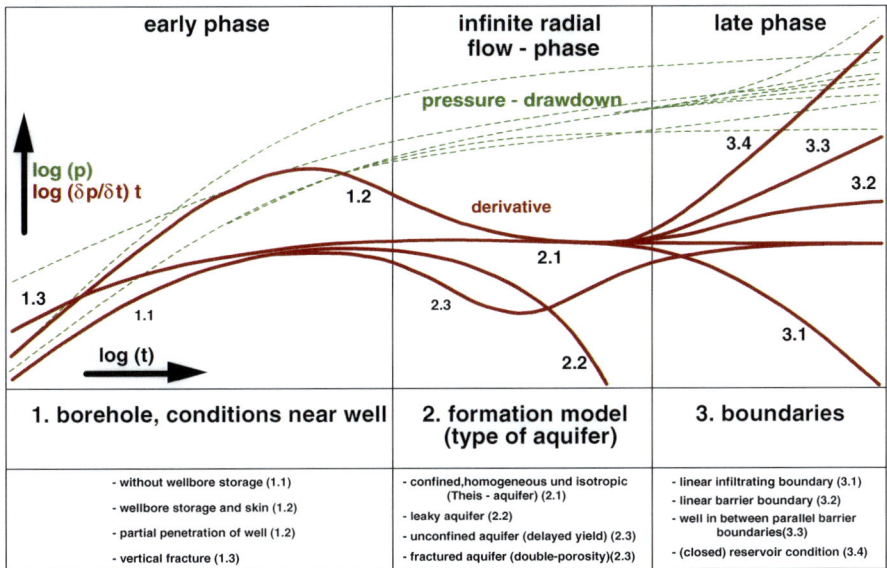

Fig. 13.6 Theoretical model *curves* for the draw down (s), the first derivative of the drawdown [($\delta s/\delta t$) t] during a pumping test at constant rate and for different formation structures (Odenwald et al. 2009)

13.2 Types of Tests, Planning and Implementation, Evaluation Procedures

Hydraulic tests must have lasted long enough to give the right answers. The correct hydraulic model can only be chosen for meaningful data interpretation if the duration of the test was sufficiently long. Implementation of pumping and injection tests follows a well-established standard procedure today (Fig. 13.7). The test is subdivided into several sub-tests, beginning with tests that explore the properties of the wellbore. The tests are being run using at least three different constant pumping (injection) rates. The future fluid production rate will be based on these test results. The following tests investigate the properties of the formation. After these testing procedures the system is left at sleep without pumping or injecting fluid. During the formation test water is pumped at a constant rate. It lasts for an extended period of time, typically substantially longer than the experiments testing the properties of the wellbore. The formation test explores the flow properties of the formation for finding the appropriate hydraulic concept model as explained in the previous section (Figs. 13.3, 13.6). The volume of the formation that responds to the pressure signal imposed by the test increases with the duration of the test. Thus the extension of a hot water reservoir and its distant hydraulic boundaries can only be investigated in tests of sufficient duration (Sect. 13.1). Short-term tests do not provide hydraulic information on the distant regions away from the wellbore. In the worst case it is not even possible to derive hydraulic parameters of the

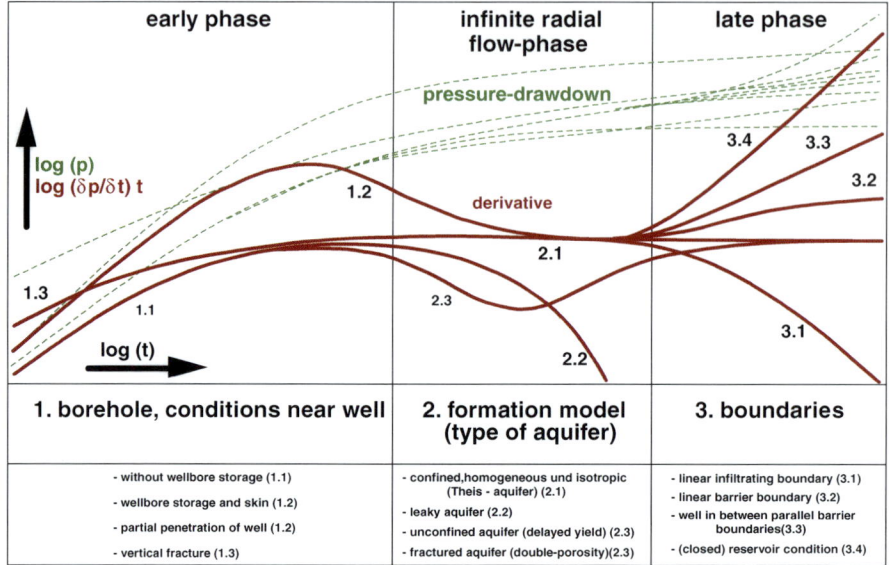

Fig. 13.7 Example of a hydraulic test arrangement. *RWL* groundwater level at-rest. *Q* extraction rate (pumping rate), the first test series pumps at 3 different rates Q1, Q2 and Q3 and investigates the properties of the well, the following test is of longer duration and pumps at a higher rate Q4 exploring the properties of the formation

formation because the tested volume remains within the wellbore storage and skin (Eq. 13.3).

Following a model concept of a homogeneous isotropic formation of infinite extension, the transmissivity and the storage coefficient of the tested formation can be derived during the radial flow period. The measured drawdown (pressure) can be graphically plotted against the logarithm of time. From the slope of the semi-log straight line (straight line on s vs. log t plots during radial flow period on Fig. 13.3) the transmissivity T [m²/s] of the formation can be computed from Eq. 13.4 (Cooper and Jacob 1946).

$$T = 2.303 Q/(4\pi \Delta s) \quad (13.4)$$

The storage coefficient S (dimensionless) can be computed from Eq. 13.5, which also consideres the skin factor s_F, the transmissivity T (Eq. 13.4), and the radius r of the well:

$$S = [2.25 \ldots T \ldots t] / \left[r^2 \ldots \left(e^{2sF} \right) \right] \quad (13.5)$$

A typical example of the evaluation of pumping test data is shown for the 4,000 m deep pilot hole of the continental deep drilling project in Germany (Fig. 13.8) (Stober and Bucher 2005a). The borehole is cased to 3,850 m; the open-hole has a length of 150 m. The open-hole is in Variscan crystalline basement with exposed amphibolites and metagabbros. The temperature at bottom hole is 120 °C. The pumping rate was held constant at nearly 1 l/s. The measured

13.2 Types of Tests, Planning and Implementation, Evaluation Procedures 245

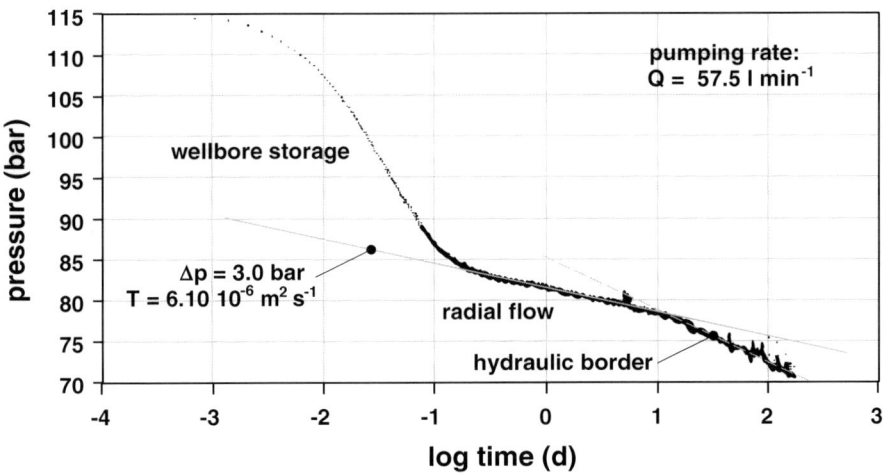

Fig. 13.8 Well test data and evaluation from the 4000 m deep pilot hole of the continental deep drilling project (KTB) in Germany (Stober and Bucher 2005a)

data, plotted on a semi-log pressure versus log time graph (Fig. 13.8), reflect the wellbore storage at about 0.2 days from test start. From then on, the data follow a straight-line relationship. This test phase signals the radial flow period and shows the hydraulic reaction of the formation, here the crystalline basement on pumping. The slope of the radial flow period can be converted to the transmissivity of the formation $T = 6.10 \times 10^{-6}$ m²/s with help of Eq. 13.4. After 12 days of pumping, pressure decreases markedly and tends to follow a linear trend with a steeper slope. This feature is caused by a hydraulic boundary with lower conductivity than the tested formation, in the example case it is caused by an impervious fault zone, the "Franconian Lineament" (Stober and Bucher 2005a).

From the computed transmissivity, the geometry of the wellbore and the observed wellbore storage (Fig. 13.8) follow the skin factor $s_F = 1.35$ from Eq. 13.3. The skin causes an additional pressure difference (drawdown) of 3.5 bar (from Eq. 13.1). The storage coefficient $S = 5 \times 10^{-6}$ can be computed from Eq. 13.5. The example shows that elaborate well testing can provide the geothermal project with a large amount of critical and important data on the hydraulic properties of the wellbore, the target formation and the hydraulic structure of the reservoir.

Long-term pumping (or injection) tests are indispensable before operation of a geothermal doublet. The hydraulic testing must be accompanied by a hydrochemical research lineup (Chap. 14). After that long-term circulation or production tests must prove the functionality of the system. Also these tests must be backed with diverse supporting experiments.

Utilization of hydrothermal reservoirs taps thermal water from deep wells almost exclusively drilled in confined hardrock aquifers. Pressure or drawdown data from these confined aquifers commonly display the effect of the tides (Fig. 13.9). This clearly indicates that the fractures and other porosity of the

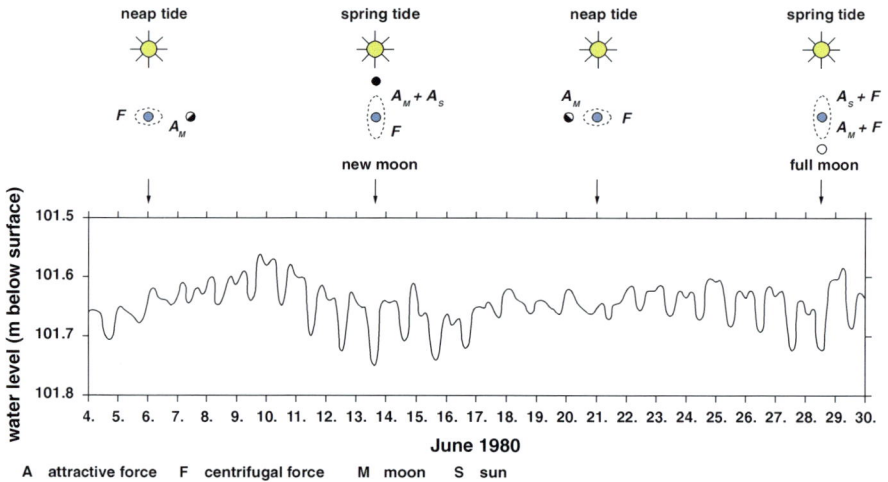

Fig. 13.9 Variation of pressure (water level) in the deep well Saulgau TB1 (in karstified upper Jurassic limestone, 650 m) caused by the tides (Example of tidal effects from: Stober 1992)

formation are interconnected and hydraulically communicate over large distances. The effect of the tides changes the shape and geometry of the voids of the formation. Thus pressure (drawdown) drops and rises in the borehole depending on the position of sun, moon and Earth (Ferris 1951; Todd 1980).

Hydraulic tests may be of little significance, despite a sophisticated test scheme, if distinct formations with different hydraulic properties are jointly tested and the properties cannot be separated. By using packer systems (Fig. 13.2) and a proper well engineering, drilled formations can be hydraulically isolated and tested separately one-by-one, so that the derived parameters refer to a distinct and well defined tested formation.

Hydraulic tests can be combined with geophysical well logging techniques. For example, data from flow meter, electrical conductivity or temperature logs can be used to assess the contribution of separate formations that have been tested together in a pumping test, permitting assignment of hydraulic conductivity values to the individual but jointly tested formations. An example for such a combined technique is the fluid-logging method (Fig. 13.10), which repeatedly measures the electrical conductivity of the pumped fluid at depth during an long-term aquifer test (Tsang et al. 1990). The hydraulic evaluation of the aquifer test provides the total transmissivity of the tested section. The relative temporal changes of the electrical conductivity in the different formations can be used for a proportional distribution of the inflow rates from which the transmissivities of the discrete inflow sections follow. Precondition for the success of the method is fluid mixing in the tested section before the start of the pumping test.

The technical equipment for performing packer tests consists of a rod assembly with a valve and one or two packers (single, double packer, Fig. 13.2). The packer is a 0.5–1 m long enforced rubber sleeve, which can be mechanically or

13.2 Types of Tests, Planning and Implementation, Evaluation Procedures

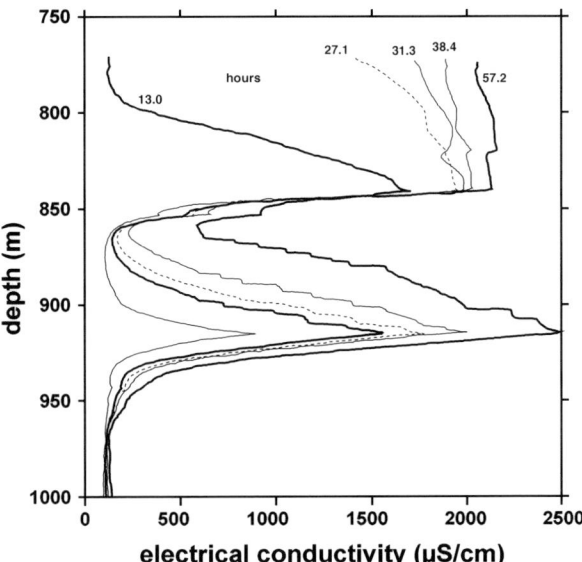

Fig. 13.10 Fluid-Logging in a 1,690 m deep drillhole, data from the test sect. 770–1,000 m (from Tsang 1987)

hydraulic-pneumatically deformed, to where the mounted and inflated device seals the section to be tested. The tested section of the borehole is normally 1.5–5 m in length. During the hydraulic test, the temperature and the pressure in the tested interval are continuously monitored for detecting leaks and infiltrating water. The principle of the test is the same like in any other hydraulic test. The pressure measured in the test section at the beginning of the test serves as reference pressure, like the water level at-rest (RWL) in open wells. The initial pressure in packer tests is measured after mounting and inflating the packers. After a so-called compliance period external disturbances decline and disappear (exception are the tides, Fig. 13.9). The first step of the test changes the pressure in the test interval by extracting or injecting water (in very dense rock: gas). Withdrawal causes a pressure decrease, injection a pressure increase. In a second step, the pump is stopped and the pressure recovers slowly to the undisturbed formation pressure (Fig. 13.7). Pre-test pressure and final formation pressure should be equal.

For packer tests there exists a large number of hydraulic testing procedures also. The selection of the appropriate method is determined by the objectives of the testing and the expected hydraulic conductivity of the formation. The fields of application of diverse test methods are primarily related to the hydraulic conductivity of the target formation (Fig. 13.1).

Slug tests are used in formations with low to intermediate hydraulic conductivity (Butler 1998). In slug tests the pressure in the borehole or the tested interval is suddenly changed and the pressure response of the system monitored. Opening the test valve of a packer test installation transfers the pressure pulse instantly to the test interval (Fig. 13.2). During the induced flow period the pressure balances by water flowing from the formation (slug-withdrawal test) or water flowing to the

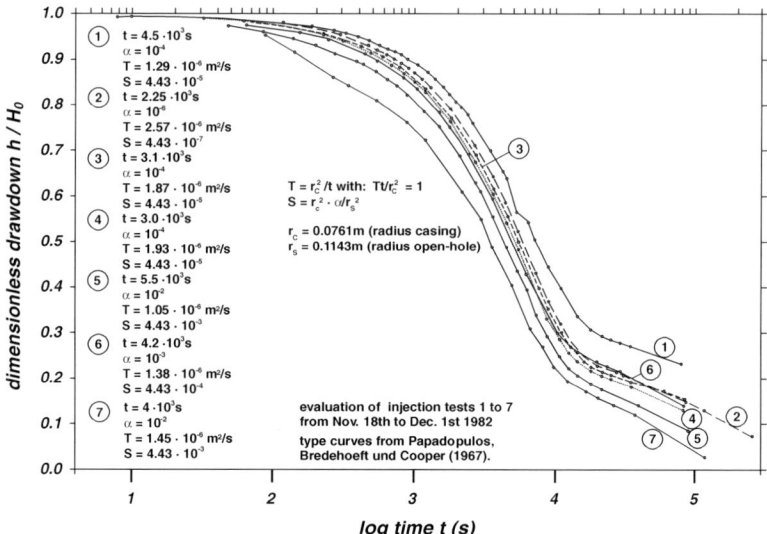

Fig. 13.11 Example of a slug test analysis from the 4,440 m deep geothermal Urach 3 (Germany) with type *curves* from Black (1985)

formation (slug-injection test) depending on the imposed pressure gradient. Slug tests can also be used in open boreholes. Slug tests of very short duration that just send a pressure pulse to the tested section are called pulse tests. The pressure signal applied in slug tests is created by very rapid withdrawal or injection of a large amount of water or mechanically inserting a displacement body. The later type of test is also called a bail-down test.

Slug tests can provide transmissivity, storage coefficient, storage and skin factor. The analysis of the pressure versus time data is typically done by means of type curves (Fig. 13.11) (e.g. Cooper et al. 1967; Ramey et al. 1975; Papadopulos et al. 1973; Black 1985). Numerical methods are also available. The derived transmissivity data can be converted to formation permeability and hydraulic conductivity (Eqs. 8.3b, 8.4a–c).

Drill Stem testing (DST) uses the drill pipe as a testing tool where the DST equipment with packer systems replace the drill bit. The packers hydraulically isolate the section to be tested. Opening a valve imposes a pressure drop in the tested interval causing water (fluid) flowing to the drillhole. Closing the valve causes the pressure to relax to the at-rest formation pressure. The standard procedure of a Drill-Stem test begins with a first short flow phase (valve open), followed by a first recovery phase (valve closed). The test continues with a long-term flow period and a long-term recovery period (Fig. 13.12). The name of the test relates to the drill stem being part of the test equipment. Depending on the test configuration, some of the test periods can be analyzed and interpreted like slug tests, the recovery periods can be evaluated with the Horner method (Fig. 13.13; Horner plot, Horner 1951). Drill Stem tests supply transmissivity, possibly also wellbore storage and skin factor.

13.3 Tracer Experiments

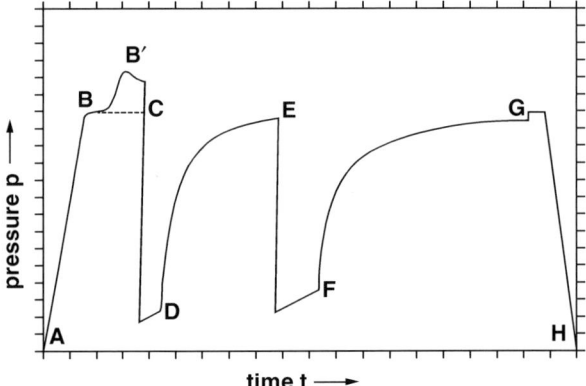

Fig. 13.12 Schematic pressure–time relationship of a drill stem test. *A–B* Mounting the DST drill pipe, *B–C* positioning and mounting the packers, *B–B'–C* expansion-related pressure response in low-permeability formations and subsequent relaxation, at *C* opening test valve, *C–D* first flow phase, *D* closing valve, *D–E* first recovery period, *E* opening valve, *E–F* second flow phase, *F* closing valve, *F–G* second recovery period. *G–H* deflating the packers and unmounting the DST equipment

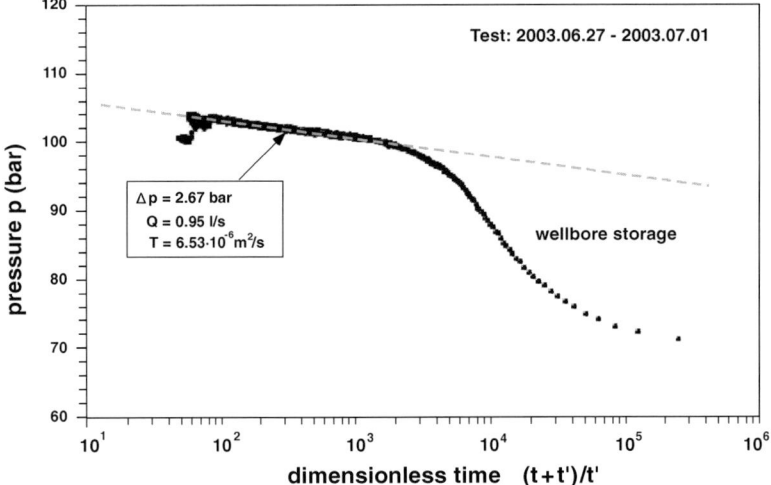

Fig. 13.13 Example of an evaluation of pressure buildup during a recovery phase with a Horner plot from test data collected in the deep geothermal well Urach 3 (Stober 2011)

13.3 Tracer Experiments

Tracers are chemicals and substances that are deposited at one location (borehole) in the underground and then their migration is traced at other locations (boreholes) in the subsurface. Tracers are substances that can be detected at very low concentrations and high-degree of dilution with confidence and low-cost routine techniques. From the travel time of the tracer between the injection and the monitoring location follows a flows velocity and the scattering of the measured data reveals mixing and distribution processes, summarized by the dispersion. Tracer tests are routinely used in groundwater engineering (hydrogeology). They are being used to gain qualitative information about water flow paths. Tracer test data can also be quantitatively analyzed and reveal flow velocity, hydraulic conductivity, flow porosity, dispersion D [m^2/s] and other parameters (Sauty 1980).

Thus tracer experiments are also very interesting and useful testing methods for developing geothermal doublets. Tracer experiments can show how re-injected cooled water from the plant spreads and migrates in the reservoir. For example, if a tracer arrives after a short period of time and with little dispersion that is with a sharp peak in the breakthrough curve at the production well then the thermal energy flux will behave in a similar way also. With the help of tracer experiments it can also be shown that geothermal wells may be hydraulically connected. Thus tracer experiments should be run in concert with the first long-term circulation tests.

The tracer substance used should be a non-reactive passive chemical that interacts with the minerals of the tested formation as little as possible. The inert behavior greatly facilitates the mathematical evaluation of tracer data and the interpretation of tracer concentration versus time plots. The ideal tracer should not be toxic; it should be stable and not decay or decompose in the formation. Ideally it has properties similar to that of water. The synthetic organic chemical fluorescein, also known as uranine, specifically the sodium salt of fluorescein is a water-soluble fluorescent tracing dye that can be detected down to 10^{-9} g/l (=1 mg Uranine in 1,000 m^3 water). Uranine comes close to the ideal tracer and is often used in groundwater engineering applications.

A quantitative rigorous and reliable evaluation of tracer data requires a careful and firm planning and realization of the test. The test should be, in view of the later mathematical description of the test data using either analytical solutions (e.g. type curves) or numerical modeling, as simple as possible. For this purpose, tracer injection should be either instantaneous (Dirac pulse) or continuous over a well defined input-period. The tracer should be injected directly into the tested formation. Samples must be taken at intervals close enough to fully cover the complete tracer transit in the observation well. Theoretically, it is necessary to sample in logarithmically equal time intervals as described in the standard tracer test text book (Käss 1998; Leibundgut et al. 2011) or general groundwater text books (Freeze and Cherry 1997; Schwartz and Zhang 2003).

Analytical solutions for tracer transport equations are available in the literature. These are solutions for the differential equations of mass transfer, for a number of different experimental arrangements. The analytical solutions can be recast in terms of dimensionless solutions that can be graphically displayed as type curves. The tracer transit on these type curves is displayed as the dimensionless tracer concentration ($C_D = C/C_{max}$) versus the logarithm of the dimensionless time

13.3 Tracer Experiments

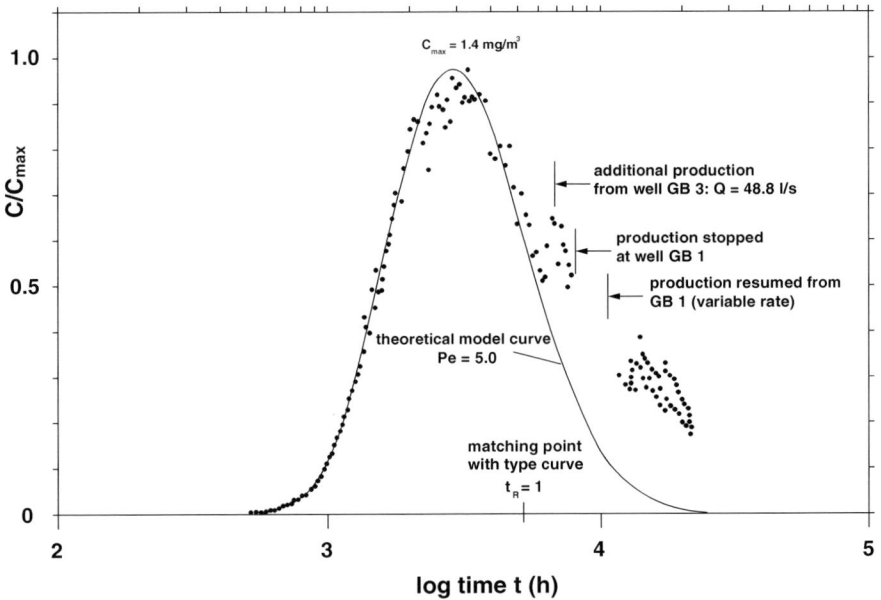

Fig. 13.14 Analysis of a tracer test with type *curves*. The data are from a tracer test implemented between the two geothermal wells Saulgau GB1 and GB2 at a distance of 430 m (Stober 1988)

($t_D = u^2 \, t/D_L$) for a series of hydraulic parameters (where u stands for the flow velocity and D_L represents the longitudinal dispersivity). For the analysis of the test data, the tracer concentration (C), normalized to the measured tracer peak concentration (C_{max}) is plotted against log time (t). The resulting data curve is then matched with the type curve that fits best. The sought-after parameters are taken from the curve with optimal fit.

An example of tracer test data are shown for the thermal aquifer of karstified upper-Jurassic limestone near Saulgau (Germany) at about 650 m depth and at 42 °C (Fig. 13.14). For the tracer test, 2 kg Uranine have been injected into the geothermal well GB3 Saulgau. During the experiment, water has been pumped at a constant rate of Q = 29 l/s from the geothermal well GB1 at a distance of 450 m from the injection well GB3. This assured a radial-convergent flow regime. The first traces of Uranine arrived at the production well GB1 already after 22 days. The maximum tracer concentration, 1.4 μg/l, was measured after 125 days. After 250 days, the pumping rate has been changed, so that only the first part of the three year long series of measurements could be analyzed with the type-curve method (Fig. 13.14). The geo-hydraulic analysis of the data yielded a value for the flow porosity of 2.7 %, a flow velocity of $u = 10^{-5}$ m/s (0.86 m/d and a longitudinal dispersion $D_L = 10^{-3}$ m^2/s (for details of the analysis see Stober 1988). During a follow-up circulation test between the two geothermal wells two additional tracer chemicals have been injected into the formation, Eosine and di-tritium oxide (very heavy water). All injected tracers could be detected in the production well, however, the temperature of the produced water was not lowered by the injection of cooled water into the injection well.

Other partially exceedingly sophisticated tracer test are occasionally being used in geothermal system development. The goal of the tests is the detailed characterization of the geothermal reservoir. The complicated tests include multi-tracer tests and dual-scale push–pull tests for the characterization of the water–rock contact surface and the change of the properties of this surface in stimulation experiments. These testing methods are under development and belong to the category of research efforts rather than being mature methods of the applied sciences (e.g. Ghergut et al. 2007).

13.4 Temperature Evaluation Methods

Large volumes of ascending or descending waters leave their thermal trace in the formation. The thermal effects of vertical fluid migration can be used for deriving hydraulic parameters of the formation simply by monitoring the thermal imprint of fluid migration. The method is called temperature at-rest monitoring. Assuming that the fluid on basement fractures thermally equilibrates with the host rock, and knowing some other thermal parameters of rock and fluid such as the density of water ρ_w, the compressibility of water, the thermal conductivity of the rock (λ) and the vertical component of the flow velocity (v_z) several parameters can be derived from temperature measurements in the wellbore (Bredehoeft and Papadopulos 1965; Mansure and Reiter 1979). Upwelling waters are displayed in vertical temperature profiles as convex curves, descending waters as concave curves in the temperature versus depth profile (Fig. 13.15). The analytical solution of the differential equation to the problem as given as Eqs. 13.6 and 13.7):

$$(T_z - T_0)/(T_H - T_0) = f(\beta, z/H) \tag{13.6}$$

$$f(\beta, z/H) = [\exp(\beta(z-z_0)/H) - 1]/[\exp\beta - 1] \tag{13.7}$$

$$\beta = \rho_W c_W / \lambda v_z H \tag{13.8}$$

where

T_z temperature at depth from z_0 bis $z_0 + H$ (measured temperatures)
T_0 temperature measured at z_0 (reference depth)
T_H temperature measured at depth $z_0 + H$
H zone thickness of upwelling or downwelling fluids.

13.3 Tracer Experiments

Fig. 13.15 Type *curves* for deriving vertical water movements from temperature logs (Bredehoeft and Papadopulos 1965)

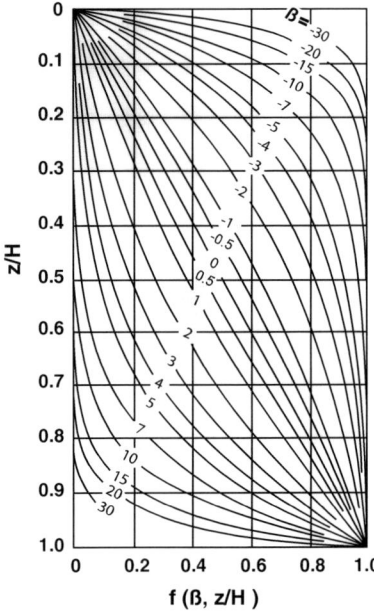

Fig. 13.16 Vertical water movements deduced from from temperature logs

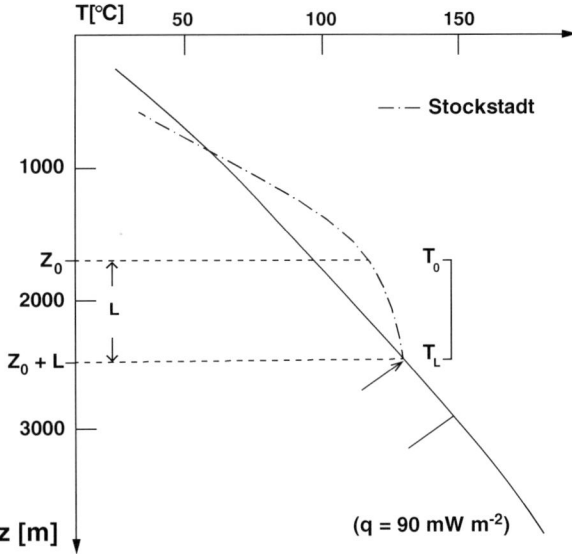

From the measured temperature profiles and with the help of type curves for the function f (Fig. 13.16) the parameter β and thus the vertical component of the flow velocity can be deduced from Eq. 13.8.

Chapter 14
The Chemical Composition of Deep Geothermal Waters and Its Consequences for Planning and Operating a Geothermal Power Plant

Silica sinter at Mammoth hot spring, Yellowstone NP, Wyo, USA

The fracture porosity of the continental crust is normally saturated with an aqueous fluid (Ingebritsen and Manning 1999; Fritz and Frape 1987; Bucher and Stober 2010; Stober and Bucher 2005b). This fluid is used to transfer thermal energy from the hot depth to the cold surface for various uses. The chemical composition of this natural heat transfer fluid depends on the predominant (reactive) rock type of the thermal reservoir and its changes along the circulation pathway. Most deep fluids are saline brines with the major components NaCl and $CaCl_2$. Typical deep fluids contain between 1 and 4 molar NaCl equivalents corresponding to a total of dissolved solids (TDS) in the range of 60–270 g/L (Emmermann and Lauterjung 1997; Möller et al. 1997; Stober and Bucher 2005a; Nordstrom et al. 1985; Pauwels et al. 1993; Kozlovsky 1984; Banks et al. 1996). The chemical composition of the fluid has a number of consequences for the operation of a power plant that will be briefly explored in this chapter.

The pre-drilling conditions are normally unknown, but the aqueous fluid residing in the fracture porosity of crystalline basement (for example) at some thousand meters depth typically has a complex composition with a high salinity and locally high amounts of dissolved gases. The natural solutes have quite different origins that can be separated into locally derived components from reaction with the rock matrix and externally derived components that have been introduced by migrating fluids (Kharaka and Hanor 2004). Natural fluid migration velocities tend to be very small at several km depth because of decreasing natural hydraulic conductivity with depth (Stober and Bucher 2007a, b, Ingebritsen and Manning 2010) and because of decreasing head gradients driving fluid migration. Consequently, natural deep fluids have a composition that is not very far from equilibrium with the host rock and thus chemical interaction between fluid and rock is slow and mild.

The situation described here does explicitly not include high-enthalpy volcanic environments where infiltrating surface fluids may chemically interact vigorously with reactive rocks. We will not describe these environments here. It is recommended that the reader consult dedicated publications that deal with high-enthalpy environments (Nicholson 1993; Giggenbach 1981).

Once the fluid becomes accessible after the first wellbore has been drilled, the fluids must be sampled and its chemical composition carefully analyzed and interpreted. Later, during operation of the plant and the associated fluid circulation, chemical changes of the fluid must be carefully monitored because they may reflect changing reservoir conditions and alteration of the reservoir structures. The hydrochemistry is a very sensitive monitor for subtle developments in the reservoir during the years of operation. Long-term operation of a system requires an excellent knowledge and understanding of chemical processes in the reservoir that are reflected in the composition of the produced fluid.

14.1 Sampling and Laboratory Analyses

Some hydrochemical parameters must be measured at the site. However, it is not a trivial matter to sample a fluid at 200 °C and 500 bar. Normally, fluids are being cooled and decompressed at the wellhead and then sampled. Electrical

14.1 Sampling and Laboratory Analyses

conductivity, pH and redox potential should be measured immediately after sampling. Alkalinity should also be titrated at the site, particularly for high-pH waters. Conservative composition parameters can be measured later after shipping to the service laboratory. Sampling and analysis of dissolved gasses requires special techniques and a certified laboratory with the appropriate expertise. Additional analysis, especially isotope composition analyses, can be helpful for solving special questions.

Because, deep fluids may contain relatively high concentrations of toxic or otherwise harmful components such as heavy metals including lead, zinc, cadmium and others, arsenic, mercury, acid, lye and consequently the pumped deep fluid must be handled with care and should not be mistaken for drinking water.

Sampling requires special techniques that isolate the fluid from contact with the highly oxidizing atmosphere. Some solutes remain stable after cooling and transport. Other components may chemically interact in various ways at low temperature and change concentration. The most common situation is that some solids may become oversaturated and start to precipitate from fluid during cooling, degassing and oxidation. Some components can be kept in solution by acidifying a part of the sampled material. This can be done with e.g. nitric acid because nitrate is not a typical component of deep fluids. However, the appropriate method of sample conservation depends on the parameters to be analyzed and the planned analytical methods, which should be known before sampling. It may not be possible to analyze for a certain helpful extra parameter if the samples have been stabilized with chemicals that precludes this (Arnórsson et al. 2006; Nicholson 1993).

The hot salty fluid is chemically fairly aggressive and tends to react with materials it comes in contact with, including steel tubing. Cooling loops from stainless steel may rapidly corrode in contact with certain brines adding metals to the solution and changing its REDOX state during sampling (Hewitt 1989; Parker et al. 1990). Sampling hose may be penetrable for oxygen and even CO_2, which may cause severe alteration of the original composition of the deep fluid.

Sampling bottles and any other equipment that has contact to the fluid to be sampled must be carefully pre-cleaned. Polybottles with tight locks are standard. Glass bottles should be avoided because they may release certain components to the fluid, although the glass contribution to high TDS fluids is probably negligible in most cases.

Flow cells are typically used to measure the on-site parameters temperature, electrical conductivity, redox potential, pH and dissolved oxygen gas. High dissolved oxygen may reflect gas leaks in the sampling devices and the flow cell. In addition to the abovementioned transient dissolved carbonate species some other solutes are unstable as well and need to be analyzed at the sampling site. This includes NH_4^+ (ammonium), NO_2^- (nitrite), HS^- (hydrogen sulfide), thiosulfate and others. Dissolved silica can cause problems if saturation with amorphous silica is exceeded during sample cooling. This can be critical in high-enthalpy systems. In low enthalpy systems it is not normally necessary to analyze for dissolved silica at the sampling site directly.

Analysis of the samples for major and trace components requires a laboratory equipped with a few specialized instruments. Most labs use ion chromatography (IC) for analyzing anions and photometry for uncharged solutes such as silica and boron. Cations can also be analyzed by ion chromatography but most labs use one of the several forms of inductively coupled plasma (ICP) spectroscopy (ICP-AES atomic emission spectroscopy = ICP-OES optical emission spectroscopy). The ICP instruments can be combined with a mass spectrometer (ICP-MS). In our lab at the University of Freiburg, the cations are being analyzed by atomic absorption spectroscopy (AAS) with a flame AAS (major components) or a graphite furnace AAS instrument (trace components). Titration methods are normally used for carbonate species and dissolved sulfide species. The quantitative analysis of elements present in a number of species and in different oxidation states such as Fe(II) and Fe(III) or As(III) and As(V), the various sulfur species, dissolved chromium, uranium and many more is analytically demanding and prone to alteration during fluid ascent from the reservoir, during sampling, handling and analysis (Arnórsson et al. 2006). Of course, knowing the precise redox state of the fluid at depth would be very helpful for making predictions about its behavior upon production, decompression and cooling (in short during later operation of the plant).

Samples that are used for cation analysis are acidified with HNO_3 and also filtered through a 45 μm acetate filter at the sampling site. Samples for anion analysis by IC are normally being filtered only. Samples for later pH and carbonate measurement, if on site analysis is not possible, need to be kept in gas-tight containers and contact to the atmosphere must be excluded.

Recommendations and technical advice for sampling and analyzing hot, highly mineralized and often gas-rich geothermal fluids including high-temperature sulfur geochemistry can be found in Ball et al. (1976), Thompson et al. (1975), Thompson and Yadav (1979), Nicholson (1993) or Cunningham et al. (1998).

The downhole-sampler is a special device that allows collection of water samples at depth thus providing in situ properties of the fluid including dissolved gasses and gas composition at depth under pressure and at reservoir temperature.

Sampling the various fluids associated with volcanic high-enthalpy fields such as fumaroles, dry-steam and wet-steam wells requires experience, expertise and special equipment. This is beyond the scope of this book (see e.g. Arnórsson et al. 2006; Sutton et al. 1992).

14.2 Deep Geothermal Waters, Data and Interpretation

A key parameter characterizing aqueous fluids is the pH value defined as the negative decimal logarithm of the activity of the hydrogen ion H^+ (pH = $-\log a_{H+}$). It is a dimension-less quantity but numerically equal to the H^+ molality of solution because of the chosen standard state $a_{H+} = 1$ for a $m_{H+} = 1$ solution. Acid solutions have a low pH (high H^+ concentration); alkaline solutions have high pH (low H^+ concentration). Neutral solutions are characterized by equal amounts of

H^+ and OH^- ions in the solution. At 25 °C pure water is neutral at pH = 7, at 108 °C the neutral point is at pH = 6.0 and at 200 °C a neutral solution has a pH = 5.5. The decrease of neutral pH is a consequence of the decreasing self-ionization constant K_w of water with temperature. Geothermal waters from crystalline basement reservoirs tend to be slightly acidic or close to neutral. pH values of deep waters at 150–200 °C tend to be in the range of 5–6 (Pauwels et al. 1993; Fritz and Frape 1987; Stober and Bucher 1999a; Bucher and Stober 2000). The parameter is difficult to assess for high-TDS fluids at high temperature and pressure. However, reliable predictions for potential risks for scaling and corrosion strongly depend on the precise knowledge of the pH of the produced fluid under reservoir conditions. Because pH is a logarithmic value the H^+ concentration in a pH = 5.5 and a pH = 5.8 solution differs by a factor of 2 (there is twice as much H^+ in a pH = 5.5 solution than in pH = 5.8 solution). Thus the numbers after the decimal point matter.

The oxidation-state of the produced fluid can be characterized with the p_e value defined as negative decimal logarithm of the activity of the electron e^- ($p_e = -\log a_{e^-}$) in analogy to the pH value. It is also a dimensionless quantity. Neutral surface waters in equilibrium atmospheric O_2 have a p_e value of about 13. Water of pH = 6 is stable in the range $p_e = 15$ (highly oxidizing) to $p_e = -5$ (very reducing). The presence of sulphate sulfur (SO_4^{2-}) in most deep geothermal fluids rather than sulfide sulfur (H_2S, HS^-) implies that p_e must be in the stable field for sulphate at the given pH of the fluids. Most deep fluids in crystalline basement reservoirs are relatively oxidized with sulphate as the dominant sulfur species in solution, with CO_2 rather than methane (CH_4) as the dominant carbon gas dissolved in the fluid and with carbonate carbon (C^{IV}) in the solution.

The solubility of many minerals depends on the oxidation state of the rock-fluid system. Particularly iron-bearing minerals may dissolve in highly reducing fluids and Fe^{2+} concentration of such fluids may be very high. On the other hand, iron-bearing minerals may not dissolve in oxidizing fluids and the concentration of Fe^{3+} in the fluids is extremely small. In most primary rock-forming Fe-minerals (e.g. biotite) in granite and gneiss iron is present in its reduced di-valent form. The mineral biotite is unstable at most reservoir conditions for geothermal applications (<350 °C). It dissolves in the pore fluid and reprecipitatates as secondary minerals such as clay. The iron is insoluble at oxidizing high p_e conditions and precipitates as iron oxide or oxide-hydroxide mineral (goethite, ferrihydrite, hematite). Thus pumped hot deep fluids may contain measurable amounts of dissolved iron only if p_e is low and the conditions reducing. This is not normally the case, however. Trivalent iron (Fe^{3+}) is soluble in very acid low pH fluids, which can be present in some volcanic environments, but the moderate pH of deep fluids in granites and gneisses typically precludes the presence of dissolved iron even under moderately oxidizing conditions. Thus very low Fe in geothermal deep-water at moderate pH indicates rather oxidizing conditions in the reservoir formation.

Generally, oxidation-reduction reactions transfer electrons from the reduced state to the oxidized state. For the important iron example: Fe^{2+} (reduced) = Fe^{3+} (oxidized) + e^- (electron). If the reaction progresses to the right-hand-side it produces electrons and divalent iron is oxidized to trivalent iron. If the reaction runs to the left it consumes electrons or reduces iron from its trivalent to the divalent state. Thus if there is a source of electrons, the environment is reducing and (like pH) p_e is low (negative). Vice versa, if p_e is high (e.g. 15), electrons are rare and the environment is oxidizing.

The REDOX state of geothermal fluids is measured as the REDOX potential using an electrode system (millivolt mV electrode). Usually the mV electrode can be connected to a pH meter. The measured REDOX potential E_H (in Volt or mV) depends on temperature, which must be recorded together with the mV measurement. The REDOX potential can be directly used as a parameter for fluid interpretation and risk prognosis. It must be converted to the parameter p_e for use in various hydrochemical software models or analyzing data on p_e versus pH diagrams:

$$p_e = E_H \; \{\Im / (2.303 \, R \, T)\} \tag{14.1}$$

where $\Im$ represents the temperature dependent Faraday constant (96,485 J V^{-1} mol^{-1}), R the universal gas constant (8.314 J K^{-1} mol^{-1}) and T temperature (K). The factor 2.303 converts to decimal log scale. The expression in { } has the dimension 1/V, together with E_H in V results in the dimensionless p_e. At 25 °C the conversion is: $p_e = 16.9 \, E_H$ (in V).

Dissolution of minerals in aqueous fluids releases true and potential electrolytes to the fluid. The ions resulting from the dissociation of the electrolytes make the solution electrically conductive. The contribution of a specific electrolyte in the solution to the total electrical conductivity (EC) depends on the charge of the ion, the degree of dissociation, the concentration of the electrolyte, TDS and other factors. The electrical conductivity (S m^{-1}) of the geothermal fluid results from the combined contributions of all ions present in the fluid. Therefore, measured EC is proportional to the TDS of the fluid. Natural near surface waters have EC in the range of 2–100 mS/m, the EC of seawater is about 4.5 S/m and the saline fluid (62 g/L TDS) from the 4,000 m deep thermal fluid (120 °C) from the pilot hole of the continental deep drilling (KTB) has a measured EC of 6.8 S/m. Because deep geothermal fluids are typically concentrated Na-(Ca)-Cl brines, EC is closely related to salinity. The EC is measured with an electrical conductivity meter, a resistance probe that can be attached to a hand-held instrument. The electrical conductivity depends on the temperature, which must be reported with the EC measurement. The EC can be measured as part of geophysical well logging. EC logs, also referred to as salinity logs, may identify and localize inflow structures of fluid with contrasting TDS (Sect. 12.2).

The measured concentration of dissolved solids is reported as mass of solute per unit volume of solution (g/L). If the amount of solute is expressed in mole (millimole) units the concentration (mole/L) is termed molarity. Another commonly used concentration unit is mass of solute per kg of water (g/kg; mg/kg). Using number of moles of solute per kg of pure water (solvent), the unit (mole/kg) is called molality. Molarity and molality should not be confused. Concentrations given in the two different units are very similar for low-TDS fluids; however, they differ for brines and other high-TDS fluids. As an example: A saturated NaCl solution at 25 °C contains 343 g NaCl per one liter solution (molarity = 5.86) and 358 g NaCl per one kg of water (molality = 6.13). Note also that solubility information is commonly given in molality that is number of moles of a substance dissolve in 1 kg of pure water.

The number and type of solutes to be analyzed in a water sample depends on the scope of investigation and the relative importance of the solute for understanding the chemical behavior of the rock-water system. Most saline deep fluids contain, in the order of importance, sodium (Na), calcium (Ca), potassium (K) and magnesium (Mg) as the dominant cations. It may be useful to analyze also for strontium (Sr), ammonium (NH_4^+) and lithium (Li). Critically important are the concentration of iron (Fe) and manganese (Mn). The two elements are present in detectable amounts only in reduced waters. Their concentration gives valuable information on the REDOX state of the fluid. Note, however, that in normal relatively oxidized waters, Fe is below 1 mg/l or even below 1 µg/l. Reported Fe concentrations above these levels must skeptically viewed. Aluminium (Al) is also present at very low concentration levels (µg/l) in most deep fluids although the fluids are in contact with Al-rich minerals, such as micas and feldspars, at depth. It should always be attempted to measure Al in the sampled deep fluids, because without Al-concentration data the saturation state of the fluid with respect to the minerals of the formation cannot be modeled.

The major anions of geothermal deep waters are chloride (Cl), carbonate or bicarbonate (CO_3^{2-}, HCO_3^-) and sulphate (SO_4^{2-}). It is recommended to analyze for fluoride (F^-) and bromide (Br^-) because Cl-rich waters and brines also contain the other halogens and because they are of diagnostic value. It is also recommended to analyze for iodide (I^-) in high-salinity fluids. The halogen data can give valuable hints about the origin of the salinity and thus the origin of the deep fluid. Other anions common in near surface waters such as nitrate (NO_3^-), nitrite (NO_2^-) and phosphate (PO_4^{3-}, HPO_4^{2-}, $H_2PO_4^-$) are less prominent in deep fluids and need not necessarily be analyzed. In reducing waters it may be necessary to analyze for sulfide (HS^-).

Because some trace elements may potentially form scales it is helpful for the assessment of the associated risks to analyze for lead (Pb), barium (Ba) and arsenic (As).

In the pH range of typical deep fluids silica (Si) and boron (B) are present as uncharged complexes in the fluid. Most thermal fluids are pumped from formations made of silicate rocks. The solubility of quartz is low at low temperature (~6 mg/kg at 25 °C) but rapidly increases with temperature. The concentration of SiO_2 in the water bears important information on the reservoir temperature and depth. Thus dissolved silica in water can be used as a geological (geochemical) thermometer. Boron may give indications on the origin of the fluids; however, it may not be necessary to analyze the parameter.

The total of dissolved solids (TDS) is the sum of all dissolved constituents (cations, anions, uncharged species). Because TDS is the total amount of solids remaining after evaporating one liter of the fluid to dryness it is common usage to convert analyzed bicarbonate to carbonate. TDS is given in g/L or mg/L.

Table 14.1 Chemical composition of water from the Muzhaerte hot spring in the Tian Shan mountains of NW China (Bucher et al. 2009b)

Date	18 August 2005		
Temp °C	55		
pH	8.29 (at 25 °C)		
EC (mS/cm)	1.38		
	mg/l	mmol/l	meq/l
Ca	38.80	0.97	1.94
Mg	0.69	0.03	0.06
Na	248.00	10.78	10.78
K	7.60	0.19	0.19
Sr	1.23	0.01	0.03
Rb	0.11	0.00	0.00
Li	0.39	0.06	0.06
Fe	<0.02		
Al	0.020	0.0007	0.0022
Alk (HCO$_3$)	57.36	0.94	0.94
SO$_4$	318.00	3.31	6.62
Cl	168.00	4.74	4.74
NO$_3$	1.52	0.02	0.02
F	7.38	0.39	0.39
Br	0.11	0.0014	0.0014
SiO$_2$	66.57	1.11	1.11
HBO$_2$	6.04	0.14	0.14
TDS	921.82		
Cl/Br	1,527		
Na/Cl		3,443	
Ca/SO$_4$		2.28	
X$_{An}$		0.29	
log a$_{SiO2}$		0.08	
		−2.98	
Cations			13.05
Anions			12.71
EN			1.33

Stable isotope composition: $\delta^{18}O = -11.72\ \%$, $\delta^2H = -82.5\ \%$

Analytical concentration data are typically reported as mg/L of the solute. The mass per volume data need to be recalculated to mmol/L and meq/L data for further research and quality tests including charge balance of the analysis. An example analysis is given in Table 14.1.

The amount of a gaseous component i dissolved in a fluid c_i (mole/L) is proportional to the partial pressure p_i (Pa) of the gas (in the gas phase). This behavior of gases is known as Henry's law:

$$c_i = K_{H_i} p_i \qquad (14.2)$$

The gas distribution coefficient K_{H_i} (Henry constant) depends on the type of gas, the temperature and the composition of the fluid (TDS). Gas solubility decreases with increasing temperature. It also decreases with increasing salinity (TDS). The most important and abundant gas in deep fluids is carbon dioxide (CO_2) followed by nitrogen (N_2). In very reducing environments methane (CH_4) and hydrogen sulfide (H_2S) may become important. The presence of high amounts of H_2S may cause serious environmental problems during plant operation. The gas must be removed from geothermal steam in high-enthalpy steam-driven power plants before releasing the steam to the atmosphere. Very gas-rich fluids have properties that deviate significantly from those of gas-poor geothermal water. These properties influence the thermal power of the system and the hydraulic properties of the reservoir including hydraulic conductivity, transmissivity, and storage coefficient (Sects. 8.2; 8.6).

Plausibility checks help to evaluate the trustworthiness and reliability of hydrochemical data purchased from a commercial laboratory. The first control parameter is the charge balance. The total of positive charges from the cations and negative charges from the anions must match in an electrically neutral solution. The calculated total of dissolved solids should be consistent with the measured electrical conductivity. The measured pH must be consistent with the analyzed concentrations of the carbonate species. Some concentration levels of specific solutes are implausible under certain conditions if the waters are reasonably close to equilibrium. As explained above, for example, high concentrations of iron in pH 6 waters are implausible if the REDOX potential or dissolved oxygen is high. Another example, if fluoride is high, calcium cannot be high at the same time because the concentrations of the two solutes are tied to the solubility of the mineral fluorite.

In the course of project development and operation of a power plant a large number of chemical fluid composition data accumulate. The data are collected in tables but for the evaluation and comparison of data graphical display and representation of data is very helpful. The type of graphical data representation primarily depends on the water composition and its variation with time but also on the aspects of fluid geochemistry one wants to illustrate. It follows from the general chemical composition of deep fluids that the diagrams must display the major component Ca, Mg, Na, K, Cl, Alkalinity and SO_4. The four major cations cannot, unfortunately, be displayed on a two dimensional diagram. Therefore, the alkalis Na and K are displayed as the sum of the alkali metals with the disadvantage that Na–K variations remain invisible on the diagrams. However, the relative proportions of the major ions in most fluids can be adequately displayed of an concentration triangle for cations (Ca-Mg-(Na+K)) and one for anions (Cl-SO_4-(carbonate alkalinity)) in % meq/L (Fig. 14.1). The cation and anion triangle is well suited to represent large amount of data since each analysis is represented by a single point on each triangle. The major disadvantage of the graphs is that TDS cannot be discriminated. It only shows the relative proportion of cations and anions. Very diluted near surface water and highly concentrated brines plot to the same point in the diagrams if the ion proportions of the two fluids are the same.

An extension of the cation and anion triangle is the so-called Piper diagram (Fig. 14.2). The extra value added by the quadrilateral that combines information already displayed on the cation and anion triangle is minimal may not be worth the effort.

264   14 The Chemical Composition of Deep Geothermal Waters and Its Consequences

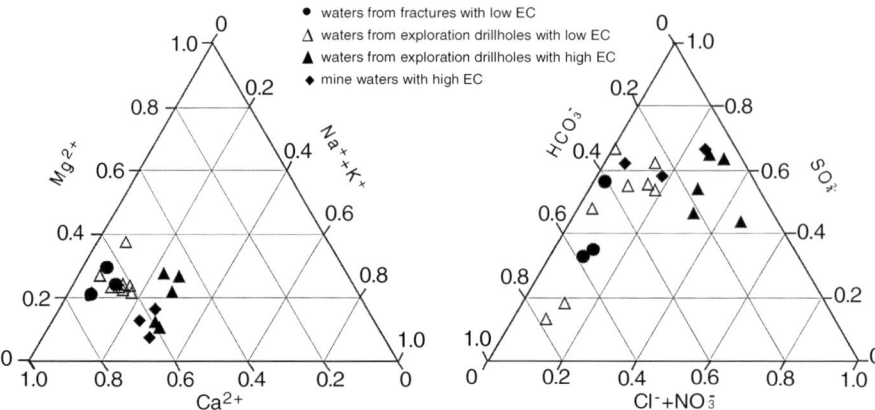

Fig. 14.1 Cation and anion *triangle* showing water samples from a deep underground mine in the crystalline basement of the Black Forest, SW Germany (Bucher et al. 2009a)

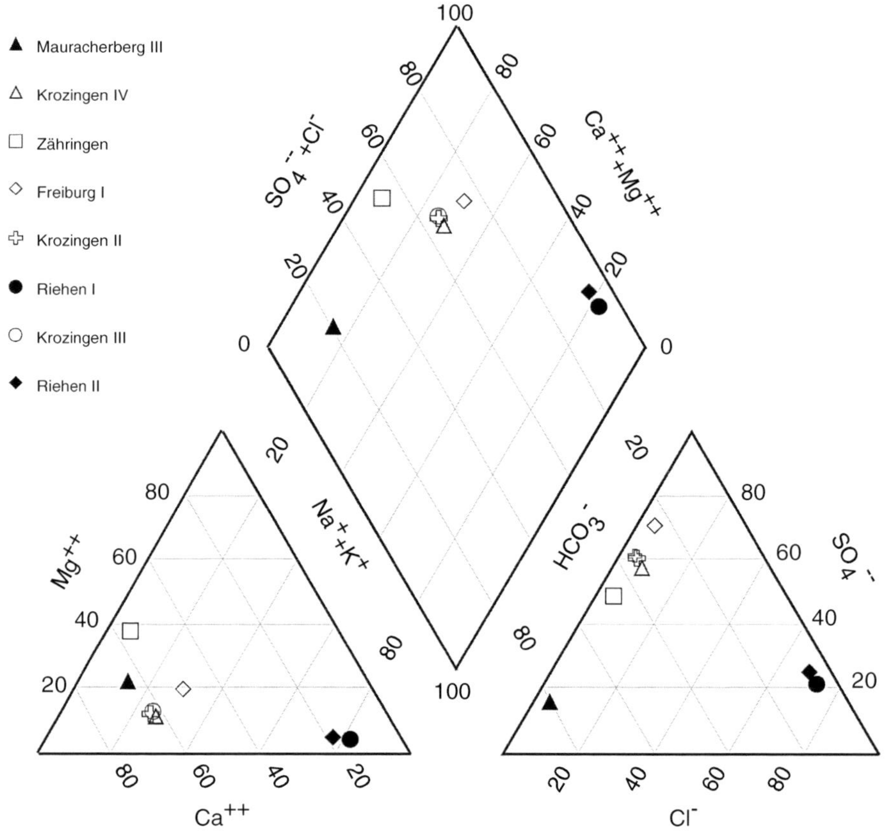

Fig. 14.2 Thermal waters from the middle Triassic Muschelkalk limestone formation in the Upper Rhine rift valley (He et al. 1999). Piper diagram

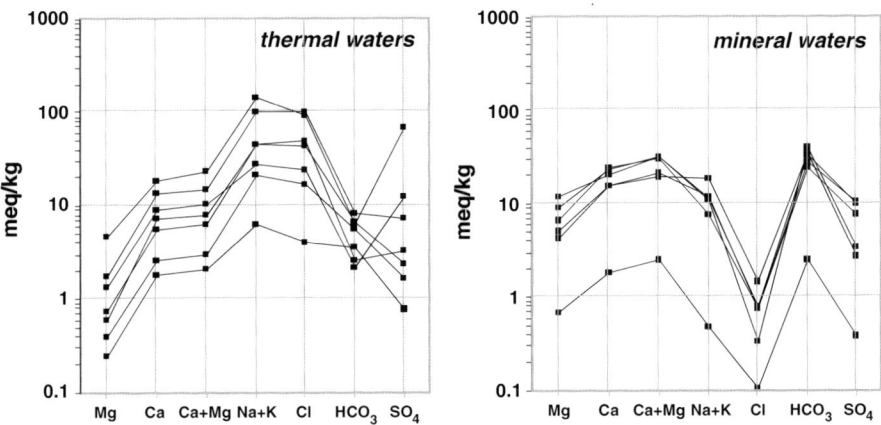

Fig. 14.3 Chemical composition of thermal and mineral waters from the crystalline basement of the Black Forest area (fractured granite and gneiss) (Stober and Bucher 1999a). Schöller diagram

Schoeller-Diagrams avoid some of the shortcomings of the ion-triangles and derivatives (Piper). The Schoeller-Diagram is a histogram-type of diagram showing the concentration in meq/l of a solute on a logarithmic scale. The solute and the order of the solutes on the diagram are arbitrary in principle. However, the great advantage of Schoeller-Diagrams is the pattern recognition potential, which only takes effect if a strict order of solutes and combination of solutes is obeyed. We strongly recommend displaying the following solutes in the strict order on Schoeller-Diagrams: Mg, Ca, Ca+Mg, Na+K, Cl, Alk, SO_4 (all in log meq/L). The pattern distinguishes high-TDS clearly from low-TDS fluid. However, the amount of data that can be reasonably displayed depends on the variability of the data. However, even if data are similar more than 20 analyses typically lead to graphically quite chaotic figures. Note that the pattern recognition potential is destroyed if, for example, Mg and Ca are put into the reversed order Ca then Mg. Schoeller-diagrams are not in common use in the US science community. However, we recommend trying this, in our opinion, useful type of diagram (Fig. 14.3).

The compositional characteristics of waters (Table 14.1) can be displayed on many different types of diagrams depending on the type of produced waters and the compositional features of the produced waters. It is important to realize that interesting compositional properties can be displayed on many different kinds of graphical diagrams including ionic-ratio diagrams (Na/K ratio, Ca/Mg ratio, Na/Cl ratio and others).

The composition of deep fluids produced by a geothermal power system has many different aspects. One aspect is the ultimate origin of the saline fluid that is pumped to the surface. A very useful tool in analyzing fluid origin is the Cl/Br ratio. Seawater has a Cl/Br mass ratio of 288. This number, therefore, is the absolute reference number of the seawater origin of salinity. Any Cl/Br ratio significantly above 288 suggests that the salinity of the fluid is derived from the

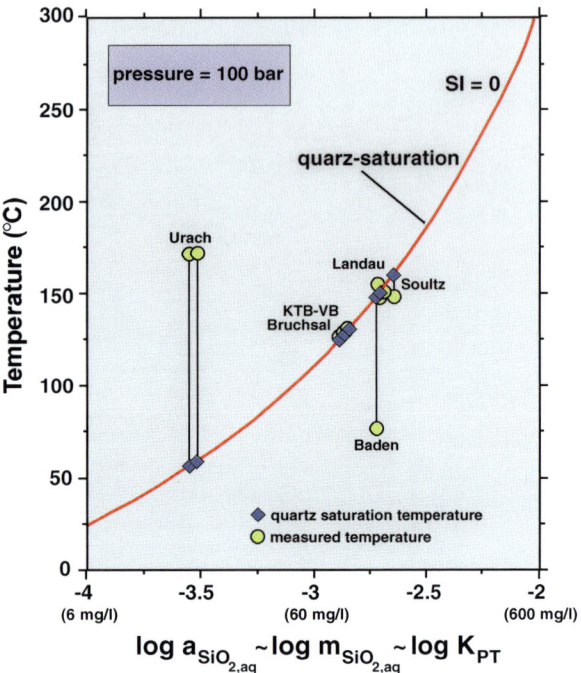

Fig. 14.4 Temperature dependence of the solubility of quartz (see text). Water data from various deep geothermal reservoirs in Germany

dissolution of evaporitic NaCl sediments. Any number massively below 300 suggests that the salinity originates from crystalline basement (Trommsdorff et al. 1985; Stober and Bucher 1999b).

The chemical composition of deep fluid reflects the temperature of the reservoir area. It is therefore worth the effort to deduce reservoir temperature and with the given geothermal gradient the depth of the geothermal reservoir. The temperature—composition relationship is referred to as a geological thermometer (geothermometer). Deducing reservoir temperatures from temperature—composition relationships is fundamentally based on the assumption of chemical equilibrium of the considered reaction.

The most important example is the solubility of quartz in water. Quartz dissolution can be described by the simple reaction:

$$SiO_{2solid} = SiO_{2aq} \quad (14.3)$$

Equilibrium of reaction 14.3 requires that:

$$\log K_{PT} = \log a_{SiO_{2aq}} \quad (14.4)$$

Because the activity—composition relationship [a = f(m)] for the uncharged silica species is close to $a_{SiO_{2aq}} = m_{SiO_{2aq}}$ and because the solubility depends on temperature and pressure $K = f(T,P)$, the solubility of quartz in water can be displayed on isobaric temperature versus log $m_{SiO_{2aq}}$ diagrams (Fig. 14.4). Figure 14.4 shows that SiO_2 dissolved in water in equilibrium with quartz rapidly increases with temperatures from 6 mg/l at 25 °C to 600 mg/l at 300 °C. The SiO_2 dissolved in the produced fluid can be analyzed and the equilibrium temperature with quartz can be read from the diagram. The influence of the pressure is very small and can be neglected. As shown on Fig. 14.4, most of the displayed deep fluids have quartz equilibrium temperatures identical to the measured bottom-hole temperatures. This is convincing evidence that the silica thermometer represents a reliable instrument for reservoir temperature estimates. However, there are two examples of thermal fluids with large mismatches between measured and silica temperatures. The Baden–Baden fluid has a measured temperature that is much lower than the quartz-saturation temperature. The silica temperature suggests that the water is at equilibrium with quartz at 150 °C. This may represent the true reservoir temperature corresponding to a depth of 5 km assuming a gradient of 30 °C/km. The water is an artesian spring and may cool on the paths to the surface. The Urach fluid has a bottom hole temperature of 170 °C but a very low concentration of dissolved silica that results in a computed temperature of only 55 °C. Since the reservoir formation is gneiss with feldspar and quartz there is plentiful of silica in the formation. The reason for the unusually low silica concentration is unknown. It shows, however, that data from geothermometers must be viewed with caution.

The stable form of SiO_2 in geothermal formations is normally quartz. Thus quartz has the lowest solubility of all forms of SiO_2. Chalcedony is a metastable form of SiO_2 and has a higher solubility than quartz. Amorphous silica has a much higher solubility than both quartz and chalcedony. A geothermal fluid in equilibrium with quartz at 150 °C contains about 150 mg/l dissolved SiO_2 (Fig. 14.4). If pumped to the surface and cooled in a power plant the fluid will be oversaturated first with respect to quartz and below 125 °C also with respect to chalcedony. In principle, the stable solid quartz should precipitate in order to maintain equilibrium. However, reaction kinetics of quartz formation is slow and the fluid remains oversaturated with Qtz and its SiO_2 concentration reflects the reservoir conditions. If the fluid would be cooled to below of about 40 °C, amorphous silica would spontaneously and rapidly precipitate. The silica crusts formed this way would age and slowly recrystallize to the more stable solid chalcedony. A 200 °C Qtz-saturated fluid can be cooled to about 70 °C before amorphous silica deposition begins. In high-enthalpy systems where 300 °C fluids may be loaded with 600 mg/l SiO_2 (Fig. 14.4) silica precipitates during extraction of the thermal energy and silica scales are a permanent and serious problem.

Note that the SiO_2 geothermometer is independent of the pH in acid and neutral fluids. In high-pH fluids the solubility of quartz rapidly increases with increasing pH. The effect must be considered if dealing with high-pH fluids.

The silica geothermometer has been calibrated and improved for equilibrium with all three SiO_2 solid phases (Fournier 1977, 1981; Fournier and Potter 1982; Arnórsson 1983; Verma and Santoyo 1997; Verma 2000; Walther and Helgeson 1977). Figure 14.4 has been computed from data of Walther and Helgeson (1977). The Verma (2000) silica geothermometer for saturation with quartz is:

$$T = \{1175.7/(4.88 - \log c_{SiO_2})\} - 273.15 \tag{14.5}$$

where T is the temperature in °C and c_{SiO_2} the concentration of dissolved SiO_2 in the fluid in mg/kg.

A number of popular geological thermometers are based on cation ratios rather than absolute concentrations. The cation ratios are controlled by exchange reactions between minerals and the fluid rather than on the solubility of a mineral in the fluid like the quartz thermometer. The exchange thermometers are either experimentally or empirically calibrated.

If the geothermal water resides in crystalline basement formations such as granite or gneiss, the rocks typically contain K-feldspar ($KAlSi_3O_8$) and plagioclase that is normally rich in Na-feldspar component ($NaAlSi_3O_8$). A fluid in contact with the two feldspars may reach equilibrium of the exchange reaction:

$$KAlSi_3O_8 + Na^+(fluid) = NaAlSi_3O_8 + K^+(fluid) \qquad (14.6)$$

The equilibrium constant of the exchange reaction (Eq. 14.6) can be written as $\log K_{PT} = \log(m_{K^+}/m_{Na^+})$ if the activity is approximated with the molality of the two cations. The value of equilibrium constant depends mostly on temperature. Thus the cation (m_{K^+}/m_{Na^+}) ratio is a function of temperature and it can be used as a geothermometer (Santoyo and Díaz-González 2010). Other cation thermometers use measured concentrations of Na and Li (Drever 2005) or Na, K and Ca (Fournier 1981) or Mg and Li (Nordstrom et al. 1985) for deriving temperature estimates. The Na-K and Na-K-Ca thermometer may not work well at temperatures below 150 °C because of slow equilibration (Fournier 1981). The empirical thermometers involving Li may give plausible and consistent results from very low to high temperatures (0–350 °C, Drever 2005).

Example calibrations of some cation thermometers are (temperature in °C, concentrations mg/kg):

$$\text{Na-K}: T = 876.3/\big(0.8775 + \log c_{Na}/c_K\big) - 273 \qquad (14.7)$$

$$\text{Mg-Li}: T = 2200/(5.47 + \log\sqrt{(c_{Mg}/c_{Li})}) - 273 \qquad (14.8)$$

$$\text{Na-Li}: T = 1590/\big(0.779 + \log c_{Na}/c_{Li}\big) - 273 \qquad (14.9)$$

Equation 14.7: Santoyo and Díaz-González (2010), Eqs. 14.8, 14.9: Drever (2005).

The ratio of the concentrations of chloride and bromide can give valuable information on the origin of the salinity and on mixing of different fluids. In the literature halogen data are given as Cl/Br or Br/Cl ratios on a mass or mole basis. We use here Cl/Br ratios on a mass basis (mg). The planetary reference fluid, standard mean ocean water has a Cl/Br = 288. If seawater mixes with low-TDS surface water the Cl/Br ration does not change and the mixtures follow the seawater dilution line (Fig. 14.5).

Granites and gneisses are the predominant reservoir rocks of the thermal waters of the Black Forest (Fig. 14.3). Leaching experiments with powders of these crystalline basement rocks readily produced leachates with considerable chloride and bromide concentrations. The average measured Cl/Br mass ratio of about 100 is significantly lower than that of seawater (Bucher and Stober 2002). The chloride and bromide in granite and gneiss is mostly released from salty deposits on the grain boundaries of the silicate minerals and from fluid and solid inclusions in the silicate

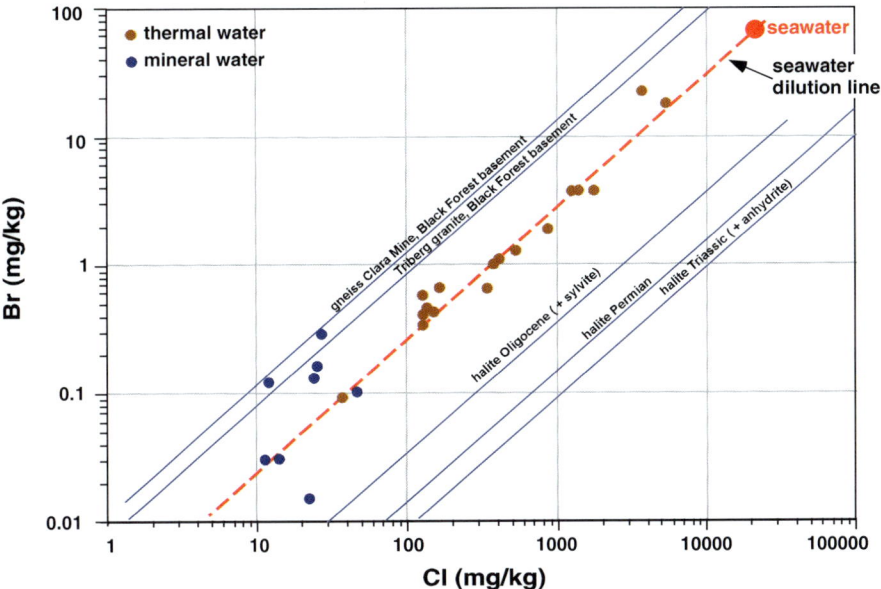

Fig. 14.5 Cl/Br ratio of deep water in the crystalline basement of the Black Forest region (Germany). Thermal waters: *filled circles*, mineral waters: *open circles*. The seawater dilution is given as *dashed line*, data from leached granite, gneiss and various evaporite halites: *full lines* (Stober and Bucher 1999b)

minerals (mostly in quartz) (Stober and Bucher 1999b). Dissolution of evaporitic salt deposits produces saline waters with extremely high Cl/Br ratios because halite cannot accommodate bromide in its structure. Dissolution of Triassic to Tertiary halite from upper Rhine valley resulted in Cl/Br ratios ranging from 2400 to 9900 (Stober and Bucher 1999b). If sampled deep geothermal fluids have Cl/Br ratios along the seawater dilution line (Fig. 14.5) fossil seawater is the probable origin of the water. The original seawater may have been diluted by near surface waters of low salinity. However, the salinity also may increase above that of seawater as a result of hydration of primary minerals a retrogression process that consumes water and desiccates the fracture porosity and passively increases the salinity of the residual water. The Cl/Br ratio does not change during the process until the brine becomes saturated with halite. Further desiccation and concurrent halite precipitation decreases Cl/Br ultimately to very low values. The salinity of waters with very high Cl/Br ratios probably originates from dissolution of salt-bearing formations. In crystalline basement Cl/Br is typically very low, which may indicate an origin of the salinity from salty fluid and solid inclusions and from alteration of Cl-bearing silicate minerals. The Cl/Br ratio may also be controlled by other processes including evaporation, cryogenic processes and others (Frape and Fritz 1982; Frape et al. 2004).

Each dissolved component, e.g. calcium, in the thermal water is distributed among a typically large number of charged and uncharged species (e.g. Ca^{2+},

$CaOH^+$, $Ca(OH)_2^\circ$, $CaCl^+$). For the evaluation of saturation states and thus the potential for scale formation it is necessary to compute the distribution of species in a specific analyzed water at the temperature (and pressure) of interest. The speciation must be computed by hydrochemical software such as WATEQ (Ball and Norstrom 1991), PHREEQC (Parkhurst and Appelo 1999), and SOLMINEQ (Kharaka et al. 1988). With the help of the software, saturation states can be explored for different operating conditions and for the reservoir itself. Meaningful models require a full analysis of the major components but also of some critical trace components. Some of these codes are powerful tools for generating transport and mixing models and allow for sophisticated numerical hydrochemical modeling. The models are, however, sensitive to input pH and temperature (Parkhurst and Appelo 1999).

The code PHREEQC, for example, can be used to compare sampled thermal water with hypothetical water that has been equilibrated with a chosen set of selected minerals. This way, it can be shown to what extent the composition of the thermal water is controlled by the solubility of the primary and secondary minerals of the reservoir rock.

Between a mineral of the rock formation and the thermal water in the fracture porosity a reaction relation exists. For example, if anhydrite ($CaSO_4$) in an evaporite formation is in contact with water the reaction can be written as:

$$CaSO_4 \text{(anhydrite)} \Leftrightarrow Ca^{2+} + SO_4^{2-} \text{ (ions in solution)} \qquad (14.10)$$

If anhydrite is a pure solid phase its activity can be defined as unity. If the anhydrite and the fluid coexist at equilibrium, the following mass-action equation must hold:

$$K_{PT} = a_{Ca^{2+}} \cdot a_{SO_4^{2-}} \qquad (14.11)$$

The dimension-less equilibrium constant is a function of pressure and, here mostly, of temperature. A sampled thermal water must be chemically analyzed (e.g. for calcium and sulphate sulfur) and the data be used for a distribution of species calculation using e.g. PHREEQC. The product of the computed ion activities is called the ion activity product (IAP) and can be written as:

$$IAP = a_{Ca^{2+}} \cdot a_{SO_4^{2-}} \qquad (14.12)$$

IAP is derived from the actual fluid of interest. It can be compared to the equilibrium condition K_{PT}. The logarithm of the quotient IAP/K is referenced to as the saturation index SI:

$$SI = \log_{10}(IAP/K) \qquad (14.13)$$

If IAP exactly matches the equilibrium condition K, $SI = 0$ and the water is at equilibrium with the considered mineral (here anhydrite). If $IAP > K$, $SI > 0$ and the water is oversaturated with the mineral and has the potential to precipitate the mineral. If IAP < K, SI < 0 and the water is undersaturated with respect to the considered mineral and has the potential to dissolve the mineral.

The models of chemical thermodynamics can also be used to study complex fluid-gas-solid reactions. However, chemical thermodynamic can predict the possibility of processes but cannot make predictions regarding when, how fast or if at all these reactions will occur. Kinetically controlled disequilibrium or metastable states may occur that are in conflict with the conditions stable equilibrium. Nevertheless, the conditions of stable equilibrium set the reference frame for all chemical processes in hydrogeothermal systems.

14.3 Mineral Scales and Materials Corrosion

The hot geothermal fluid interacts at depth with the minerals of the reservoir formation. Typically, the fluid is not in an overall chemical equilibrium with the reservoir rock because of slow reaction kinetics and slow diffusion under most reservoir conditions and temperatures of 200 °C and lower. However, the fluids are normally not far from equilibrium and reaction progress of fluid-rock interaction is small. This means the fluid composition does not change much over long periods of time.

The installation and operation of a geothermal power plant fundamentally changes this situation. The deep fluid is pumped to the surface, partly decompressed and substantially cooled. This may, and normally does change the saturation state of the fluid. As a consequence the fluid may become oversaturated with respect to one or several minerals. The minerals may be deposited as scales in the borehole and in the surface installations. Particularly exposed to mineral scales and crusts are heat exchangers, steam separators, filter systems and pipes. However, the re-injected cool fluid may cause increased chemical interaction in the reservoir formation with potential consequences for the porosity and the hydraulic conductivity of the formation.

Another important chemical aspect is the generally high corrosion potential of the hot saline deep fluids. The corrosive fluids chemically attack the casing of the borehole, the submersible pump and all materials of the surface installations it comes in contact with.

The thermal water (fluid) system includes the surface installations that come in contact with the pumped deep fluid also comprising the fluid pump and the re-injection pipe. Dimensioning the system typically considers physical parameters such as pressure, temperature and production rate. The parameters define pipe diameters, pressure ratings and thermal properties of the materials. However, the thermal water system must be operated at a minimum pressure that prevents outgassing and unmixing dissolved gaseous components from the fluid. CO_2-loss from the decompressed fluid is the major cause of carbonate (calcite) scales in the installations. The potentially created gas-fluid mixture may result in a two-phase flow system in the installations with difficult to handle pressure variations in the system. In special environments chemical inhibitors may help to prevent carbonate

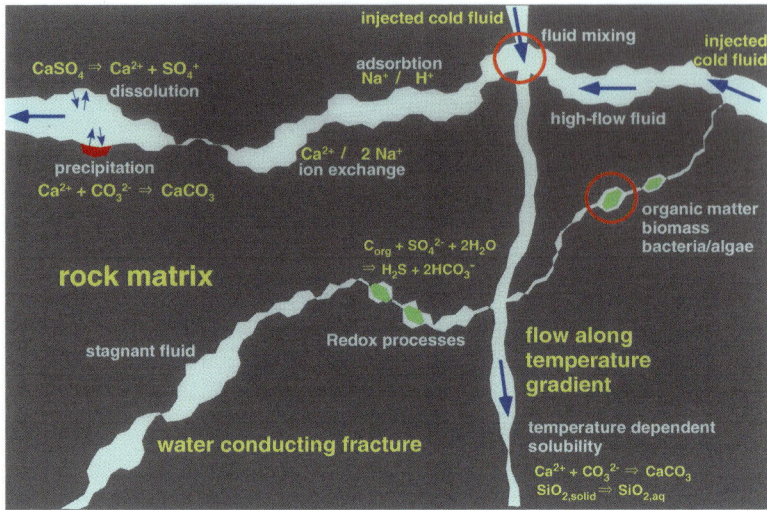

Fig. 14.6 Chemical processes along water conducting fractures of a geothermal reservoir resulting from injection of cool fluid in a hot reservoir formation (see text)

scales and still run the system at degassing conditions if degassing can be prevented at uneconomical high pressures only. Dimensioning and design of the heat exchanger must consider supply and return temperature, pressure gradient between primary and secondary loop, temperature and pressure of the secondary loop fluid, gas-content and composition of the produced thermal fluid and the heat capacity and viscosity of the fluids of the primary and secondary loop.

At the injection side of geothermal doublet systems boiling must be prevented. Ideally, the return flow pipe to the injection borehole is placed clearly below the dynamic water table. Filter systems at the production and injection well prevent solid particles from entering the installations of the power plant. This reduces abrasion and may restrict scales to the filter units and protect heat exchangers and pumps from scale formation. The most typical scales include carbonates (calcite, aragonite) and sulfates (anhydrite).

Significant precipitation reactions in the reservoir will reduce the hydraulic conductivity of the fractured rock system. The resulting increase of the necessary injection pressure may require more powerful (and more costly) pumps. The uptake capacity of the return flow by the injection well can be greatly reduced with disastrous economic consequences.

It is therefore highly recommended to consider, predict and model possible chemical processes in the reservoir as soon as chemical data on the composition of the deep fluid become available. Injection of cold fluid into the fracture system of the heat reservoir causes a number of chemical processes to run simultaneously and for the entire duration of the plant operation (Fig. 14.6). The chemical effects (Fig. 14.6) of re-injecting cooled fluids into the reservoir include dissolution/precipitation reactions (typically involving carbonate and sulfate), ion-exchange and adsorption reactions, chemical consequences of mixing of fluids of different composition and temperature, solubility effects of fluid flow along temperature gradients (vertical flow), REDOX processes in low-flow fractures with stagnant fluids (that also may be coated with biomass at low temperature < 110 °C) (Fig. 14.6).

The produced fluid comes in contact with many different installations and materials used in these installations including the casing of the production and injection borehole, the pumps installed in each of these boreholes and the filters, heat-exchangers, steam separators, turbines, pipes and other surface installations of the power plant. Many of components of the produced fluid may however cause severe corrosion of the materials used in the geothermal system. Corrosion-relevant substances in the geothermal fluid include: Oxygen, hydrogen sulfide, carbon dioxide, sulfate, and chloride. The origin of the trouble makers can be related to contamination with near surface fluids (leaks), the presence of abundant sulfide minerals in the geothermal reservoir formation, deep thermal CO_2 sources (metamorphic, magmatic), oxidation of primary sulfides by oxygen-rich surface fluids and resident fossil seawater at depth (Lund et al. 1976; Ellis and Conover 1981).

It is important to realize that corrosion is not an exclusive problem of deep geothermal systems but that also near surface systems can be affected by serious material corrosion problems. The origin of dissolved gases in near-surface groundwater is normally directly the atmosphere or modified soil gasses.

If a geothermal fluid equilibrates with the atmosphere its dissolved gasses, readjusts their concentrations according to the partial pressure of the gas in the atmosphere (Sect. 14.2). A fluid that has been saturated with CO_2 at high temperature and pressure releases CO_2 to the atmosphere until an equilibrium concentration has been established. The fluids will essentially lose all gases with very low partial pressures in the atmosphere (H_2S, H_2, CH_4). However, it will gain gasses that have a very low partial pressure in the reservoir formation (e.g. O_2). Dissolved gasses in the fluid must be analyzed reliably and regularly. The reported concentration units must be clearly specified (recalculation to other units must be possible, clear and easy) and the analytical procedures must be documented.

Decreasing the concentration of the dissolved carbon dioxide in the fluid by degassing CO_2 as a result of contact to the atmosphere or of decompression of fluid may cause precipitation of carbonate (calcite). The process of carbonate scale formation can be understood from the reaction:

$$Ca^{2+} + 2HCO_3^- \Rightarrow CaCO_3 (calcite) + H_2O + CO_2\uparrow \qquad (14.14)$$

If the CO_2 on the right hand side of Eq. 14.14 escapes (indicated by arrow), the reaction will progress to the right hand side according to Le Catelier's principle and precipitate the mineral calcite or aragonite (Fig. 14.7). Carbonate scales can be massive and must be prevented or minimized. The most important measure follows from Eq. 14.14, namely that CO_2 loss must be barred. This can be realized by strictly operating in a closed system under about 10–20 bar pressure and under isolation from the atmosphere (geothermal doublet systems). The necessary

Fig. 14.7 Carbonate scales from a pipe used in a geothermal plant: *Aragonite and calcite on top, pure calcite below* at the pipe wall

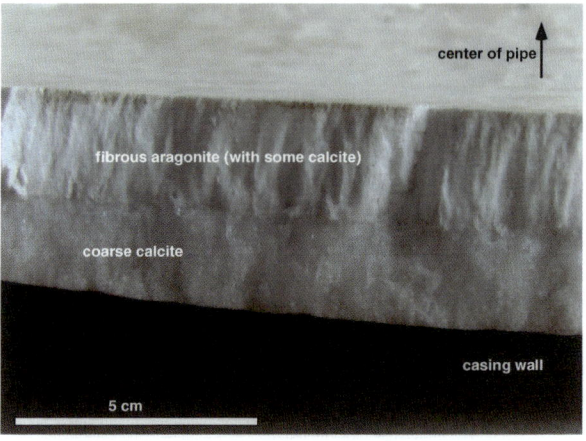

minimum pressure for preventing carbonate scales for a specific system and fluid composition can be computed from thermodynamic models or it can be experimentally determined. In both methods, exact data on the composition of the produced geothermal fluid are needed.

Other solids such as many scale-forming sulphates cannot be stopped from precipitating from cooled fluids by supporting measures on the gas-side. A good example is the solubility of anhydrite as described by Eq. 14.10. It follows from the inverse temperature dependence of the reaction that anhydrite may not precipitate from the fluid if the fluid is saturated with anhydrite at high-T and then being cooled in the heat exchanger. However, from Eq. 14.10 and from Chatelier's principle it also follows that anhydrite will precipitate from an anhydrite-saturated fluid if Ca^{2+} is increased (e.g. by on-going calcite dissolution) or SO_4^{2-} is increased (e.g. by sulfide to sulphate oxidation due to contact to the atmosphere). Other frequently observed sulphate scales such as barite ($BaSO_4$) form by similar mechanisms. Scale prevention is difficult in the case of sulphates. Inhibitor chemicals may help. If not, mechanical cleaning of the system as long as the scales are soft is the only remedy.

Some scales may contain high concentrations of toxic or radioactive substances (e.g. As, Cd, Pb, Hg, ^{210}Pb, ^{224}Ra). Therefore, exposed equipment and the scales must be handled with great care and disposed of in a legally correct manner.

The occasionally high concentrations of CO_2 or H_2S in deep waters may corrode C-steel pipes. The degree of corrosion damage depends strongly on pH of the fluid. Therefore, pH and fluid composition must be known for a reliable evaluation of the corrosion risk. If corrosion resistance of C-steels is insufficient at a particular site, Cr-Ni steels or Ni-based materials need to be considered.

Steel pipes and casing can be corroded by direct interaction with hot low-pH fluids according to the overall reaction that dissolves the steel into the aqueous fluid:

$$\text{Fe(metal pipe)} + 2H^+ = Fe^{2+} + H_2(\text{gas}) \qquad (14.15a)$$

$$2\text{Fe}(\text{metal pipe}) + 3\text{H}_2\text{O} = \text{Fe}_2\text{O}_3(\text{hematite, iron ochre}) + 3\text{H}_2(\text{gas}) \quad (14.15b)$$

Reaction 14.15a depends on pH and is favored by acid fluids. The produced Fe(II) in solution is no major problem as long as no atmospheric oxygen gains access to the fluid, in which case Fe(III) oxides, hydroxides and sulfates may cause severe scaling problems (iron ochre sedimentation). This situation is described by reaction 14.15b. If H_2 can be measured in the produced fluid, casing corrosion is likely to be in progress.

The iron oxidation process can be described by the reaction:

$$2\text{Fe}^{2+}(\text{iron in solution}) + 2\text{H}_2\text{O} + \text{O}_2 = 4\text{H}^+ + 2\text{FeO(OH)} \text{ (iron ochre scale)} \quad (14.16)$$

It oxidizes dissolved iron in the fluid (e.g. from pipe corrosion) by dissolved O_2 of typically atmospheric origin. Other oxidizing agents may also play an important role at some sites. CO_2 for example is a potential oxidizer that will be reduced to elementary carbon or to methane gas in the process. Because Fe^{3+} is essentially insoluble at moderate pH it precipitates as various minerals including goethite (FeO(OH)), hematite, schwertmanite, ferrihydrite and many others. It follows from Eq. 14.16 that the oxidation process produces protons and decreases pH, which in turn promotes the corrosion reaction 14.15a.

Reaction 14.15b dissolves metal from casing and other pipes and gradually reduces wall thickness. However, the co-produced hydrogen gas diffuses into the metallic materials and causes embrittlement of the steel (pitting corrosion). Corrosion velocity strongly depends on pH (Eq. 14.15a) and increases dramatically with decreasing pH. If pH is below 4 steel corrosion becomes a severe problem. Increasing dissolved CO_2 also decreases pH, however, carbonic acid is a relatively mild acid.

Hydrogen sulfide (H_2S) occurs as gas dissolved in reduced geothermal fluids. The gas, if present, may cause corrosion and scaling problems combined. The gas may react with the metals of casing and pipes according to the reaction (14.17):

$$\text{Fe}(\text{casing and pipes}) + 2\text{H}_2\text{S} = \text{FeS}_2(\text{pyrite scale}) + 2\text{H}_2 \quad (14.17)$$

The reaction corrodes the steel of the casing and produces difficult to remove pyrite scales and hydrogen gas that worsens corrosion.

If H_2S meets an oxidizing environment (e.g. atmospheric oxygen), sulfide sulfur is transferred to sulfate sulfur by the reaction:

$$2\text{H}_2\text{S} + 2.5\text{O}_2 = 2\text{H}^+ + \text{SO}_4^{2-} + \text{H}_2\text{O} \quad (14.18)$$

The reaction produces sulfuric acid thereby decreasing pH and promoting the corrosion reaction 14.15b. So H_2S in the geothermal fluid is a notorious troublemaker. Continued corrosion at a hole or crack (welded connections) is self-accelerating. Pitting corrosion can be best prevented by selecting (expensive) corrosion resistant materials. Other measures include increasing the pH, decreasing fluid temperature, increasing flow velocity, adding scaling inhibitors and installing cathodic protection equipment.

The scaling-potential can be predicted from the composition of the produced fluid and its temperature changes during cycling through the surface installations, quantitative data on gas unmixing in the operation cycle and from thermodynamic or experimental models. The evaluated scaling-potential indicates a tendency or possibility that mineral scales may form during operation of the plant. The thermodynamic analysis also suggests sensible procedures for scale prevention. However, the actual active scaling processes that must be expected at a given site are very difficult to predict reliably. This is because thermodynamics states if a precipitation reaction is possible but it cannot predict the velocity of its progress (kinetics). It may, in fact, not run at all because no crystal nuclei can be formed. In this context, the texture and roughness of surfaces of the used materials is important for nucleation and thus also influences scale formation. Along the fluid flow path, with its abrupt changes in flow direction and flow velocity, local zones of supersaturation for a certain mineral may develop even if the fluid is undersaturated with respect to that mineral in the reservoir or in most other zones on the system.

Laboratory testing of materials may help to select the best corrosion-resistant steels for the particular site. This requires, of course, knowledge of the composition of the produced fluids. Corrosion and scales can also be minimized or prevented by avoiding sharp 90° turns in the flow direction and massive fluid flow velocity variations in the pipe system by using large radius pipe bends (minimizing zones of turbulent flow). Thus choosing the materials used for components of the thermal water loop must be made with great care. Pressure control, fluid filtration, and optimizing the operation management are further aspects of corrosion and scaling control.

At temperatures below about 120 °C microbial processes may bring additional complexity to the scaling problem. Microbial biomass may form at the low-temperature side of the heat exchanger and may pose a particular problem at the injection borehole. Microbial metabolism may produce scales and may promote, support and accelerate inorganic scaling. Dissolved organic carbon (DOC) in the fluid can be analyzed in the laboratory. DOC can originate from the decay of biomass or from its metabolic processes. DOC may also originate from technical products used in the system such as lubricants, grease, oil and other organic technical substances. The biomass may use these technical organic substances as nutrients.

The described chemical aspects of geothermal plant development show that chemical studies on the composition of the fluids and the secondary interaction products are indispensable. The data should be collected as early as possible in order to develop strategies for scaling prevention. Later, during operation of the plant, a chemical monitoring program should be established that recognizes upcoming troubles and failures rapidly. This will enable the plant management to devise appropriate protective measures.

Once mineral scales have formed in wellbores chemical and mechanical removal measures will be the appropriate remedy (Crabtree et al. 1999). Scale removal techniques must be applied as soon as possible. The techniques should cause no damage to the pipes and wellbore and should preferably hinder future scale formation. Carbonate scales can be removed with acids, other soluble scales may also be removed with a variety of inorganic and organic dissolver chemicals. Mechanical removal of scales is performed with a large variety of workover tools such as casing scrapers.

References

Water level indicator

References

Aadony, B. S. (1999). *Modern well design* (p. 240). Rotterdam: Balkema.
Acuña, J., & Palm, B. (2009). Local conduction heat transfer in U-pipe borehole heat exchangers. In *Excerpt from the Proceedings of the COMSOL Conference* (p. 6), Milan.
Agarwal, R. G., Al-Hussainy, R., & Ramey, H. J. Jr. (1970). An investigation of wellbore storage and skin effect in unsteady liquid flow: I. Analytical treatment. *SPE Journal, 10*(3), 279–290.
Ahrens, T. J. (1995) *Global Earth physics: A handbook of physical constants*. Washington: American Geophysical Union.
Amann, R., Glöckner, F. -O., & Neef, A. (1997). Modern methods in subsurface microbiology: In situ identification of microorganisms with nucleic acid probes. *FEMS Microbiology Reviews, 20*(3/4), 191–200.
Antics, M., & Sanner, B. (2007). Status of geothermal energy use and resources in Europe. In *Proceedings European Geothermal Congress* (p. 1–8). Germany: Unterhaching.
Antics, M., Papachristou, M., & Ungemach, P. (2005). Sustainable heat mining, a reservoir engineering approach. In *Proceedings of 13th Workshop on Geothermal Reservoir Engineering* (pp. 14). Stanford: Stanford University.
Armstead, H. C. H. (1983). *Geothermal energy.* (p. 404). London: E. & F. N. Spon.
Armstead, H. C. H., & Tester, J. W. (1987). *Heat mining*. London: E. & F. N Spon.
Arnórsson, S. (1983). Chemical equilibria in Iceland geothermal systems. Implications for chemical geothermometry investigations. *Geothermics, 24*, 603–629.
Arnórsson, S., Bjarnason, J. Ö., Giroud, N., Gunnarsson, I., & Stefánsson, A. (2006). Sampling and analysis of geothermal fluids. *Geofluids, 6*, 203–216.
Australian Drilling Industry. (1997). *Drilling: The manual of methods, applications, and management* (p. 624). Boca Raton: CRC Press.
Avseth, P., Mukerji, T., & Mavko, G. (2010). *Quantitative Seismic interpretation: Applying rock physics tools to reduce interpretation risk* (p. 408). Cambridge: Cambridge University Press.
Baisch, S., Weidler, R., Vörös, R., Wyborn, D., & de Graaf, L. (2006). Induced seismicity during the stimulation of a geothermal HFR reservoir in the Cooper Basin, Australia. *Bulletin of the Seismological Society of America, 96*, 2242–2256.
Ball, J. W., & Nordstrom, D. K. (1991). WATEQ4F, current version 4.00 2012: First release 1991 as user's manual for WATEQ4F, with revised thermodynamic data base and test cases for calculating speciation of major, trace, and redox elements in natural waters. *Open-File Report 91-183; U.S. Geological Survey*, 10 pp.
Ball, J. W., Jenne, E. A., & Burchard, J. M. (1976). Sampling and preservation techniques for waters in geyers and hot springs, with a section on gas collection by A.H. Truesdell. In *Workshop on sampling geothermal effluents, 1st, proceedings, environmental protection agency 600/9-76-011* (pp. 218–234).
Banks, D., Odling, N. E., Skarphagen, H., & Rohr-Trop, E. (1996). Permeability and stress in crystalline rocks. *Terra Nova, 8*, 223–235.
Barenblatt, G. E., Zeltov, J. P., & Kochina, J. N. (1960). Basic concepts in the theory of homogeneous liquids in fissured rocks. *Journal of Applied Mathematics and Mechanics (USSR), 24*(5), 1286–1303.
Baria, R. A., & Green, S. P. (1989). Microseismics: A key to understanding reservoir growth. In E. R. Baria (Ed.), *Hot Dry Rock Geothermal Energy, Proceedings of Camborne School of Mines International Hot Dry Rock Conference* (pp. 363–377). London: Robertson Scientific Publications.
Baria, R., Michelet, S., Baumgärtner, J., Dyer, B., Gerard, A., & Nicholls, J., et al. (2004). Microseismic monitoring of the world largest potential HDR reservoir. In *Proceedings of the 29th Workshop on Geothermal Reservoir Engineering*. California: Stanford University.
Baria, R., Jung, R., Tischner, T., Nicholls, J., Michelet, S., & Sanjuan, B., et al. (2006). Creation of an HDR/EGS reservoir at 5000 m depth at the European HDR project. In *Proceedings 31st Workshop on Geothermal Reservoir Engineering*, Stanford, California.
Bassetti, S., Rohner, E., Signorelli, S., & Matthey, B. (2006). *Documentation of cases of damage of geothermal probes* (in German) (p. 65). Zürich: Schlussbericht Energie Schweiz.

References

Batchelor, A. S. (1977). Brief summary of some geothermal related studies in the United Kingdom. In *2nd NATO/CCMS Geothermal Conference 22–24 June, Section 1.21* (pp. 27–29). Los Alamos.

Bear, J. (1979). *Hydraulics of groundwater*. New York: McGraw Hill Book Comp.

Bencic, A. (2005). Hydraulic Fracturing of the Rotliegend Sst. in N-Germany—Technology, Company History and Strategic Impotance. In *SPE Technology Transfer Workshop*, Suco, Zeit Bay Field.

Benfield, A. E. (1939). Terrestrial heat flow in great Britain. In *Proceedings of the Royal Society of London Series A: Containing Papers of A mathematical and Physical Character, 173*(955), 428–450.

Berkaloff, E. (1967). Interprétation des pompages d'essai. Cas de nappes captives avec une strate conductrice d'eau privilégiée. *Bull. B.R.G.M. (deuxième série)*, section III: 1, 33–53.

Bertani, R. (2007). World geothermal generation in 2007. *GHC Bulletin, R 7*, 19 pp.

Bertani, R. (2010). Geothermal power generation in the world: 2005–2010 Update Report. In *Proceedings of the World Geothermal Congress in Bali*.

Bertleff, B., Joachim, H., Koziorowski, G., Leiber, J., Ohmert, W., Prestel, R., et al. (1988). Data from geothermal wells in Baden-Württember (Germany) (in German). *Jh. geol. Landesamt Baden-Württemberg, 30*, 27–116.

Binder, J. (2007). New technology drilling rig. In *Proceedings of the European Geothermal Congress* (p. 4).

BINE. (2009). Geothermal electricity generation combined with a heating network. *BINE Information*, 7.

Bjelm, L. (2006). Underbalanced drilling and possible well bore damage in low-temperature geothermal environment. In *Proceedings of the 31st Workshop on Geothermal Reservoir Engineering, Stanford* (p. 6).

Black, J. H. (1985). The interpretation of slug tests in fissured rocks. *Quarterly Journal of Engineering Geology and Hydrogeology, 18*(2), 161–171.

Bommer, J. J., Oates, S., Cepeda, J. M., Lindholm, C., Bird, J., Torres, R., et al. (2006). Control of hazard due to seismicity induced by a hot fractured rock geothermal project. *Engineering Geology, 83*(4), 287–306.

Bondor, P. L., & De Rouffignac, E. (1995). Land subsidence and well failure in the Belridge diatomite oil field, Kern county, California. Part II. Applications. *IAHS Publications-Series of Proceedings and Reports-Intern Assoc Hydrological Sciences*, (Vol. 234, pp. 69–78).

Bourdet, D., Ayoub, J. A., & Pirard, Y. M. (1989). Use of pressure derivative in well-test Interpretation. *The Society of Petroleum Engineers*, 293–302.

Bourgoyne, A. T., Millheim, K. K., Chenevert, M. E., & Young, F. S. (1986). Applied drilling engineering. In S. T. Series (Ed.), *Society of Petroleum Engineers* (p. 502). Richardson: SPE

Bredehoeft, J. D., & Papadopulos, I. S. (1965). Rates of vertical groundwater movement estimated from the earth's thermal profile. *Water Resources Research, 1*, 325–328.

Brown, D. W. (2009a). Hot dry rock geothermal energy: Important lessons from Fenton Hill. In *Proceedings of Thirty-Fourth Workshop on Geothermal Reservoir Engineering, Stanford University, Stanford* (p. 4).

Brown, D. W. (2009b). Hot dry rock geothermal energy: Important lessons from Fenton Hill. In *Proceedings of Thirty-Fourth Workshop on Geothermal Reservoir Engineering* (p. 4). Stanford University, Stanford.

Brown, D. W., Duchane, D. V., Heiken, G., & Hriscu, V. T. (2012). *Mining the earth heat: Hot dry rock geothermal energy* (p. 657). Heidelberg: Springer.

Bucher, K., & Grapes, R. (2011). *Petrogenesis of metamorphic rocks* (p. 428). Berlin: Springer.

Bucher, K., & Stober, I. (2000). The composition of groundwater in the continental crystalline crust. In I. Stober & K. Bucher (Eds.), *Hydrogeology in crystalline rocks* (pp. 141–176). Dordrecht: Kluwer Academic Publishers.

Bucher, K., & Stober, I. (2002). Water–rock reaction experiments with Black Forest gneiss and granite. In I. Stober & K. Bucher (Eds.), *Water–rock interaction, water science and technology library* (pp. 61–96). Dordrecht: Kluwer Academic Publishers.

Bucher, K., & Stober, I. (2010). Fluids in the upper continental crust. *Geofluids, 10*, 241–253.
Bucher, K., Zhu, Y., & Stober, I. (2009a). Groundwater in fractured crystalline rocks, the Clara mine, Black Forest (Germany). *International Journal of Earth Sciences (Geologische Rundschau), 98*, 1727–1739.
Bucher, K., Zhang, L., & Stober, I. (2009b). A hot spring in granite of the Western Tianshan, China. *Applied Geochemistry, 24*, 402–410.
Budi Kesuma Adi Putra, I. M. (2008). Drilling practice with aerated drilling fluid: Indonesian and Icelandic geothermal fields. Geothermal Training Programme (Vol. 11, pp. 77–100). Iceland: UN University Reykjavik.
Butler, J. J., Jr. (1998). The design, performance, and analysis of slug tests (p. 252). New York: Lewis Publishers.
Câmara, G., Souza, R. C. M., Freitas, U. M., Garrido, J., & Ii, F. M., 1996. SPRING: Integrating remote sensing and GIS by object-oriented data modelling. Image Processing Division (DPI), National Institute for Space Research (INPE). *Computers and Graphics, 20*(3), 395–403.
Caine, J. S., & Tomusiak, S. R. A. (2003). Brittle structures and their role in controlling porosity and permeability in a complex Precambrian crystalline-rock aquifer system in the Colorado Rocky Mountain Front Range. *GSA Bulletin, 115*(11), 1410–1424.
Carslaw, H. S., & Jaeger, J. C. (1959). *Conduction of heat in solids* (p. 342). Oxford: Oxford at the Clarendon Press.
Cataldi, R. (1992). Review of historiographic aspects of geothermal energy in the mediterranean and mesoamerican areas prior to the modern age. *Geo-Heat Centre Quarterly Bulletin, 18*, 13–16.
Cinco, L. H., Ramey, H. J., & Miller, F. G. (1975). Unsteady-state pressure distribution created by a well with an inclined fracture. *Society of Petroleum Engineers of AIME*(SPE 5591), 18.
Clauser, C. (2003). *Numerical simulation of reactive flow in hot aquifers using SHEMAT and processing SHEMAT*. Heidelberg: Springer.
Clauser, C. (2009). Heat transport processes in the Earth's crust. *Surveys in Geophysics, 30*, 163–191.
Cook, N. G. W. (1976). Seismicity associated with mining. *Engineering Geology, 10*, 99–122.
Cooper, H. H., & Jacob, C. E. (1946). A Generalized graphical method for evaluating formation constants and summarizing well-field history. *Transactions, American Geophysical Union, 27*, 526–534.
Cooper, H. H. J., Bredehoeft, J. D., & Papadopulos, I. S. (1967). Response of a finite-diameter well to an instantaneous charge of water. *Water Resources Research, 3*(1), 263–269.
Cornet, F., Helm, J., Poitrenaud, H., & Etchecopar, A. (1997). Seismic and aseismic slips induced by large-scale fluid injections. *Pure and Applied Geophysics, 150*, 563–583.
Crabtree, M., Eslinger, D., Fletcher, P., Miller, M., Johnson, A., & King, G. (1999). Fighting scale—Removal and prevention. *Oilfield Review*, 30–45 (Autumn).
Cunningham, K. M., Nordstrom, D. K., Ball, J. W., Schoonen, M. A. A., Xu, Y., & DeMonge, J. M. (1998). Water-chemistry and on-site sulfur-speciation data for selected Springs in Yellowstone National Park, Wyoming, 1994–1995. *U.S. Department of the Interior, Open-File Report 98*, 40.
Dash, Z. V., Murphy, H. D., & Cremer, G. M. (1981). Hot dry rock geothermal reservoir testing: 1978 to 1980. Los Alamos National Laboratory Report LA-9080-SR.
Davis, S. D., & Pennington, W. D. (1989). Induced seismic deformation in the Cogdell oil field of West Texas. *Bulletin of the Seismological Society of America, 79*, 1477–1495.
Devereux, S. (2012). *Drilling technology in nontechnical language* (2nd ed., pp. 270–370). Oklahoma: PennWell Corp.
Dezayes, C., Gentier, S., & Genter, A. (2005). Deep Geothermal Energy in Western Europe: The Soultz Project (Final Report). BRGM/RP-54227-FR, 51 pp.
Diersch, H. J. (1994). FEFOLW, finite element subsurface flow & transport simulation system. *Reference Manual*.
Dingh, H. T., Kuever, J., Mussmann, M., Hassel, A. W., Stratmann, M., & Widdel, F. (2004). Iron corrosion by novel anaerobic microorganisms. *Nature, 427*, 829–832.

References

DiPippo, R. (2012). *Geothermal power plants: Principles, applications, case studies and environmental impact* (3rd ed., 600 pp). Washington: Butterworth Heinemann.

Drever, J. I. (2005). *Water, weathering, and soil* (p. 626). Oxford: Elsevier.

Duchane, D. V., & Brown, D. W. (2002). Hot dry rock (HDR) geothermal energy research and development at Fenton Hill, New Mexico. *GHC Bulletin, 32*, 13–19.

Duffield, R. B., Nunz, G. J., Smith, M. C., & Wilson, M. G. (1981). *Hot dry rock, geothermal energy development program* (p. 211). Los Alamos National Laboratory Report.

Dyes, A. B., Kemp, C. E., & Caudle, B. H. (1958). Effect of fractures on sweep-out pattern. *Transactions AIME, 213*, 245–249.

EIA (2012). International Energy Outlook. *US energy information administration*. http://www.eia.gov.

Ellis, P. F., & Conover, M. F. (1981). Material selection guideline for geothermal energy systems. *NTIS Code DOE/RA/27026-1*. Radian Corporation, Austin, TX.

Ellis, D. V., & Singer, J. M. (2007). *Well logging for earth scientists* (2nd ed., p. 708). Heidelberg: Springer.

Emmermann, R., & Lauterjung, J. (1997). The German continental deep drilling program KTB: Overview and major results. *Journal of Geophysical Research, 102*, 18179–18201.

Ernst, P. L. (1977). *A hydraulic fracturing technique for dry hot rock experiments in a single borehole* (p. 7). Dallas: Soc. Petrol. Engineers of AIME.

Eugster, W. J. (1998). Longterm behavior of the geothermal probes at Elgg (Zurich, Switzerland) (in German). In: *Projekt 102, Polydynamics* (p. 38). Zürich: Schlussbericht PSEL.

Evans, K., Genter, A., & Sausse, J. (2005). Permeability creation and damage due to massive fluid injections into granite et 3.5 km at Soultz: 1- Borehole observations. *Journal of Geophysical Research, 110*, 1–19.

Faure, G. (1986). *Principles of Isotope Geology* (2nd ed., p. 608). New York: Wiley.

Ferris, J. G. (1951). Cyclic fluctuations of water level as a basis for determining aquifer transmissivity. *International Association of Science Hydrology Publ., 33*, 148–155.

Fielding, E. J., Blom, R. G., & Goldstein, R. M. (1998). Rapid subsidence over oil fields measured by SAR interferometry. *Geophysical Research Letters, 25*(17), 3215–3218.

Forrer, S., Mégel, T., Rohner, E., & Wagner, R. (2008). Better planning security for geothermal probe projects (in German). *bbr Fachmagazin für Brunnen- und Leitungsbau, 5*, 42–47.

Fournier, R. O. (1977). Chemical geothermometers and mixing models for geothermal systems. *Geothermics, 5*, 41–50.

Fournier, R. O. (1981). Application of water geochemistry to geothermal exploration and reservoir engineering. In L. Rybach & L. I. P. Muffler (Eds.), *Geothermal systems: Principles and case histories* (pp. 109–143). New York: Wiley.

Fournier, R. O., & Potter, R. W. (1982). An equation correlating the solubility of quartz in water from 25°C to 900°C at pressures up to 10,000 bar. *Geochimica et Cosmochimica Acta, 46*, 1969–1973.

Frape, S. K., & Fritz, P. (1982). The chemistry and isotopic composition of saline waters from the Sudbury Basin, Ontario. *Canadian Journal of Earth Sciences, 19*, 645–661.

Frape, S. K., Blyth, A., Blomqvist, R., McNutt, R. H., & Gascoyne, M. (2004). Deep fluids in the continents: II. Crystalline rocks. In J. I. Drever., H. D. Holland & K. K. Turekian, (Eds.), *Surface and ground water, weatherin, and soils. Treatise on Geochemistry* (pp. 541–580). Elsevier, Amsterdam.

Freeze, K. A., & Cherry, J. A. (1997). *Groundwater* (p. 604). Englewood Cliffs: Prentice Hall.

Fridleifsson, I. B., Bertani, R., Huenges, E., Lund, J. W., Ragnarsson, A., & Rybach, L., (2008). The possible role and contribution of geothermal energy to the mitigation of climate change. In O. H Trittin (Ed.) *PCC Scoping Meeting on Renewable Energy Sources, Proceedings*, pp. 59–80, Luebeck, Germany.

Fritz, P., & Frape, S. K. (1987). *Saline water and gases in crystalline rocks* (p. 259). Ottawa: The Runge Press Limited. GAC Special Paper 33.

GEA (2012a). *Annual US geothermal power production and development report*, April 2012.

GEA. (2012b). *Annual US geothermal power production and development report*, p. 35.
Gehlin, S. (2002). *Thermal response test, method development and evaluation*. Unpublished Doctoral Theses Thesis, University of Technology, Luleå, Sweden.
Gehlin, S., & Nordell, B. (1997). Thermal response test—a mobile equipment for determining thermal resistance of boreholes. In *Proceedings 7th International Conference on Thermal Energy Storage Megastock '97*.
Genter, A., Keith, E., Cuenot, N., Fritsch, D., & Sanjuan, B. (2010). Contribution to the exploration of deep crystalline fractured reservoir of Soultz of the knowledge of enhanced geothermal systems (EGS). *Comptes Rendus Geoscience, 342*, 502–516.
Genter, A., Cuenot, N., Goerke, X., Melchert, B., Sanjuan, B., & Scheiber, J. (2012). Status of the Soultz geothermal project during explotation between 2010 and 2012. In *Proceedings of Thirty-Fourth Workshop on Geothermal Reservoir Engineering, Stanford University, Stanford, Cal, USA* (p. 11).
Gérard, A., Genter, A., Kohl, T., Lutz, P., Rose, P., & Rummel, F. (2006). The deep EGS (enhanced geothermal system) project at Soultz-sous-Forêts (Alsace, France). *Geothermics, 35*, 473–483.
Ghergut, I., Sauter, M., Behrens, H., Rose, P., Licha, T., & Lodemann, M. et al. (2007). Tracer-assisted evaluation of hydraulic stimulation experiments for geothermal reservoir candidates in deep crystalline and sedimentary formations. In *EGC Proceedings European Geothermal Congress* (pp. 1–12). Unterhaching.
Giardini, D. (2009). Geothermal quake risks must be faced. *Nature, 462*, 848–849.
Giggenbach, W. F. (1981). Geothermal mineral equilibria. *Geochimica et Cosmochimica Acta, 45*, 393–410.
Giroud, N. (2008). *A Chemical Study of Arsenic*. Boron and Gases in High-Temperature Geothermal Fluids in Iceland: Dissertation at the Faculty of Science, University of Iceland. 110 p.
Grasso, J. R. (1992). Mechanics of seismic instabilities induced by the recovery of hydrocarbons. *Pure and Applied Geophysics, 139*, 507–534.
Gringarten, A. C., & Ramey, H. J. (1974). Unsteady-state pressure distributions created by a well with a single horizontal fracture, partial penetration, or restricted entry. *Society of Petroleum Engineers Journal 257*, 413–426.
Gupta, H. K., & Rastogi, B. K. (1976). *Dams and Earthquakes* (p. 229). Amsterdam: Elsevier.
Gustafsson, A. M. (2006) *Thermal response test—Numerical simulations and analysis*. Unpublished Licentiate Theses Thesis, University of Technology, Luleå, Sweden.
Mlcak, H., & Mirolli, M. (2002). Notes from the North: A Report on the Debut Year of the 2 MW Kalina Cycle® Geothermal Power Plant in Husavik. Iceland: Exergy, Hreinn Hjartarson, Orkuveita Húsavíkur, Marshall Ralph, Power Engineers, Inc.
Harbaugh, A. W. (2005). MODFLOW-2005; The U.S. geological survey modular ground-water model—The ground-water flow process. Techniques and methods. *U.S. Geological Survey*, 6–A16.
Hasnaina, S. M. (1998a). Review on sustainable thermal energy storage technologies, Part I: Heat storage materials and techniques. *Energy Conversion and Management, 39*, 1127–1138.
Hasnaina, S. M. (1998b). Review on sustainable thermal energy storage technologies, part II: Cool thermal storage. *Energy Conversion and Management, 39*, 1139–1153.
Hawkins, M. F. (1956). A note on the skin effect. *Transactions AIME, 207*, 356–357.
He, K., Stober, I., & Bucher, K. (1999). Chemical evolution of thermal waters from limestone aquifers of the Southern Upper Rhine Valley. *Applied Geochemistry, 14*, 223–235.
Hekel, U. (2011). Hydraulische Tests. In K. Bucher., A. Gautschi., T. Geyer., U. Hekel., M. Mazurek., & I. Stober (Eds.) *Hydrogeologie der Festgesteine* (p. 15). FH-DGG: Freiburg.
Hellström, G. (1998). Thermal performance of borehole heat exchangers. In *The second Stockton international Geothermal Conference*.
Hellström, G., & Sanner, B. (2000). *EED earth energy designer, Computer program for borehole heat exchangers*. Sweden: Lund University.

References

Hewitt, A. D. (1989). Leaching of metal pollutants from four well casings used for groundwater monitoring. In *Special Report 89-32*, USA Cold Regions Research and Engineering Laboratory.

Hole, H. (2006). Lectures on geothermal drilling and direct uses. *UNU-GTP, Iceland, report* (Vol. 3, p. 32).

Horner, D. R. (1951). Pressure build-up in wells. In E. Bull (Ed.), *Proceedings of 3rd World Petroleum Congress* (pp. 503–521). Leiden, Netherlands.

Hsieh, P. A., & Bredehoeft, J. S. (1981). A reservoir analysis of the Denver earthquakes-A case of induced seismicity. *Journal of Geophysical Research, 86*, 903–920.

Huber, A. (2008). Code EWS, Computing Geothermal Probes (in German). Huber Energietechnik AG.

Huenges, E. (2010). *Geothermal Energy Systems: Exploration, Development, and Utilization* (p. 486). Berlin: Wiley-VCH Verlag GmbH & Co. KGaA.

Hurtig, E., Großwig, S., & Kasch, M. (1997). Fiberoptical temperature measurement: Montoring the temperature field at geothermal probe sites. *Geothermische Energie, 5*(18), 31–34.

Husen, S., Bachmann, C., & Giardini, D. (2007). Locally triggered seismicity in the central Swiss Alps following the large rainfall event of August 2005. *Geophysical Journal International, 171*(3), 1126–1134.

Ibrahim, O. M. (1996). Design Considerations for Ammonia-Water Rankine Cycle. *Energy, 21*, 835–841.

Ichikawa, S., Yasuga, H., Tosha, T., & Karasawa, H. (2000). Development of downhole pumps for binary cycle power generation using geothermal water. In *Proceedings World Geothermal Congress 2000, Kyushu–Tohoku, Japan* (pp. 1283–1288).

IEA-GIA. (2012). Trends in geothermal applications. *Publication on the IEA Geothermal Implementing Agreement*, p. 41.

IGA. (2009). News. *Newsletter of the International Geothermal Association, 76*, 1–18.

Ingebritsen, S. E., & Manning, C. E. (1999). Geological implications of a permeability-depth curve for the continental crust. *Geology, 27*, 1107–1110.

Ingebritsen, S. E., & Manning, C. E. (2010). Permeability of the continental crust: Dynamic variations inferred from seismicity and metamorphism. *Geofluids, 10*, 193–205.

Ingerle, K. (1988). Computation of aquifer cooling by heat pumps (in German). *Österreichische Wasserwirtschaft, 40*(11/12).

International Energy Agency (IEA). (2009). *Geothermal energy 12th annual report 2008*. Wairakei, New Zealand, p. 19.

International Energy Agency (IEA) (2012). *Key world energy statistics*, p. 80.

Jodocy, M., & Stober, I. (2008). Development of a geothermic information system for Germany; State of Baden-Württemberg (in German). *Erdöl-Erdgas-Kohle, 10*, 386–393.

Jodocy, M., & Stober, I. (2009). Geologisch-geothermische Tiefenprofile für den südwestlichen Teil des Süddeutschen Molassebeckens. *Z. dt. Ges. Geowiss, 160*(4), 359–366.

Johnson, A. I. (1991). Land Subsidence. *IAHS Publication, 200*, 680.

Kalfayan, L. (2008). *Production enhancement with acid stimulation* (2nd ed., p. 270). Tulsa: PennWell Corporation.

Kalina, A. L. (1984). Combined-cycle system with novel bottoming cycle. *Journal of Engineering for Gas Turbines and Power, 106*, 737–742.

Kappelmeyer, O., & Haenel, R. (1974). *Geothermics with special reference to application* (p. 238). Stuttgart: E. Schweizerbart Science Publishers.

Kappelmeyer, O., & Rummel, F. (1980). Investigations on an artificially created frac in a shallow and low permeable environment. In *Proceedings of the 2nd International Seminar EC Geoth* (1048–1053). Energy Res, Strasbourg.

Käss, W. (1998). *Tracing techniques in geohydrology* (p. 581). Rotterdam: A. A. Balkema.

Kharaka, Z. K., & Hanor, J. S. (2004). Deep fluids in the continents: I. Sedimentary basins. In J. I. Drever, H. D. Holland & K. K. Turekian (Eds.), *Surface and ground water, weatherin, and soils. Treatise on Geochemistry* (pp. 499–540), Elsevier, Amsterdam.

Kharaka, Y. K., Gunter, W. D., Aggarwal, P. K., Perkins, E. H., & DeBraal, J. D. (1988). SOLMINEQ.88: A computer program for geochemical modeling of water-rock interactions. *U.S. Geological Survey, Water-Resources Investigations Report 88-4227*, 420 pp.

Kipp, K. L. J. (1997). Guid to the revised heat and solute transport simulator HST3D. In *Water-Resources Investigations* (p. 149), U.S. Geological Survey.

Kobus, H. (1992). *Schadstoffe im Grundwasser / DFG, Deutsche Forschungsgemeinschaft, Band 1. Wärme- und Schadstofftransport im Grundwasser* (p. 480). Wiley, ISBN-13:978-3527271313

Kobus, H., & Mehlhorn, H. (1980). Approximative computation of the continuous operation of geothermal installations (in German). In *GWF 121*.

Kohl, T., & Hopkirk, R. J. (1995). "FRACTURE" a simulation code for forced fluid flow and transport in fractured porous rock. *Geothermics, 24*, 345–359.

Kohl, T., Evans, K. F., Hopkirk, R. J., Jung, R., & Rybach, L. (1997). Observation and simulation of non-Darcian flow transients in fractured rock. *Water Resources Research, 33*, 407–418.

Kovach, R. L. (1974). Source mechanisms for Wilmington oil field, California subsidence earthquakes. *Bulletin of the Seismological Society of America, 64*, 699–711.

Kozlovsky, Y. A. (1984). The world's deepest well. *Scientific American, 251*, 106–112.

Kraft, T., Mai, M. P., Wiener, S., Deichmann, N., Ripperger, J., Kästli, P., et al. (2009). Enhanced geothermal systems: Mitigating risk in urban areas. *EOS, Transactions American Geophysical Union, 90*(32(11)), 273–274.

Kruseman, G. P., & de Ridder, N. A. 1994. Analysis and evaluation of pumping test data (p. 377). Wageningen, Netherlands: International Institute for Land Reclamation and Improvement ILRI.

Ladner, F., Schanz, U., & Häring, M. O. (2008). Deep-heat-mining-project Basel: First insights from the development of an enhanced geothermal system (EGS) (in German). *Bulletin Angewandte Geologie, 13*(1), 41–54.

Landolt-Börnstein (1992). Numerical data and functional relationships in science and technology. In *Physical properties of rocks*. Berlin-Heidelberg-New York: Springer.

Langenbruch, C., & Shapiro, S. A. (2010). Decay rate of fluid-induced seismicity after termination of reservoir stimulations. *Geophysics, 75*(6), MA53–MA62.

Leibundgut, C., Moaloszewski, P., & Külls, C. (2011). Tracer in hydrology (p. 432). New York: John Wiley & Sons.

Lemmelä, R., Sucksdorff, Y., & Gilman, K. (1981). Annual variation of soil temperature at depth 20–700 cm in an experimental field in Hyrylä, South-Finland during 1969–1973. *Geophysica, 17*, 143–154.

Lund, J. W. (2007). Characteristics, development and utilization of geothermal resources. *Geo-Heat Centre Quarterly Bulletin, 28*, 1–9.

Lund, J. W., Silva, J. F., Culver, G., Lienau, P. J., Svanevik, L. S., & Anderson, S. D. (1976). Corrosion of downhole heat exchangers, Appendix A. *DOE Contract E(10-1)-1548, Oregon*, Institute of Technology, Klamath Falls, OR.

Majer, E., Baria, R., & Stark, M., (2008). Protocol for induced seismicity associated with enhanced geothermal systems. In *Report produced in Task D Annex I (9 April 2008), International Energy Agency-Geothermal Implementing Agreement* (incorporating comments edited by Bromley C., Cumming W., Jelacic A., & Rybach L.).

Mansure, A. J., & Reiter, M. (1979). A vertical groundwater movement correction for heat flow. *Journal of Geophysical Research, 84*(7), 3490–3496.

Mareschal, J. -C., & Jaupart, C. (2013). Radiogenic heat production, thermal regime and evolution of the continental crust. Tectonophysics 11 p. http://dx.doi.org/10.1016/j.tecto.2012.12.001

Matthews, C. S., & Russel, D. G. (1967). Pressure buildup and flow tests in wells. In *AIME Monograph 1* (p. 167). New York: H.L. Doherty Series SPE of AIME.

Mazurek, M. (2000). Geological and hydraulic properties of water-conducting features in crystalline rocks. In I. Stober & K. Bucher (Eds.), *Hydrogeology of crystalline rocks* (pp. 3–26). Dordrecht: Kluwer Academic Publisher.

McGarr, A. (1991). On a possible connection between 3 major earthquakes in California and oil production. *The Bulletin of the Seismological Society of America 81*, 948–970.

Midgley, D. (1990). A review of pH measurement at high temperatures. *Talanta, 37*, 767–781.

MIT (2007). The future of geothermal energy, impact of enhanced geothermal systems (EGS) on the United States in the 21st Century, Massachusetts Institute of Technology, U.S.A.

Mogensen, P. (1983). Fluid to duct wall heat transfer in duct system heat storages. In *Proceedings International Conference Subs Heat Storage*, pp. 652–657.

Möller, P., Weise, S. M., Althaus, E., Bach, W., Behr, H. J., Borchardt, R., et al. (1997). Paleo- and recent fluids in the upper continental crust—Results from the German Continental deep drilling program (KTB). *Journal of Geophysical Research, 102*, 18245–18256.

Mottaghy, D., & Pechnig, R. (2009). Numerical 3-D model and prediction of the temperature evolution of thermal reservoirs (in German). *BBR—Fachmagazin für Brunnen- und Leitungsbau, 60–10*, 44–51.

Narayanan, K. R., & Shankara, (2004). What is a Heat Pipe?" The Chemical Engineers' Resource Page.

Nicholson, K. (1993). *Geothermal fluids: Chemistry and exploration techniques* (p. 263). Berlin: Springer.

Nicholson, C., & Wesson, R. L. (1990). *Earthquake hazard associated with deep well injection—a report to the U.S. environmental protection agency* (p. 74). Florida: U.S. Geological Survey Bulletin.

Nielsen, K. A. (2007). *Fractured aquifers: Formation evaluation by well testing* (p. 229). Victoria, BC, Canada: Trafford Publishing.

Nordstrom, D. K., Andrews, J. N., Carlsson, L., Fontes, J. -C., Fritz, P., Moser, H., & Olsson, T. (1985). Hydrogeological and hydrogeochemical investigations in boreholes. In *Final report of the phase I geochemical investigations of the Stripa groundwaters* (pp. 85–106), Technical Report STRIPA Project, Stockholm.

Odenwald, B., Hekel, U., & Thormann, H. (2009). Groundwater flow—groundwater storage (in German). In K. J. Witt (Hrsg.) (Ed.),*Grundbau-Taschenbuch, Teil 2: Geotechnische Verfahren* (p. 950). Ernst & Sohn.

Omori, F. (1894). On the aftershocks of earthquakes. *Journal of Colloid Science, 7*, 111–200.

Owens, S. R. (1975). Corrosion in disposal wells. *Water and Sewage Works, RN 75*, 10–12.

Pahud, D. (1998). *PILESIM: Simulation tool of heat exchanger pile system*. Lausanne: Laboratory of Energy Systems, Swiss Federal Institute of Technology.

Papadopulos, S. S., Bredehoeft, J. D., & Cooper, H. H, Jr. (1973). On the analysis of 'slug test' data. *Water Resources Research, 9*(4), 1087–1089.

Park, Y. M., & Sonntag, R. E. (1990). A Preliminary Study of the Kalina Power Cycle in Connection with a Combined Cycle System. *International Journal of Energy Research, 14*, 153–162.

Parker, L. V., Hewitt, A. D., & Jenkins, T. F. (1990). Influence of casing materials on trace-level chemicals in well water. *Ground Water Monitoring Review, 10*(2), 146–156.

Parkhurst, D. L., & Appelo, C. A. J. (1999). User's guide to PHREEQC (version 2)—a computer program for speciation, batchreaction, one dimensional transport, and inverse geochemical calculations. In *Water-Resources Investigations Report 99–4259*, U.S. Geological Survey, Denver, Colorado, p. 312.

Pauwels, H., Fouillac, C., & Fouillac, A. M. (1993). Chemistry and isotopes of deep geothermal saline fluids in the Upper Rhine Graben: Origin of compounds and water-rock interactions. *Geochimica et Cosmochimica Acta, 57*, 2737–2749.

Pearson, C. (1981). The relationship between microseismicity and high pore pressures during hydraulic stimulation experiments in low permeability granitic rocks. *Journal of Geophysical Research, 86*(B9), 7855–7864.

Pine, R. J., & Batchelor, A. S. (1984). Downward migration of shearing in jointed rock during hydraulic injections. *International Journal of Rock Mechanics Mining Sciences and Geomechanical Abstracts, 21*(5), 249–263.

Poletto, F. B., & Miranda, F. (2004). *Seismic While Drilling: Fundamentals of Drill-Bit Seismic for Exploration* (p. 546). Amsterdam: Elsevier.

Pollack, H. N., Hurter, S. J., & Johnson, J. R. (1993). Heat flow from the Earth's interior: Analysis of the global data set. *Reviews of Geophysics, 31*, 267–280.

Poppei, J., Mayer, G., & Schwarz, R. (2006). Groundwater energy designer (GED): Software for utilization of groundwater for heating and cooling. In *Colenco Power Engineering AG Report for the Swiss Federal Energy Agency* (p. 70), Baden, Switzerland.

Portier, S., André, L., & Vuataz, F. -D. (2007). Review on chemical stimulation techniques in oil industry and applications to geothermal systems. In *Engine* (p. 32). Neuchatel: CREGE.

Pruess, K. (1987). TOUGH2 transport of unsaturated groundwater and heat. In *User's Guide, Version 2.0 (1999)*, Lawrence Berkeley Laboratory Report LBL-43134.

Putra, I. M. B. K. A. (2008). Drilling practice with aerated drilling fluid: Indonesian and Icelandic geothermal fields. *Geothermal Training Programme* (Vol. 11, pp. 77–100). Iceland: UN University Reykjavik.

Puttagunta, S., Aldrich, R. A., Owens, D., & Mantha, P. (2010). Residential ground-source heat pumps: In field system performance and energy modeling. *GRC Transactions, 34*, 941–948

Ramey, H. J. Jr. (1962). Wellbore heat transmission. *JPT 435 Trans AIME*, p. 225.

Ramey, H. J. J., Agarwal, R. G., & Martin, I. (1975). Analysis of 'slug test' or DST flow period data. *Journal of Canadian Petroleum Technology, 3*(37), 47.

Rauch, W. (2009). EGON. In *User manual*. Austria: University of Innsbruck.

Reich, M. (2011). Grundlagen der Richtbohrtechnik. *Erdöl Erdgas Kohle, 127* (1), 35–40.

Reynolds, J. M. (2011). *An Introduction to Applied and Environmental Geophysics* (2nd ed., p. 712). New York: Wiley-Blackwell.

Ruck, W., Adinolfi, M., & Weber, W. (1990). Chemical and environmental aspects of heat storage in the subsurface. *Zeitschrift für angewandte Geowissenschaften, 9*, 119–129.

Russell, D. G., & Truitt, N. E. (1964). Transient pressure behavior in vertically fractured reservoirs. *Journal of Petroleum Technology, 16*(10), 1159–1170.

Rutledge, J. T., Phillips, W. S., & Mayerhofer, M. J. (2004). Faulting induced by forced fluid injection and fluid flow forced by faulting. *Bulletin of the Seismological Society of America, 94*, 1817–1830.

Rybach, L. (1976). Radioactive heat production in rocks and its relation to other petrophysical parameters. *Pageoph, 114*, 309–317.

Rybach, L. (2004). EGS—State of the art. In *Tagungsband der 15. Fachtagung der Schweizerischen Vereinigung für Geothermie*, Basel.

Rybach, L., Wilhelm, J., & Gorhan, H. (2003). Geothermal use of tunnel waters—A Swiss speciality. *International Geothermal Conference*, Reykjavík, Sept. 2003, S05 Paper 051, pp. 17–23.

Sanner, B., & Chant, V. G. (1992). Seasonal cold storage in the ground using heat pumps. *Newsletter IEA Heat Pump Center, 10*(1), 4–7.

Sanner, B., Reuss, M., & Mands, E. (2000). Thermal response test—Experiences in Germany. In *Proceedings Terrastock 2000, 8th International Conference on Thermal Energy Storage* (pp. 177–182), Stuttgart, Germany.

Santoyo, E., & Díaz-González, L. (2010). Improved proposal of the Na/K-geothermometer to estimate deep equilibrium temperatures and their uncertainties in geothermal systems. In *Proceedings World Geothermal Congress* (p. 7). Bali, Indonesia.

Sauty, J. P. (1980). An analysis of hydrodispersive transfer in aquifers. *Water Resources Research, 16*(1), 145–158.

Sauty, J. P., Gringarten, A. C., Landel, P. A., & Menjoz, A. (1980). Lifetime optimization of low enthalpy geothermal doublets. In A. S. Strub & P. Ungemach (Eds.), *Advances in European geothermal research* (pp. 706–719). The Netherlands: D. Reidel Publications Co.

Schädel, K., & Dietrich, H. -G. (1979). Results of the fracture experiments at the geothermal research borehole Urach 3. In R. Haenel (Ed.), *The Urach Geothermal Projekt (Swabian Alb, Germany)* (pp. 323–344). Stuttgart: Schweizerbart'sche Verlagsbuchhandlung.

References

Schädel, K., & Stober, I. (1984). The thermal anomaly of Urach seen from a geological perspective (in German). *h. geol. Landesamt Baden-Württemberg, 26*(2), 19–25.
Schmidt, T., Mangold, D., & Müller-Steinhagen, H. (2003). Central solar heating plants with seasonal storage in Germany. *Solar Energy, 76*(1–3), 165–174.
Schön, J. (2004). *Physical properties of rocks* (p. 600). Amsterdam: Elsevier.
Schwartz, F. W., & Zhang, H. (2003). Fundamentals of ground water (p. 592). New York: Wiley.
Segall, P. (1989). Earthquakes triggered by fluid extraction. *Geology, 17*, 942–946.
Segall, P., Grasso, J. R., & Mossop, A. (1994). Poroelastic stressing and induced seismicity near the Lacq gas field, Southwestern France. *Journal of Geophysical Research, 99*, 15423–15438.
Shapiro, S. A., & Dinske, C. (2009). Fluid-induced seismicity: Pressure diffusion and hydraulic fracturing. *Geophysical Prospecting, 57*, 301–310.
Shapiro, S. A., Dinske, C., & Kummerow, J. (2007). Probability of a given-magnitude earthquake induced by a fluid injection. *Geophysical Research Letters, 34*, L22314.
Shaw, J. H., Connors, C. D., & Suppe, J. (Eds.). (2005). *Seismic interpretation of contractional fault-related folds*. American Association of Petroleum Geologists.
Sheriff, R. E., & Geldart, L. P. (2006). *Exploration Seismology* (p. 592). Cambridge, UK: Cambridge University Press.
Signorelli, S., (2004). *Geoscientific investigations for the use of shallow low-enthalpy systems*. In *ETH No. 15519*. Dissertation of the Swiss Federal Institute of Technology Zurich, Zurich, p. 157.
Smith, M. C., Aamodt, R. L., Potter, R. M., & Brown D. W. (1975). Man-made geothermal reservoirs. *Proceedings of UN Geothermal Symposium* (Vol. 3, pp. 1781–1787).
Smolczyk, H. G., (1968). Chemical reactions of strong chloride solutions with concrete. In *5th International. Symposium on Chemistry Cement*, Tokyo (pp. 274–280).
Stark, A. (2008). *Seismic Methods and Applications: A Guide for the Detection of Geologic Structures, Earthquake Zones and Hazards, Resource Exploration, and Geotechnical Engineering* (p. 592). Boca Raton: Brown Walker Press.
Stober, I. (1986). Strömungsverhalten in Festgesteinsaquiferen mit Hilfe von Pump- und Injektionsversuchen (The Flow Behaviour of Groundwater in Hard-Rock Aquifers—Results of Pumping and Injection Tests) in German. In *Geologisches Jahrbuch, Reihe C*, p. 204.
Stober, I. (1988). Geohydraulic results from tests in hydrothermal wells in Baden-Württemberg (in German). In B. Bertleff., H. Joachim., G. Koziorowski., J. Leiber., W. Ohmert., & R. Prestel et al. (Eds.) (pp. 27–116) Jh. geol. Landesamt Baden-Württemberg, Freiburg i.Br.
Stober, I. (1992). The tides and their hydraulic effects on groundwater (in German). *DGM, 36*(4), 142–147.
Stober, I. (2011). Depth- and pressure-dependent permeability in the upper continental crust: Data from the Urach 3 geothermal borehole, Southwest Germany. *Hydrogeology Journal, 19*, 685–699.
Stober, I., & Bucher, K. (1999a). Deep groundwater in the crystalline basement of the Black Forest region. *Applied Geochemistry, 14*, 237–254.
Stober, I., & Bucher, K. (1999b). Origin of salinity of deep groundwater in crystalline rocks. *Terra Nova, 11*(4), 181–185.
Stober, I., & Bucher, K. (2005a). The upper continental crust, an aquifer and its fluid: Hydraulic and chemical data from 4 km depth in fractured crystalline basement rocks at the KTB test site. *Geofluids, 5*, 8–19.
Stober, I., & Bucher, K. (2005b). Deep-fluids: Neptune meets Pluto. In C. Voss (Ed.), *The future of hydrogeology. Hydrogeology Journal* (pp. 112–115).
Stober, I., & Bucher, K. (2007a). Hydraulic properties of the crystalline basement. *Hydrogeology Journal, 15*, 213–224.
Stober, I., & Bucher, K. (2007b). Erratum to: Hydraulic properties of the crystalline basement. *Hydrogeology Journal, 15*, 1643.

Stober, I., Richter, A., Brost, E., & Bucher, K. (1999). The ohlsbach plume: Natural release of deep saline water from the crystalline basement of the Black Forest. *Hydrogeology Journal, 7,* 273–283.

Stober, I., Fritzer, T., Obst, K., & Schulz, R. (2009). Nutungsmöglichkeiten der Tiefen Geothermie in Deutschland. In *Bundesministerium für Umwelt, Naturschutz und Reaktorsicherheit* (p. 73) Berlin.

Sun, H., Feistel, R., Koch, M., & Markoe, A. (2008). New equations for density, entropy, heat capacity and potential temperature of a saline thermal fluid. *Deep-Sea Research I,.* doi:10.1016/j.dsr.2008.05.011.

Sutton, A. J., McGee, K. A., Casadevall, T. J., & Stokes, B. J. (1992). Fundamental volcanic-gas-study techniques: An integrated approach to monitoring. In J. W. Ewert & D. A. Swanson (Eds.), *Monitoring volcanoes: Techniques and strategies used by the staff of the cascades volcano observatory, 1980–90* (pp. 181–188). U.S. Geological Survey.

Talwani, P., Chen, L., & Gahalaut, K. (2007). Seismogenic permeability, ks. *Journal of Geophysical Research, 112,* doi:10.1029/2006JB004665.

Telford, W. M., Geldart, L. P., & Sheriff, R. E. (2010). *Applied Geophysics* (2nd ed., p. 792). Cambridge, UK: Cambridge University Press.

Teodoriu, C., & Falcone, G. (2009). Comparing completion design in hydrocarbon and geothermal wells: The need to evaluate the integrity of casing connections subject to thermal stress. *Geothermics, 38,* 238–246.

Theis, C. V. (1935). The relation between the lowering of the piezometric surface and the rate and duration of discharge of a well using groundwater storage. *Transactions AGO,* pt. 2, 519–524.

Thompson, J. M., & Yadav, S. (1979). Chemical analysis of waters from Geysers, Hot Springs, and pools in Yellowstone National Park, Wyoming, from 1974 to 1978. In *U.S. Geological Survey Open-File Report 79-704* (p. 49).

Thompson, J. M., Presser, T. S., Barnes, R. B., & Bird, D. B. (1975). Chemical analysis of the water of Yellowstone National Park, Wyoming from 1965 to 1973. In *U.S. Geological Survey Open-File Report 75-25* (p. 59).

Tischner, T., Schindler, M., Jung, R., & Nami, P. (2007). HDR project Soultz: Hydraulic and seismic observations during stimulation of the 3 deep wells by massive water injections. In *Proceedings, thirty-second workshop on geothermal engineering, Stanford University* (p. 7). California: Stanford.

Tiwari, G. N., & Ghosal, M. K. (2005). Renewable energy resources: Basic principles and applications. *International Journal of Industrial Engineering Computations, 3,* 649–662.

Tobler, D. J., Stefánsson, A., & Benning, L. G. (2008). In-situ grown silica sinters in Icelandic geothermal areas. *Geobiology, 6,* 481–502.

Todd, D. K. (1980). *Groundwater hydrology* (2nd edn., p. 535). New York: John Wiley & Sons.

Trefry, M. G., & Muffels, C. (2007). FEFLOW: A finite-element ground water flow and transport modeling tool. *Ground Water, 45*(5), 525–528.

Trommsdorff, V., Skippen, G., & Ulmer, P. (1985). Halite and sylvite as solid inclusions in high-grade metamorphic rocks. *Contributions to Mineralogy and Petrology, 89,* 24–29.

Tsang, C. -F. (1987). A borehole fluid conductivity logging method for the determination of fracture inflow parameters. In *Report of the Earth Science Division* (p. 53). California: Lawrence Berkley Laboratory, University of California.

Tsang, C. -F., Hufschmied, P., & Hale, F. V. (1990). Determination of fracture inflow parameters with a borehole fluid conductivity logging method. *Water Resources Research, 26*(4), 561–578.

Ungemach, P. (2001). Insight into geothermal reserhoir management district heating in the Paris Basin, France. *GHC Bulletin, 22,* 3–13.

Ungemach, P., Antics, M., & Papachristou, M. (2005). Sustainable Geothermal Reservoir Management. In *Proceedings World Geothermal Congress 2005* (pp. 24–29). Antalya, Turkey.

Ungemach, P., & Turon, R. (1988). Geothermal Well Damage in the Paris Basin: A Review of Existing and Suggested Workover Inhibition Procedures.*SPE Formation Damage Control Symposium* (17 pp), 8–9 February 1988, Bakersfield, California.

Van Everdingen, A. F. (1953). The skin effect and its influence on the productive capacity of a well. *Petroleum Transactions AIME, 198,* 171–176.

Verma, M. P. (2000). Revised quartz solubility temperature dependence equation along the water-vapor saturation curve. In *Proceedings World Geothermal Congress* (pp. 1927–1932). Kyushu-Tohoku, Japan.

Verma, S. P., & Santoyo, E. (1997). Improved equations for Na/K, Na/Li, and SiO_2 geothermometers by outlier detection and rejection. *Journal of Volcanology and Geothermal Research, 79,* 9–23.

VDI (2001). *Use of subsurface thermal resources* (in German) (p. 4640). Richtlinienreihe: Union of German Engineers (VDI).

Wagner, R., & Clauser, C. (2005). Evaluating thermal response tests using parameter estimation for thermal conductivity and thermal capacity. *Journal of Geophysics and Engineering, 2,* 349–356.

Wagner, W., & Kretschmar, H. -J. (2008). *International steam tables, properties of water and steam.* Heidelberg: Springer.

Walther, J. V., & Helgeson, H. C. (1977). Calculation of the thermodynamic properties of aqueous silica and the solubility of quartz and its polymorphs at high pressures and temperatures. *American Journal of Science, 277,* 1315–1351.

Weast, R. C., & Selby, S. M. (1967). *CRC Handbook of chemistry and physics* (48th edn). Cleveland: CRC Press.

Wells, D. L., & Coppersmith, K. J. (1994). New empirical relationships among magnitude, rupture length, rupture width, rupture area and surface displacement. *Bulletin of the Seismological Society of America, 84,* 974–1002.

Williams, B. B., Gidley, J. L., & Schechter, R. S. (1979). Acidizing fundamentals. *Society of Petroleum* (p. 273).

Witt, K.-J. (2009). *Grundbau-Taschenbuch, Teil 2, Geotechnische Verfahren.* Berlin: Ernst & Sohn.

Wyss, M. (1979). Estimating maximum expectable magnitude of earthquakes from fault dimensions. *Geology, 7,* 336–340.

Wyss, R. (2001). The blowout at the geothermal probe wellbore at Wilen (Obwalden, Switzrland) (in German). *Bulletin Angewandte Geologica, 6*(1), 25–40.

Zapp, F. J., & Rosinski, C. (2007). Effects of heat transfer fluid parameters on the transfer of thermal energy of thermal ground probes (in German). Bochum: Der Geothermiekongress.

Zobak, M. D., Barton, C. A., Brudy, M., Castillo, D. A., Finkbeiner, T., Grollimund, B. R., et al. (2003). Determination of stress orientation and magnitude in deep wells. *International Journal of Rock Mechanics and Mining Sciences, 40,* 1049–1076.

Printed in Great Britain
by Amazon